Edited by
Günter Obe, Burkhard Jandrig,
Gary E. Marchant, Holger Schütz,
and Peter M. Wiedemann

Cancer Risk Evaluation

Related Titles

Hsu, C.-H., Stedeford, T. (eds.)

Cancer Risk Assessment

Chemical Carcinogenesis, Hazard Evaluation, and Risk Quantification

2010

ISBN: 978-0-470-23822-6

Wiedemann, P. M., Schütz, H. (eds.)

The Role of Evidence in Risk Characterization

Making Sense of Conflicting Data

2008

ISBN: 978-3-527-32048-6

Edited by
Günter Obe, Burkhard Jandrig, Gary E. Marchant,
Holger Schütz, and Peter M. Wiedemann

Cancer Risk Evaluation

Methods and Trends

The Editors

Prof. Dr. Günter Obe
Ret. from University Duisburg-Essen
Present address:
Gershwinstrasse 33
14513 Teltow
Germany

Dr. Burkhard Jandrig
Max-Delbrück-Center
for Molecular Medicine (MDC)
Robert-Rössle-Str. 10
13125 Berlin
Germany

Prof. Dr. Gary E. Marchant
S. Day O'Connor College of Law
Arizona State University
Tempe, AZ 85287-796
USA

Dipl. Päd. Holger Schütz
Research Center Jülich, Inst.
of Neuroscience and Medicine (INM-8)
52425 Jülich
Germany

Prof. Dr. Peter M. Wiedemann
Karlsruher Institut für Technologie (KIT)
TAB – Büro für Technikfolgen-
Abschätzung beim Deutschen Bundestag
Neue Schönhauser Straße 10
10178 Berlin
Germany

Cover
Microscopic Cancer Cell © PhotoDisc/Getty Images

Library of Congress Card No.: applied for

British Library Cataloguing-in-Publication Data
A catalogue record for this book is available from the British Library.

Bibliographic information published by the Deutsche Nationalbibliothek
The Deutsche Nationalbibliothek lists this publication in the Deutsche Nationalbibliografie; detailed bibliographic data are available on the Internet at http://dnb.d-nb.de.

Wiley-Blackwell is an imprint of John Wiley & Sons, formed by the merger of Wiley's global Scientific, Technical, and Medical business with Blackwell Publishing.

Cover Design Adam-Design, Weinheim
Typesetting Thomson Digital, Noida, India
Printing and Binding betz-druck GmbH, Darmstadt

Printed in the Federal Republic of Germany
Printed on acid-free paper

ISBN: 978-3-527-32753-9
ePDF ISBN: 978-3-527-63462-0
ePub ISBN: 978-3-527-63463-7
Mobi ISBN: 978-3-527-63464-4
oBook ISBN: 978-3-527-63461-3

Contents

Preface

Both the modern biomedical research such as genomics and proteomics and the rapid advances in high-throughput screening molecular technologies have revolutionized the knowledge about functional and regulatory genomics, which is beginning to make an immense impact on our understanding of human health and disease. These developments have also brought great hope to improve cancer risk assessment, even to solve scientific controversies about cancer risk claims, such as the debate whether electromagnetic fields from mobile telephony cause cancer in humans.

During the past few years, we were able to focus on this question in an integrated multidisciplinary research project on the implications of modern biomedicine on risk assessment (IMBA), sponsored by the Helmholtz Association of German Research Centres. As a health technology assessment project, IMBA analyzed how new developments in biomedicine, which are often summarized under the term "toxicogenomics," will transform the present risk management framework. IMBA looked into a wide range of scientific and social challenges that deserve careful attention, particularly on issues related to risk assessment, risk perception, and risk communication.

In 2008, we organized an international workshop in Berlin as part of the IMBA project. The aim of the workshop was to compare the potential of genomics and traditional approaches used in cancer risk assessment, particularly genotoxicity studies, with regard to their potential to inform assessment of unclear risks, that is, risks where evidence is insufficient for a conclusive risk assessment. The unclear risks chosen for discussion were radio frequency electromagnetic fields. Topics such as the validity and reliability of genotoxic research for cancer risk assessment, the prospects of toxicity testing and risk assessment, and the implications for policy making were critically reviewed and evaluated by experts in the fields of ionizing and nonionizing radiation, genotoxicity, molecular medicine, and epidemiology.

The discussions during the workshop motivated us to plan a publication on these topics. Further impetus came from the ongoing societal debate on the health implications of electromagnetic fields, which seems not to be solved but stimulated by new molecular biomarker studies and high-throughput technologies in this field. We think that in a climate of excitement about the promises of molecular medicine, it is crucial to explore the validity of molecular biomarkers and evaluate their added value for risk assessment. We hope that this book will contribute to effective interdisciplinary communication and collaboration in the fields of molecular biology, cancer research, risk assessment, and public health policy.

We are grateful to all authors of the book for investing their valuable time in writing their contributions and participating in the review process in order to make the book valuable for all readers. Last but not least, we appreciative the support of the Helmholtz Association of German Research Centres.

Berlin, December 2010

Günter Obe, Burkhard Jandrig,
Gary E. Marchant, Holger Schütz,
Peter M. Wiedemann

List of Contributors

Robert A. Baan
WHO–International Agency for Research on Cancer
The IARC Monographs Programme
150, cours Albert Thomas
69372 Lyon Cedex 08
France

Gabriele Berg-Beckhoff
Unit for Health Promotion Research
Institut of Public Health
University of Southern Denmark
Niels Bohrs Vej 9
6700 Esbjerg
Denmark

Maria Blettner
Johannes Gutenberg-University of Mainz
Institute of Medical Biostatistics, Epidemiology, and Informatics
Obere Zahlbacher Straße 69
55101 Mainz
Germany

Jochen Buschmann
Fraunhofer Institute for Toxicology and Experimental Medicine
Department of Toxicology & Environmental Hygiene
Nikolai-Fuchs-Strasse 1
30625 Hannover
Germany

Vincent J. Cogliano
Acting Director, Integrated Risk Information System (IRIS)
National Center for Environmental Assessment
U.S. Environmental Protection Agency
1200 Pennsylvania Ave NW (8601P)
Washington DC 20460
USA

Clemens Dasenbrock
Fraunhofer Institute for Toxicology and Experimental Medicine
Department of Toxicology & Environmental Hygiene
Nikolai-Fuchs-Strasse 1
30625 Hannover
Germany

Marco Durante
GSI Helmholzzentrum für
Schwerionenforschung
Biophysics Department
Planckstrasse 1
64291 Darmstadt
Germany

and

Technical University of Darmstadt
Department of Condensed Matter
Physics
Hochschulstraße 3
46289 Darmstadt
Germany

Bernd Epe
University of Mainz
Institute of Pharmacy and Biochemistry
Staudingerweg 5
55128 Mainz
Germany

Markus Fußer
University of Mainz
Institute of Pharmacy and Biochemistry
Staudingerweg 5
55128 Mainz
Germany

Burkhard Jandrig
Max Delbrück Center for Molecular
Medicine
Robert-Rössle-Str. 10
13125 Berlin
Germany

Wolfgang Kemmner
Max Delbrück Center for Molecular
Medicine
Experimental and Clinical Research
Center (ECRC)
Research Group Surgical Oncology
Robert-Rössle-Str. 10
13125 Berlin
Germany

Jürgen Kiefer
Universität Giessen
Am Dornacker 4
35435 Wettenberg
Germany

Alexander Lerchl
Jacobs University Bremen
School of Engineering and Science,
Research II
Campus Ring 6
28759 Bremen
Germany

Dariusz Leszczynski
STUK – Radiation and Nuclear Safety
Authority
Laippatie 4
00881 Helsinki
Finland

David C. Lloyd
Health Protection Agency
Chilton
Didcot OX11 0RQ
UK

Gary E. Marchant
Arizona State University Sandra
Day O'Connor College of Law
P.O. Box 877906
Tempe, AZ 85287-796
USA

David L. McCormick
IIT Research Institute
10 West 35th Street
Chicago, IL 60616
USA

Meike Mevissen
University of Bern
Vetsuisse Faculty
Department of Clinical Research and Veterinary Public Health
Division Veterinary Pharmacology and Toxicology
Länggassstrasse 124
3012 Bern
Switzerland

William F. Morgan
Pacific Northwest National Laboratory
Cell Biology and Biochemistry
P.O. Box 999, MSIN P7-56
Richland, WA 99354
USA

Günter Obe
Ret. from University Duisburg-Essen
Present address:
Gershwinstrasse 33
14513 Teltow
Germany

Christopher J. Portier
National Center for Environmental Health/Agency for Toxic Substances and Disease Registry
Centers for Disease Control and Prevention
1600 Clifton Road
Atlanta, GA 30333

Thomas J. Prihoda
University of Texas Health Science Center
Department of Pathology
San Antonio, TX 78229
USA

Brigitte Schlehofer
German Cancer Research Centre
Unit of Environmental Epidemiology
Im Neuenheimer Feld 280
69120 Heidelberg
Germany

Alexander Schramm
Universitätsklinikum Essen,
Pädiatrie III
Onkologisches Labor
Hufelandstr. 55
45122 Essen
Germany

Holger Schütz
Research Center Jülich
Institute of Neuroscience and Medicine (INM-8)
52425 Jülich
Germany

Joachim Schüz
International Agency for Research on Cancer (IARC)
Section of Environment and Radiation
150, cours Albert Thomas
69372 Lyon Cedex 08
France

Michal R. Schweiger
Max-Planck Institute for Molecular Genetics
Department of Vertebrate Genomics
Ihnestrasse 63–73
14195 Berlin
Germany

Marianne B. Sowa
Pacific Northwest National Laboratory
Cell Biology and Biochemistry
P.O. Box 999, MSIN P7-56
Richland, WA 99354
USA

Reuben Thomas
National Institute of Environmental
Health Sciences
Laboratory of Toxicology and
Pharmacology
Environmental Systems Biology
P.O. Box 12233, MD B2-08
Research Triangle Park, NC 27709
USA

Bernd Timmermann
Max-Planck Institute for Molecular
Genetics
Ihnestrasse 63–73
14195 Berlin
Germany

Vijayalaxmi
University of Texas Health Science
Center
Department of Radiology
San Antonio, TX 78229
USA

Richard Wakeford
The University of Manchester
Dalton Nuclear Institute
Pariser Building – G Floor
P.O. Box 88, Sackville Street
Manchester M60 1QD
UK

Peter M. Wiedemann
Karlsruher Institut für Technologie
(KIT)
TAB – Büro für Technikfolgen-
Abschätzung beim Deutschen
Bundestag
Neue Schönhauser Straße 10
10178 Berlin
Germany

1
Introduction

Cancer is one of the leading causes of human mortality. Over the past 30 years, the global burden of cancer has more than doubled. According to the recent World Cancer Report, published by the World Health Organization (WHO) and the International Agency for Research on Cancer (IARC), in 2008 there were 7 million deaths from cancer. Affected by the still growing and aging world population, this figure is expected to increase to 17 million annually by 2030 [1]. While many environmental cancer risk factors, such as exposures to ionizing radiation or tobacco smoke, alcohol consumption, or excessive sun exposure, have been established [2], assessments of cancer hazards and risks are difficult and often highly uncertain. Of the more than 900 agents that have been evaluated by IARC, only 12% have been classified as being clearly carcinogenic to humans [3]. And even if an agent has been identified as a carcinogen, the risk it poses to a given population is often hard to estimate. The reasons for these difficulties are manifold. First of all, there are different types of cancer that differ in their etiology. Another reason – and that is the focus of this book – is that cancer causation is hard to investigate. Experimental studies in humans are for obvious ethical reasons not possible, thus cancer risk assessment has to rely on indirect evidence.

At present, assessments of carcinogenicity are based on three pillars: epidemiological studies in humans, studies in experimental animals, and genotoxicity studies. Epidemiological studies aim at identifying the causes of cancer by studying the covariation between exposure to an agent and cancer incidence. Although there is a long debate on if and when epidemiology actually can provide causal evidence [4], there is little disagreement that epidemiological studies are the most important source of knowledge for cancer risk assessment [2, 5]. In studying the carcinogenicity of agents, epidemiological studies have to rely on given exposures to the respective agents, for instance, radon emanating from the soil or electromagnetic fields emitted from mobile communication devices. These conditions are usually not under control of the investigators, and although epidemiologists have developed an elaborate methodology to match specific study demands [6], problems such as bias and confounding frequently limit the conclusiveness of their results.

Cancer Risk Evaluation: Methods and Trends,
Edited by Günter Obe, Burkhard Jandrig, Gary E. Marchant, Holger Schütz, and Peter M. Wiedemann.

Compared to epidemiology, animal studies have the advantage of permitting experimental designs, where (at least in principle) everything can be controlled. This allows the most stringent test of a causal relationship between the exposure to an agent and an adverse effect. At least for chemical agents, there is a kind of "gold standard" that is used for carcinogenicity testing, which includes 2-year studies with rodents [7]. However, these studies are time consuming and expensive, limiting the number of agents that are tested [2]. Beside ethical considerations regarding the use of animals in research, the appropriateness of animal models for investigating and predicting human diseases has been disputed [8]. It should also be noted that this gold standard is not so well established for some physical agents. For example, many animal studies investigating the potential carcinogenicity of radio frequency electromagnetic fields (RF EMF) use only one type of animals and often for a short period [9]. An important limitation of using animal studies for carcinogenicity testing is that the experimental results always have to be extrapolated to humans, which is of course acknowledged in evaluations of evidence for cancer risk assessment [2, 5].

Basically, the same holds for genotoxicity studies, where experimental findings also have to be evaluated with regard to their implications for humans. Their value lies in the fact that cancer results primarily from genetic changes in single cells. Therefore, agents that are able to damage cellular DNA lead to mutations and then possibly to cancer. For instance, people exposed to ionizing radiation have both an elevated cancer risk and elevated frequencies of chromosomal aberrations in their peripheral lymphocytes, showing the mutagenic activity of ionizing radiation. Mutations are initiating events for the development of cancer and therefore testing of various agents for their possible mutagenicity is an important part of cancer risk assessment [10].

Over the past years, new technologies have been developed that promise new insight into cancer risk assessment by focusing on the role of the genome for understanding cancer initiation and development [11, 12]. These so-called omics technologies include genomics for DNA variations, transcriptomics for messenger RNA, proteomics for peptides and proteins, and metabolomics for intermediate products of metabolism. Technological breakthroughs allow simultaneous examination of thousands of genes, transcripts, proteins, and metabolites with high-throughput techniques and analytical tools to extract information. These new technologies are expected to provide a highly sensitive detection of low-dose effects, more reliable extrapolation of risk estimates across doses, routes, and species, and valuable insight into the mechanism of action of toxicants. Overall, the ability to classify chemicals and other stressors based on their effects at omics level would permit the development of new testing strategies in cancer risk assessment. At present, genomics- and transcriptomics-based approaches are most promising, while metabolomics, though in principle quite potent, is quite nascent in its development, as present techniques and the methodology are far away from inspecting the whole metabolome. High-throughput screening technologies have their own technical limitations and uncertainties. The transcriptome and proteome are highly dynamic and change rapidly and dramatically in response to perturbations or even during normal cellular events. The modern screening technologies still have the problem of reproducibility and variability between studies and are prone to produce false positive results [13, 14].

An important aspect here is quality control of scientific investigations. Although in general not limited to the omics field, the huge amount of data produced with microarray experiments and the extensive data processing required for analysis make open data accessibility to allow independent reevaluation of findings an important claim, which is increasingly acknowledged in the scientific community [15, 16]. Another aspect of quality control is how to evaluate the reliability of controversial scientific results. As said before, it is difficult to rule out errors in high-throughput screening research. Even more complicated is the proper dealing with fraud suspicions. Although fraud in science is by no means a new phenomenon, recent scandals in highly prestigious scientific journals have also called the public's attention to this issue [17]. Thus, the highly welcome new approaches to cancer risk assessment also call for the establishment of rules that allow a careful evaluation of study results. Furthermore, better risk communication is required for informing health professionals, the media, and the general public about the meaning of omics findings for risk assessment [18]. A particular problem here is if and when uncertainties in risk assessment should be communicated to a nonexpert audience. On a more general level, the question arises how these uncertainties should be addressed in risk management. This is likely to intensify the current debate about the application of the precautionary principle. Of course, these problems are not specific to omics; however, apart from providing new knowledge for risk assessment, omics is also likely to introduce new uncertainties [19–21].

The following chapters of this book provide insight into new developments of cancer risk assessment and their accompanying scientific discussions. While the focus is on cancer and radiation, especially nonionizing radiation, the various chapters provide the reader with a comprehensive view on cancer biology, cancer assessment methods including epidemiology, animal research, and genotoxicity studies as well as omics approaches and applications. Furthermore, it covers the comparative assessment of radiation risks and addresses policy considerations such as risk communication and application of the precautionary principle.

The book is organized in seven parts. Part One gives an overview of the current understanding of cancer development and approaches to cancer risk assessment. Jandrig (Chapter 2) shows that, apart from mutations, other cellular changes have to be taken into account to understand the complex biology of cancer. Epe and Fußer (Chapter 3) describe the various determinants of generation, repair, and steady-state levels of endogenous DNA modifications. Baan and Cogliano (Chapter 4) provide insight into cancer hazard identification as the first step in cancer risk assessment and cancer prevention, as outlined in the IARC Monographs Programme.

The role of epidemiology in cancer risk assessment is addressed in Part Two. Schüz, Berg-Beckhoff, Schlehofer, and Blettner (Chapter 5) consider the particularly challenging possible adverse health effects of exposure to electromagnetic fields (EMF) that have remained a scientific and political controversy until today. Their first example is the relationship between extremely low-frequency (ELF) fields from power lines and the risk of childhood leukemia. Their second example is the relationship between RF EMF, specifically those emitted from mobile phones, and the risk of brain tumors. Wakeford (Chapter 6) presents data for cancer risk assessment of ionizing radiation. Among others, he provides cancer risk figures based on epidemiology from Hiroshima survivors and children exposed during and after the Chernobyl accident.

Animal studies are indispensable for cancer hazard identification and results of this type of research are presented in Part Three. Buschmann and Dasenbrock (Chapter 7) refer to recent advances in animal studies on RF EMF testing the possible carcinogenic effects related to cell phones and base stations. On the basis of a comprehensive discussion of the PERFORM-A project, they demonstrate how existing data gaps relevant for risk assessment can be closed. Pointing to the strengths and limitations of epidemiological cancer studies of ELF fields, McCormick (Chapter 8) shows how laboratory animal research can fill gaps in EMF cancer risk assessment. The author discusses the findings of various types of experimental animal studies and comes to the conclusion that available animal data do not support an elevated cancer risk.

Part Four highlights the importance of studying chromosomal damage, which is a highly reliable endpoint for cancer hazard and risk assessment. Obe, Lloyd, and Durante (Chapter 9) outline current approaches to investigating chromosomal aberrations. They argue that elevated frequencies of chromosomal aberrations in peripheral lymphocytes of human populations are associated with elevated cancer frequencies and allow calculation of cancer risks in persons exposed to ionizing radiation, such as astronauts. Vijayalaxmi and Prihoda (Chapter 10) show how meta-analysis as a tool for statistical data synthesis can be used to systematically summarize evidence from cytogenetic studies in mammalian somatic cells that have been exposed to radio frequency radiation. They conclude that exposure to radio frequency radiation does not increase frequencies of chromosomal aberrations and micronuclei, which are two endpoints for chromosomal damage.

The potential of omics technologies as new tools for cancer risk assessment are discussed in Part Five. Technological breakthroughs allow simultaneous examination of thousands of genes, transcripts, proteins, and metabolites with high-throughput techniques and analytical tools to extract information. Modern screening technologies speed up the discovery process and give a broader insight into biochemical events that follow the exposure to potentially harmful agents, such as chemical substances, ionizing radiation, or electromagnetic fields. The different methodologies and techniques are discussed in this part with respect to actual applications and future developments. Schweiger and Timmermann (Chapter 11) explain the huge potential that whole genome approaches afford for understanding complex genetic diseases such as cancer. They provide an overview of the advancement of genome analysis technologies and illustrate how these are used for investigating the mechanisms underlying cancer development. The authors close with an outlook on how the genomics approach might ultimately lead to an individualized cancer treatment. Kemmner (Chapter 12) outlines the use of transcriptomics, or gene expression profiling, in cancer risk assessment, for instance, with regard to classification of human cancers and prediction of cancer recurrence and metastasis. The author discusses technical challenges of gene expression profiling, such as sample preparation and data analysis, and gives examples of microarray applications in cancer research. Proteomics, the analysis of proteins, and its relevance to cancer risk assessment, is discussed by Schramm (Chapter 13). While proteomics comprises a variety of technical disciplines, its application to cancer risk assessment can be described as a multistep process including sample preparation, separation, quantitation, and protein identification. The author discusses particular challenges of these steps and concludes with an outlook on future developments of proteomics for individualized cancer therapy.

Examples of using omics technologies for risk assessment are described in Part Six. Portier and Thomas (Chapter 14) provide a critical discussion of omics and high-throughput screening strategies concerning cancer risk assessment. First, they discuss the difficulties of traditional cancer risk assessment, in particular with animal studies, and then describe how omics might be used to overcome these problems. They conclude that while there is little doubt that omics will be of major importance for future risk assessment, there is still much research needed, before it finds regulatory approval in risk assessment. Morgan and Sowa (Chapter 15) show how omics might be used for risk assessment of exposure to low-level ionizing radiation. So far, risk assessment had to rely mainly on epidemiological data, for instance, from Japanese A-bomb survivors, but here epidemiology clearly reaches its limits. The authors discuss studies that used gene expression profiling, proteomic profiling, and metabolomic profiling to investigate the effects of low-level ionizing radiation. Their conclusion is that while significant progress has been made in using omics for cancer risk assessment, the future challenge is to integrate the various omics technologies to allow a "systems level" approach. The next two chapters then address how transcriptomics and proteomics can be used for cancer risk assessment of RF EMF. Mevissen (Chapter 16) provides an overview of studies investigating the effects of RF EMF exposure on gene expression. She makes it clear that these studies differ strongly in scientific quality and focus, and are insufficient for drawing conclusions regarding effects the RF EMF exposure has on organisms. A similar picture emerges from the review of proteomics studies that is given by Leszczynski (Chapter 17). So far, only few studies have investigated the effects of RF EMF exposure on the proteome, and many of them have methodological shortcomings.

The last part of the book addresses challenges for risk management. Lerchl (Chapter 18) reports recent examples of apparent scientific misconduct and discusses heuristics that can help detect data fabrication. He also offers some advice how to handle such misconduct appropriately. Kiefer (Chapter 19) offers a comparative risk assessment across the electromagnetic spectrum based on the Bradford Hill criteria. He argues that at present only ionizing radiation fulfils all requirements for cancer hazard identification. Wiedemann and Schütz (Chapter 20) discuss the challenges of communicating about uncertainty in cancer risk assessments to nonexperts. They offer ample evidence that, in contrast to common beliefs, informing about uncertainties might create misperceptions and misunderstandings of risk. Furthermore, they discuss how to explain inconclusive scientific evidence, a task particularly important for hazard assessment. Finally, Marchant (Chapter 21) considers the role of the precautionary principle in risk management. Weighing the pros and cons, he concludes that despite its rhetorical appeal, the precautionary principle remains problematic in its practical application, which in large part is due to the ambiguity and arbitrariness of the principle.

References

1 Boyle, P. and Levin, B. (eds) (2008) *World Cancer Report 2008*, International Agency for Research on Cancer, Lyon.

2 Fontham, E.T.H., Thun, M.J., Ward, E., Balch, A.J., Delancey, J.O.L., and Samet, J.M.,on behalf of ACS Cancer and the Environment Subcommittee (2009) American Cancer Society perspectives on environmental factors and cancer. *CA Cancer J. Clin.*, **59**, 343–351.

3 IARC (2010) *Agents Classified by the IARC Monographs*, vols. 1–100, International Agency for Research on Cancer, Lyon. Available at http://monographs.iarc.fr/ENG/Classification/index.php (Accessed July 19, 2010).

4 Rothman, K.J. and Greenland, S. (2005) Causation and causal inference in epidemiology. *Am. J. Public Health*, **95**, S144–S150.

5 International Agency for Research on Cancer (2006) *Preamble to the IARC Monographs. IARC Monographs Programme on the Evaluation of Carcinogenic Risks to Humans*, International Agency for Research on Cancer. Available at http://monographs.iarc.fr/ENG/Preamble/CurrentPreamble.pdf (Accessed July 26, 2010).

6 Rothman, K.J., Greenland, S., and Lash, T.L. (eds) (2008) *Modern Epidemiology*, 3rd edn, Lippincott Wilkins & Wilkins, Philadelphia, PA.

7 Fung, V.A., Barrett, J.C., and Huff, J. (1995) The carcinogenesis bioassay in perspective: application in identifying human cancer hazards. *Environ. Health Perspect.*, **103**, 680–683.

8 Hackam, D.G. and Redelmeier, D.A. (2006) Translation of research evidence from animals to humans. *JAMA*, **296**, 1731–1732.

9 Dasenbrock, C. (2005) Animal carcinogenicity studies on radiofrequency fields related to mobile phones and base stations. *Toxicol. Appl. Pharmacol.*, **207**, 342–346.

10 Parsons, B.L., Myers, M.B., Meng, F., Wang, Y., and McKinzie, P.B. (2010) Oncomutations as biomarkers of cancer risk. *Environ. Mol. Mutagen.* doi: 10.1002/em.20600.

11 Bishop, W.E., Clarke, D.P., and Travis, C.C. (2001) The genomic revolution: what does it mean for risk assessment? *Risk Anal.*, **21**, 983–987.

12 Simmons, P.T. and Portier, C.J. (2002) Toxicogenomics: the new frontier in risk analysis. *Carcinogenesis*, **23**, 903–905.

13 Troester, M.A., Millikan, R.C., and Perou, C.M. (2009) Microarrays and epidemiology: ensuring the impact and accessibility of research findings. *Cancer Epidemiol. Biomarkers Prev.*, **18**, 1–4.

14 Vlaanderen, J., Moore, L.E., Smith, M.T., Lan, Q., Zhang, L., Skibola, C.F., Rothman, N., and Vermeulen, R. (2010) Application of omics technologies in occupational and environmental health research: current status and projections. *Occup. Environ. Med.*, **67**, 136–143.

15 Kaye, J., Heeney, C., Hawkins, N., de Vries, J., and Boddington, P. (2009) Data sharing in genomics: re-shaping scientific practice. *Nat. Rev. Genet.*, **10**, 331–335.

16 Field, D., Sansone, S.-A., Collis, A., Booth, T., Dukes, P., Gregurick, S.K., Kennedy, K., Kolar, P., Kolker, E., Maxon, M., Millard, S., Mugabushaka, A.-M., Perrin, N., Remacle, J.E., Remington, K., Rocca-Serra, P., Taylor, C.F., Thorley, M., Tiwari, B., and Wilbanks, J. (2009) 'Omics data sharing. *Science*, **326**, 234–236.

17 Science (2006) Special Online Collection: Hwang et al. Controversy. Available at http://www.sciencemag.org/sciext/hwang2005/ (Accessed September 10, 2010).

18 McBride, C.M., Bowen, D., Brody, L.C., Condit, C.M., Croyle, R.T., Gwinn, M., Khoury, M.J., Koehly, L.M., Korf, B.R., Marteau, T.M., McLeroy, K., Patrick, K., and Valente, T.W. (2010) Future health applications of genomics: priorities for communication, behavioral, and social sciences research. *Am. J. Prev. Med.*, **38**, 556–565.

19 Adelman, D.E. (2005) The false promise of the genomics revolution for environmental law. *Harv. Environ. Law Rev.*, **29**, 117–177.

20 Battershill, J.M. (2005) Toxicogenomics: regulatory perspective on current position. *Hum. Exp. Toxicol.*, **24**, 35–40.

21 Boverhof, D.R. and Zacharewski, T.R. (2006) Toxicogenomics in risk assessment: applications and needs. *Toxicol. Sci.*, **89**, 352–360.

Part One
Models and Approaches

Cancer Risk Evaluation: Methods and Trends,
Edited by Günter Obe, Burkhard Jandrig, Gary E. Marchant, Holger Schütz, and Peter M. Wiedemann.

2
Models of Cancer Development: Genetic and Environmental Influences

Burkhard Jandrig

2.1
Introduction

The past decade has brought enormous advances in the understanding of the molecular pathogenesis of cancer. Cancer research has generated a prodigious amount of knowledge revealing cancer as a disease characterized by dynamic changes in the genome, at the expression level, and influenced by environmental factors. Deep insight could especially be achieved in the area of cancer genetics where the explosion of sequence and molecular profiling data elucidated the complexity of human malignancies. Several familial cancer genes with high-penetrance mutations could be identified. However, multigenic models suggest that a high proportion of cancers may arise as a consequence of the combined effects of common low-penetrance alleles and rare disease-causing variants that pose moderate cancer risks.

At the genetic level, oncogenes and tumor suppressor genes play prominent roles. Mainly mutations produce oncogenes with dominant gain of function and tumor suppressor genes with recessive loss of function. These genes have been intensely studied in human and animal cancer cells and in experimental models. It turned out that these genes are involved in a molecular machinery regulating proliferation, differentiation, and death. Similar mechanisms govern the transformation of normal cells into malignant cancers.

Tumorigenesis is a multistep process reflected by stochastic genetic alterations that drive a progressive transformation of normal cells via a series of premalignant states into highly malignant derivatives [1]. Genomes of tumor cells are altered at multiple sites, ranging from subtle point mutations to gross changes in chromosome number, size, or structure [2]. Experimental transformation of cultured cells into tumorigenic ones and transgenic animal models of tumorigenesis have repeatedly shown multiple rate-limiting steps, each conferring some kind of growth advantage [3].

Cancer cells often show defects in the so-called regulatory circuits that normally govern cell proliferation and homeostasis. A number of essential circuits dictate malignant growth (Figure 2.1) [4, 5]. Tumors are characterized by acquired novel capabilities such as self-sufficiency in growth signals, insensitivity to growth inhibitory (antigrowth) signals, evasion of programmed cell death (apoptosis), limitless

Cancer Risk Evaluation: Methods and Trends,
Edited by Günter Obe, Burkhard Jandrig, Gary E. Marchant, Holger Schütz, and Peter M. Wiedemann.

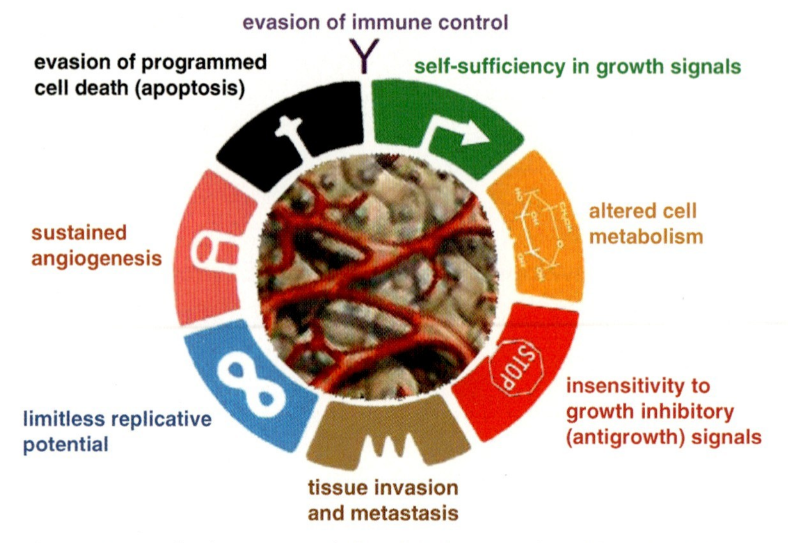

Figure 2.1 Specific characteristics (hallmarks) of cancer (adapted from Refs [4, 5]).

replicative potential, altered cell metabolism, sustained angiogenesis, evasion of immune response, tissue invasion, and metastasis.

2.2 Specific Characteristics of Tumors

Normal cells require stimulatory growth signals before they can switch from a quiescent into a proliferative state. Signaling molecules such as diffusible growth factors, extracellular matrix components, or cell-to-cell adhesion molecules bind to transmembrane receptors leading to conformational changes and activation of signal transduction pathways. Many of the so far known oncogenes have a place in these pathways and foster unregulated growth signaling. Tumor cells often show a greatly reduced dependence on exogenous stimulation by generating many of their own growth signals, thereby creating an autocrine stimulation and disrupting important homeostatic mechanisms [6–8].

Cell surface receptors that transduce growth stimulatory signals into the cell are other targets of deregulation during tumor pathogenesis. In many tumors, growth factor receptors are overexpressed or truncated enabling the cancer cells to become hyperresponsive to growth factors or to elicit ligand-independent signaling [9].

In addition, tumor cells can also switch the expression of integrin receptors in favor of progrowth signals [10]. These receptors physically link cells to the extracellular matrix. Successful binding to specific parts of the matrix can influence cell behavior, resistance to apoptosis, or activation of the cell cycle. In addition, signaling between the diverse cell types within a tumor, especially from the stromal cell components of the tumor mass, may enhance the tumor growth potential.

Signals emitted by ligand-activated growth factor receptors and integrins can activate different cytoplasmic pathways, most prominent the ras-raf-MAP kinase pathway. The signal transduction cascades are often linked to other pathways creating a cross-talking connection network.

Tissue homeostasis is mainly maintained by growth inhibitory signals blocking proliferation and by controlling the cell cycle clock. Tumor cells have to switch off such impediments by specifically altered components that govern the transit of the cell through the G1 phase of its growth cycle. At the molecular level, almost all antiproliferative signals function through the retinoblastoma protein (pRb) and its relatives p107 and p130. Disruption of the pRb pathway leads to an insensitivity to antigrowth factors, *inter alia* activates E2Fs, and thus allows cell proliferation. The pRb signaling circuit can be disrupted in a variety of ways; for example, functional pRb may be lost through mutation of its gene [11] originally defining the concept of tumor suppressor loss in cancer.

Some tumor cells use various strategies to defend themselves from an irreversible switch into postmitotic differentiated states and to avoid terminal differentiation. For example, in colon cancer, mutations both in the Apc gene and in genes that modify or interact with Apc result in an escape of enterocytes in the colonic crypts from going into a terminal differentiated state [12]. Again, this eventually disturbs tissue homeostasis massively.

Furthermore, tumor cell populations are often able to expand in number by avoiding cellular safety devices, for example, multiple cell death mechanisms. Programmed cell death or apoptosis and necrosis are not the only cell death programs involved in the regulation of tissue homeostasis and the removal of unwanted cells [13]. However, the precise mechanisms of apoptotic steps and the cross-talk of physiologic signals are relatively well established. Special proteins check the extra- and intracellular environment for conditions of normality or abnormality and regulate execution pathways. Many of the signals affect the mitochondria, thereby releasing cytochrome c and influencing members of the bcl-2 family. The p53 tumor suppressor protein has a prominent role as a cell cycle checkpoint and in regulating apoptosis [14]. The ultimate effectors of apoptosis are proteolytic caspases. In the end, the chromosomes degrade, the nucleus is fragmented, the cytoplasmic and the nuclear skeletons break down, and the cells are digested.

In tumors, the apoptotic program can be influenced by overexpressed oncogenes and circumvented by inactivated tumor suppressor proteins, which can dramatically affect the dynamics of tumor progression.

Growth signal autonomy, insensitivity to antigrowth signals, and resistance to apoptosis all lead to uncoupling of the cellular growth program from signals in its environment. Nevertheless, normal mammalian cells carry an intrinsic program of a finite replicative potential that limits their multiplication and has to be disrupted in tumors.

Most types of tumor cells cultivated *in vitro* are immortalized. At some point during the course of the multistep tumor progression, these cells breach the mortality barrier and acquire an unlimited replicative potential. Tumor cells are able to maintain the telomeres at a length above a critical threshold by upregulated expression of the telomerase or through a mechanism known as alternative lengthening of telomeres [15]. Inactivation of the tumor suppressor proteins p53 and Rb, for example, can circumvent senescence and eventually lead to immortalization.

Tumors are characterized by a rampant growth that requires a perpetual supply of oxygen and nutrients. A prerequisite, however, is that the process of angiogenesis is initiated and new blood vessels are developed. Angiogenesis is regulated by a series of positive and negative signals. Most studied are the angiogenesis-initiating vascular endothelial growth factor (VEGF) and fibroblast growth factors (FGFs) that are under complex transcriptional control [16].

Using increasingly sophisticated animal models will make it possible to ascribe specific roles to each of the regulators and to distinguish the molecular mechanisms that control their production and activity.

The above-mentioned specific characteristics of tumors are often accompanied by alterations in cell metabolism. In this case, tumor cells use elevated amounts of glucose as a carbon source for anabolic reactions. An increase in glycolysis that is maintained under conditions of high oxygen tension leads to enhanced lactate production [17] and to acidic conditions in the environment, thereby favoring tumor invasion and suppressing anticancer immune effectors. Enhanced glucose uptake for glycolytic ATP generation or anabolic reactions gives a competitive edge for tumor growth. The hypoxia-inducible transcription factor HIF-1 plays a key role in the metabolic reprogramming of cancer cells by activating genes encoding glucose transporters and glycolytic enzymes [18]. In addition, HIF-1 downregulates E-cadherin required for the maintenance of intercellular contacts within epithelia.

During the development of most types of solid tumors, cells move out, invade adjacent tissues, and travel to distant sites in the body to establish new clusters of cells. These metastases consist of cancer cells and normal supporting cells incorporated from the host tissue.

Several classes of proteins are involved in the invasion and metastatic process, among them cell–cell adhesion molecules (e.g., N-CAM), members of the cadherin families, and integrins, which link cells to extracellular matrix substrates [19]. The most frequent alteration in cell–environment interactions in epithelial cancer involves E-cadherin. E-cadherin function is lost in a majority of carcinomas by mutational inactivation of the E-cadherin or β-catenin genes, transcriptional repression, or proteolysis of the extracellular cadherin domain. In addition, a switch in the expression of N-CAM from a highly to a poorly adhesive form plays a critical role in the processes of invasion and metastasis [20].

Extracellular proteases represent another general player in invasive and metastatic processes. Protease genes are upregulated, protease inhibitor genes are downregulated, and inactive precursor forms of proteases are converted into active enzymes. Cancer cells are able to invade surrounding stroma, permeate blood vessel walls, and infiltrate normal epithelial cell layers by active proteases on the cell surface. However, matrix-degrading proteases are often not produced by the carcinoma cells themselves, but by the adjacent stromal and inflammatory cells. In summary, the activation of extracellular proteases and the altered binding specificities of cadherins, CAMs, and integrins are of pivotal importance for tumors to acquire invasive and metastatic capabilities, to metastasize, and eventually to kill the host.

2.3 Tumorigenesis as a Multistep Process

A rising cancer incidence with increasing age indicates that the formation of tumor is a complex process that usually proceeds over a period of decades. Tumor progression is mainly driven by a sequence of randomly occurring mutations and epigenetic alterations of DNA that affect the genes controlling cell proliferation, survival, and other specific properties associated with the malignant cell phenotype (see above). Normal cells possess multiple independent mechanisms that regulate their growth and differentiation potential; therefore, several separate events are necessary to override these control mechanisms. Many mutations are caused by a repeated exposure to carcinogens that can increase the rate of tumor progression by many orders of magnitude above the spontaneous background rate. Examples of well-studied environmental carcinogens include γ-irradiation, X-ray, UV-B, polycyclic aromatic hydrocarbons, heterocyclic amines, aflatoxin B1, and alkylating agents. The likelihood of developing a detectable tumor is determined by the cumulative exposure to a carcinogenic stimulus rather than the age at which this exposure begins. In addition, viruses might contribute to the cancer phenotype by several mechanisms. Multistep tumor formation can be clearly seen on the basis of histopathological alterations in a variety of organ sites, most prominent in colon cancer, where a carcinoma can sometimes be observed to grow out from an adenomatous polyp.

The accumulation of genetic alterations as colon tumor progression proceeds involves both the activation of oncogenes and the apparent inactivation of at least three distinct tumor suppressor genes (Figure 2.2). Inactivation of the APC gene is one of the first incidents, but the precise order of the subsequent changes may vary from tumor to tumor. Barrett's esophagus, which is a precursor lesion to esophageal carcinoma, is also characterized by a number of alternative genetic paths ranging

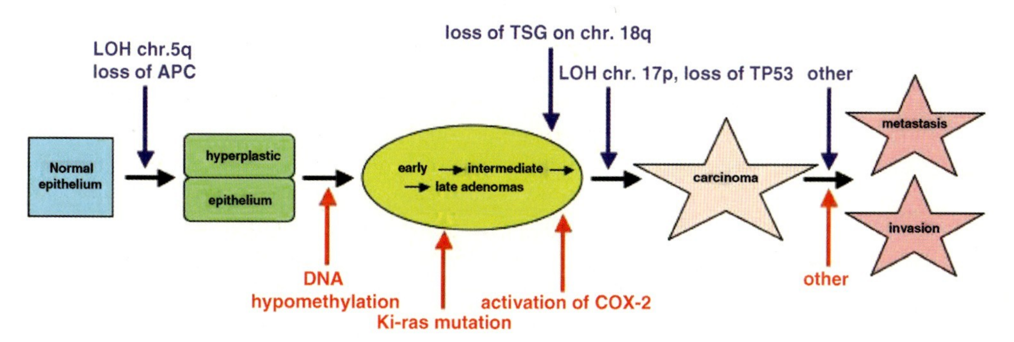

Figure 2.2 Accumulation of genetic alterations in colon carcinoma progression (adapted from Ref. [2]).

from the initial metaplasia and dysplasia to cancer clones [21]. However, in other organ sites such as breast or prostate cancer, similar pathways could not be described. Each tumor type seems to have its own genetic bibliography involving its own particular set of mutated oncogenes and tumor suppressor genes.

Oncogene collaboration experiments provided an *in vitro* model of multistep transformation *in vivo* and suggested a rationale for the complex genetic steps that accompany and cause tumor formation in human beings. Cancer phenotypes can be achieved by collaborative action of several genes or genetic alterations in experimental transformation of various human cell types *in vitro*, but does this happen in all human cell types? And does this happen in (all) spontaneously arising human tumors? In humans, it is still not possible to measure the kinetics of individual steps of tumor progression. Therefore, experimental protocols were developed to induce carcinomas in rodents using genetically engineered (ras transgenics, APC knockout) or chemically induced tumor models (azoxymethane/dextran sodium sulfate, 7,12-dimethylbenz[*a*]anthracene/12-*O*-tetradecanoylphorbol-13-acetate). In these models, the functional role of genes in tumorigenesis during initiation, promotion, and progression and the intracellular pathways involved can be studied in precise detail. In addition, these models can also offer insight into the mechanisms by which different chemical or physical modulators influence the outcome of early-stage carcinogenesis.

2.4
Epigenetic Changes in Cancer Development

Epigenetics encompasses the interaction of genetic material with its surroundings to produce a phenotype and can be seen as the basis of cellular differentiation. It explains that in a multicellular organism development generates a vast number of cell types with distinct functions and diverse but relatively stable gene expression profiles despite the same genotype of the cells. Epigenetic mechanisms coordinate important biological processes such as reprogramming of genomes during differentiation and development, RNA interference leading to posttranscriptional gene silencing, genomic imprinting, or X-chromosome inactivation [22].

Besides mutations in the DNA, the mechanisms that regulate the interpretation of the genetic code play a growing role in understanding carcinogenesis. Examples of such mechanisms include covalent and noncovalent modifications to the DNA, such as DNA methylation of cytosine dinucleotides (CpG), and to chromosomal proteins such as posttranslational modifications of histones and nonhistones (acetylation, methylation, ubiquitinylation, phosphorylation, sumoylation, or ADP ribosylation). Regulation of gene transcription can thus be affected in the short or long term. Epigenetic processes play an important role in the normal homeostasis of stem and progenitor cells in tissues and disruption can result in cancer. In tumors, DNA is often globally hypomethylated, contributing to genome instability and activation of oncogenes, and locally hypermethylated causing silencing of tumor suppressor genes. In addition, changes in histone modifications and chromatin organization can also be observed. In recent years, methylation of DNA and histone tail modifications have emerged as the most critical players of transcriptional regulation. DNA methylation is conferred by DNA methyltransferases (DNMTs) forming CpG islands especially enriched in gene promoters or the first exon (in about 70% of protein-coding mammalian genes). These CpG islands are normally unmethylated in transcriptionally active genes such as housekeeping or tumor suppressor genes, whereas developmental and tissue-specific genes mostly appear to be methylated and silenced in differentiated tissues. Meanwhile, many tumor suppressor genes have been identified to be hypermethylated in tumorous tissues compared to their normal counterparts, for example, RB1, VHL, CDKN2A, BRCA1, or ST18 [23, 24]. However, the basic mechanisms underlying the aberrant DNA methylation and the selection of genes that become methylated are barely understood. DNMTs are ubiquitously expressed at distinct levels in normal human tissues, but are overexpressed in various tumor types.

Histones are not solely DNA-packaging proteins, but store epigenetic information. Acetylation of histone lysines is generally associated with transcriptional activation, whereas methylation of certain histone residues is associated with transcriptional repression. The presence of the hypoacetylated and hypermethylated histones H3 and H465 silences certain genes with tumor suppressor-like properties, such as p21WAF1, despite the absence of hypermethylation of the CpG island [25]. Expression patterns of histone-modifying enzymes such as histone acetyltransferases distinguish cancer tissues from their normal counterparts and differ according to the tumor type.

DNA methylation and histone modifications interact with each other in the regulation of gene expression. It is generally believed that DNA methylation is the initiating event that marks certain genomic sites for the establishment of a transcriptionally inactive chromatin state [26]. However, there is increasing complexity of the relations between various epigenetic repression systems.

2.5 miRNAs and Cancer

Since the discovery of a class of small noncoding RNAs (called microRNA or miRNA) and their role in tumorigenesis, a multitude of studies have established a lot of evidence that supports the increased and accelerated progression in oncological research. A global reduction in miRNA levels is emerging as a common hallmark of cancer. MicroRNAs are regulatory RNAs of 17–30 nucleotides in length. They perfectly match 3′ untranslated regions of target messenger RNAs (mRNAs) resulting in its degradation or inhibition of mRNA translation. It is the function of the target mRNA that determines a miRNA acting either as a tumor suppressive if directed against protooncogene transcripts or as an oncogenic if directed against tumor suppressor gene transcripts. It should be noted that some miRNAs can have dual oncogenic and tumor suppressive roles depending on the cell type and pattern of gene expression. Prominent members of miRNAs include the let-7 family whose depletion in breast, lung, and colon cancer causes enhanced tumorigenicity [27].

Some miRNAs are key regulators of multiple genes that regulate different processes in cancer biology. Other miRNAs confer increased invasion capacities and promote tumor metastasis. The number of transcripts known to be regulated by miRNAs is growing rapidly.

In cancer, miRNAs are found to be massively deregulated. Several miRNAs reside in chromosomal regions that are either frequently deleted or amplified or affected by copy number variations in tumors. Another mechanism responsible for this deregulation is the epigenetic silencing of miRNA genes. In turn, some miRNAs directly repress enzymes of the epigenetic machinery, including DNA methyltransferases, histone deacetylases, and histone methyltransferases. In addition, a failure of posttranscriptional regulation may also lead to impaired miRNA maturation. The production of mature miRNAs underlies a complex process of subsequent modifications of the primary transcript. After export to the cytoplasm, the precursor miRNA is further processed. Repeatedly, discrepancies between the levels of primary transcript, precursor, and mature miRNA have been reported in tumorous cells.

Signatures of deregulated miRNAs can be useful in subtyping different carcinomas or determining their aggressiveness. Genome-wide profiling of miRNAs in different cancer types identified differentially expressed candidates including predictive miRNAs able to distinguish between normal and tumor tissue [28]. In addition, several correlations between downregulation of certain miRNAs and clinicopathological features such as tumor size, lymph node status, the expression of p53, and others were found. Not only signatures of miRNA expression may be used as tumor markers for diagnosis and patient risk stratification, but deregulated miRNAs may also represent novel targets for anticancer therapies.

2.6
Cancer Stem Cells

The cancer stem cell hypothesis suggests that cancers are derived from a stem cell compartment in a multistep process involving the accumulation of mutations in a variety of oncogenes and tumor suppressor genes. Cancer stem cells (CSCs) are a subset of cancer cells within a tumor that have the ability both to self-renew and to differentiate. CSCs are long-term residents of exposed epithelial tissues and uniquely susceptible to the accumulation of oncogenic lesions by carcinogens. CSCs share many characteristics with carcinoma cells, such as immortality, the absence of contact inhibition, and the ability to undergo self-renewal. It has been estimated that as many as 25% of the cancer cells within certain tumors have the properties of CSCs [29]. The actual proportion of CSCs within a tumor depends on the cell type of origin, stromal microenvironment, accumulated mutations, and stage of malignant progression reached by a tumor. Growing evidence suggests that signaling pathways that regulate self-renewal in normal stem cells are deregulated in cancer-initiating cells, resulting in uncontrolled expansion, aberrant differentiation, and formation of tumors with heterogeneous phenotypes. The existence of CSCs may also help to understand many of the properties of carcinomas, such as their clonal origin, their heterogeneity, and their plasticity. Increased plasticity may be present within tumor populations, enabling bidirectional interconvertibility between CSCs and non-CSCs [30].

Multipotency of lineage differentiation is likely to be a frequent but not an essential property of CSCs. CSCs with high initiating and tumor growth driving capacity give rise to more differentiated nontumorigenic progeny. Selection pressure can act at this level. Acquired additional genetic alterations can be beneficial for special clones. The genomes of populations of tumor cells often become increasingly unstable. Consequently, the resulting tumor mass is composed of an increasing number of distinct sectors, each dominated by a genetically distinct subclone. As the tumor progresses, genetic and epigenetic alterations may result in the emergence of a self-renewing metastatic CSC that can enter the blood stream and seed a secondary tumor in a distinct organ.

2.7
Cancer and the Environment

The behavior of tumor cells can also be influenced by various environmental factors that play a dominant role in the majority of sporadic cancers (see Chapters 4 and 6). Cancer development is not only due to exogenous or endogenous carcinogens but also due to their interactions with genes that are involved in the detoxification of these carcinogens, repair of DNA damage, and control of cell signaling and cell cycle. The cells in the body are constantly exposed to carcinogens from the macroenvironment (chemical carcinogens, radiation, viruses, etc.) or microenvironment (reactive oxygen species such as superoxide anions, hydroxyl radicals, etc.).

The microenvironment surrounding the tumor cell (stromal fibroblasts, adipocytes, and endothelial cells, as well as the extracellular matrix) and the immune system are known to play important roles in cancer progression. There are a lot of reciprocal interactions between cancer cells and their microenvironment or niche. Lesions in the surrounding mesenchymal tissue can enormously modulate the risk of epithelial malignancy.

The epithelial to mesenchymal transition plays an important role in metastasis and is defined as the switch from nonmotile, polarized epithelial cells to motile, nonpolarized mesenchymal cells, with the potential to migrate from a primary tumor to distant organs. Loss of E-cadherin and a simultaneous gain of mesenchymal N-cadherin allow cells to lose adhesive affinity for other epithelial cells and become more migratory and invasive [31]. In addition to transcriptional repressors such as Snail or Twist, one of the most potent inducers of an epithelial to mesenchymal transition is the transforming growth factor-β (TGF-β).

A key environmental stressor associated with tumor progression is cell oxygen deficiency or hypoxia. Induction of the hypoxia-inducible factor (HIF) family of transcription factors regulates cellular processes including glucose metabolism, angiogenesis, cell proliferation, and tissue remodeling in response to low oxygen levels. Insufficient oxygen limits tumor cell division while at the same time causing malignant cells to switch to anaerobic metabolism, increasing genetic instability, promoting angiogenesis, and inducing cell adaptations allowing for more invasive behavior [32].

Chronic inflammatory conditions are often associated with tumor development. Dysregulation of tissue repair, for example, can lead to abnormalities in the inflammatory response and ultimately tumorigenesis. Inflammation in carcinomas involves different processes such as an influx of proinflammatory cytokines including tumor necrosis factor-α and TGF-β, cytotoxic mediators, proteases, matrix metalloproteinases, interleukins, and interferons and produces potent lymphangiogenic growth factors allowing tumor growth and metastatic spread to the lymph nodes [33]. Tumor cells themselves produce cytokines that attract neutrophils, macrophages, lymphocytes, and dendritic cells. Tumor-associated macrophages produce proangiogenic growth factors such as VEGF contributing to tumorigenic growth and metastatic potential.

In multistage epithelial carcinogenesis in the skin driven by transgenic expression of HPV16, humoral antibodies are produced against extracellular matrix components [34]. Stromal accumulation of autoantibodies in premalignant skin regulates recruitment, composition, and bioeffector functions of leukocytes in tumor tissue, which in turn promotes progression and subsequent carcinoma development.

In vitro, individual tumor lines have RNA expression patterns that clearly define them from other lines even when grown in different environments. However, gene expression remains constant in individual subcutaneous tumors as the tumors increase in size [35, 36]. In the past few years, models have been developed that rely on Cre recombinase-mediated deletion of floxed sequences to activate an oncogene or inactivate a tumor suppressor gene in only a subset of cells of a tissue at a defined point in time. Cells bearing such mutation are surrounded by nonmutated competitors and may better reflect the situation during spontaneous tumorigenesis.

2.8 Systems Analysis of Cancer

Tumors are not only highly heterogeneous with respect to cell type and tissue origin but also involve dysregulation of multiple pathways controlling fundamental cell processes such as proliferation, differentiation, migration, and apoptosis. The activities of molecular networks that execute metabolic or cytoskeletal processes and regulate them by signal transduction are altered in a complex manner by diverse genetic mutations in concert with the environmental context. It is therefore necessary to develop actionable understanding of this multivariate dysregulation. High-throughput experimental platform technologies ranging from genomic sequencing to transcriptomic, proteomic, and metabolomic profiling are used for characterization of tumor cells and surrounding tissues at the molecular level (see Chapters 11–13). Even tumors of a particular tissue type comprise highly heterogeneous sets of mutations in a great number of different genes and a large number of gene products contribute to the tumor cell phenotype.

Signal transduction pathways are organized in networks, an alteration of one pathway can lead to changes in others directly via protein–protein interactions or indirectly via transcriptional or translational influences. These networks connect to components beyond the tumor cells themselves, including other cells in the environment along with the extracellular matrix [37]. Therefore, characterization of a cellular dysregulation will need to be multivariate and quantitative because single molecular biomarkers or qualitative constituent lists will be inadequate. Systems genetic analysis of experimental animal cancer models, in which extensive control over the environment and the initiating lesion is possible, has enormously expanded the understanding of how genetic alterations affect the development of tumors [38]. These artificial cancer models approximate human disease to varying degrees and provide an opportunity to perform interventional and gene–environment interaction studies that are difficult or impossible to perform in human populations. In addition, significant advances can be obtained by applying computational modeling approaches to elucidate the pathways most critically involved in tumor formation and progression, the impact of particular mutations on signal transduction, and consequences of altered cell behavior in tissue environments. Although individual genes or environmental factors may be a critical component in the pathogenesis of cancer, it is ultimately the modulation of underlying pathways that determines the resultant phenotype.

New techniques are required to examine dysregulated networks and to identify pathways within the context of a cellular network. Studying the interactomes and especially protein–protein and protein–DNA associations provides a framework for analysis of empirical data of various types, such as transcriptomic, phosphoproteomic, and phenotypic assessments [39]. Functional proteomics based on optical, spectroscopic, and microarray methods is one example of a systematic analysis of biochemical networks to provide data for a better understanding of networks and to interpret the action of chemical or physical agents [40].

The development from a single cell with a disturbed network to a metastatic tumor is exceptionally complex. However, there exist several key processes common to most cancers including uncontrolled excessive proliferation, angiogenesis, resistance to apoptosis, and metastasis. By using systems biology modeling techniques, it will be possible to understand each of these processes and how they interact to drive tumor progression.

2.9 Outlook

Genomic, transcriptomic, proteomic, and metabolomic studies have significantly advanced our understanding of carcinogenesis. Despite an impressive progress in the development of methods in all omics fields, substantial hurdles remain. Challenges especially are to bring omics technologies to clinical applications and to use them in risk assessment strategies. Mainly current proteomics technologies are too slow, too complex, and too expensive to be used in a clinical laboratory and, in general, the existence of many different experimental approaches leaves a deficit in standardization. However, great progress is to be expected from further developed array-based methods, optical methods, and microengineering approaches. For instance, the increasing use of advanced bioinformatics and systems biology tools has led to the identification of cancer-associated phosphorylation networks [41] and a further stream of data can be expected to understand the regulatory interplay between individual molecules and in networks. In addition, these investigations may bridge the gap between mechanistic understanding and mainly phenomenological markers (see Chapter 14).

Further development of more cost-effective high-throughput technologies able to comprehensively assess DNA, RNA, protein, and metabolites will visualize the interconnected events within a cell that determine how inputs from the environment and the network rewiring influences cell behavior. An interdisciplinary systems biology effort integrating engineering, physics, and mathematical approaches with biological and medical insight into an iterative process will be necessary to convert the information contained in multidimensional data not only to classify tumors and advance clinical therapies but also to get better prognosis concerning the carcinogenic power of unknown agents.

References

1 Renan, M.J. (1993) How many mutations are required for tumorigenesis? Implications from human cancer data. *Mol. Carcinog.*, **7**, 139–146.

2 Kinzler, K.W. and Vogelstein, B. (1996) Lessons from hereditary colorectal cancer. *Cell*, **87**, 159–170.

3 Karakosta, A., Golias, C., Charalabopoulos, A., Peschos, D., Batistatou, A., and Charalabopoulos, K. (2005) Genetic models of human cancer as a multistep process. Paradigm models of colorectal cancer, breast cancer, and chronic myelogenous and acute lymphoblastic leukaemia. *J. Exp. Clin. Cancer Res.*, **24** (4), 505–514.

4 Hanahan, D. and Weinberg, R.A. (2000) The hallmarks of cancer. *Cell*, **100**, 57–70.

5 Kroemer, G. and Pouyssegur, J. (2008) Tumor cell metabolism: cancer's Achilles' heel. *Cancer Cell*, **13**, 472–482.

6 Ikushima, H. and Miyazono, K. (2010) TGFbeta signalling: a complex web in cancer progression. *Nat. Rev. Cancer*, **10** (6), 415–424.

7 Pasquale, E.B. (2010) Eph receptors and ephrins in cancer: bidirectional signalling and beyond. *Nat. Rev. Cancer*, **10** (3), 165–180.

8 Turner, N. and Grose, R. (2010) Fibroblast growth factor signalling: from development to cancer. *Nat. Rev. Cancer*, **10** (2), 116–129.

9 Saif, M.W. (2010) Colorectal cancer in review: the role of the EGFR pathway. *Expert Opin. Investig. Drugs*, **19** (3), 357–369.

10 Stupack, D.G. (2007) The biology of integrins. *Oncology (Williston Park, NY)*, **21** (9 Suppl. 3), 6–12.

11 Poznic, M. (2009) Retinoblastoma protein: a central processing unit. *J. Biosci.*, **34** (2), 305–312.

12 Taketo, M.M. and Edelmann, W. (2009) Mouse models of colon cancer. *Gastroenterology*, **136** (3), 780–798.

13 Zhivotovsky, B. and Orrenius, S. (2010) Cell death mechanisms: cross-talk and role in disease. *Exp. Cell Res.*, **316** (8), 1374–1383.

14 Amaral, J.D., Xavier, J.M., Steer, C.J., and Rodrigues, C.M. (2010) The role of p53 in apoptosis. *Discov. Med.*, **9** (45), 145–152.

15 Cesare, A.J. and Reddel, R.R. (2010) Alternative lengthening of telomeres: models, mechanisms and implications. *Nat. Rev. Genet.*, **11** (5), 319–330.

16 Korc, M. and Friesel, R.E. (2009) The role of fibroblast growth factors in tumor growth. *Curr. Cancer Drug Targets*, **9** (5), 639–651.

17 Warburg, O., Posener, K., and Negelein, E. (1924) Über den Stoffwechsel der Tumoren. *Biochem. Z.*, **152**, 319–344.

18 Semenza, G.L. (2010) HIF-1: upstream and downstream of cancer metabolism. *Curr. Opin. Genet. Dev.*, **20** (1), 51–56.

19 Makrilia, N., Kollias, A., Manolopoulos, L., and Syrigos, K. (2009) Cell adhesion molecules: role and clinical significance in cancer. *Cancer Invest.*, **27** (10), 1023–1037.

20 Zecchini, S. and Cavallaro, U. (2010) Neural cell adhesion molecule in cancer: expression and mechanisms. *Adv. Exp. Med. Biol.*, **663**, 319–333.

21 Graham, T.A. and McDonald, S.A. (2010) Genetic diversity during the development of Barrett's oesophagus-associated adenocarcinoma: how, when and why? *Biochem. Soc. Trans.*, **38** (2), 374–379.

22 Veeck, J. and Esteller, M. (2010) Breast cancer epigenetics: from DNA methylation to microRNAs. *J. Mammary Gland Biol. Neoplasia*, **15** (1), 5–17.

23 Esteller, M. (2008) Epigenetics in cancer. *N. Engl. J. Med.*, **358** (11), 1148–1159.

24 Jandrig, B., Seitz, S., Hinzmann, B., Arnold, W., Micheel, B., Koelble, K., Siebert, R., Schwartz, A., Ruecker, K., Schlag, P.M., Scherneck, S., and Rosenthal, A. (2004) ST18 is a breast cancer tumor suppressor gene at human chromosome 8q11.2. *Oncogene*, **23** (57), 9295–9302.

25 Richon, V.M., Sandhoff, T.W., Rifkind, R.A., and Marks, P.A. (2000) Histone deacetylase inhibitor selectively induces p21WAF1 expression and gene-associated histone acetylation. *Proc. Natl. Acad. Sci. USA*, **97** (18), 10014–10019.

26 Jaenisch, R. and Bird, A. (2003) Epigenetic regulation of gene expression: how the genome integrates intrinsic and environmental signals. *Nat. Genet.*, **33** (Suppl.), 245–254.

27 Boyerinas, B., Park, S.M., Hau, A., Murmann, A.E., and Peter, M.E. (2010) The role of let-7 in cell differentiation and cancer. *Endocr. Relat. Cancer*, **17** (1), F19–F36.

28 Iorio, M.V. and Croce, C.M. (2009) MicroRNAs in cancer: small molecules with a huge impact. *J. Clin. Oncol.*, **27** (34), 5848–5856.

29 Quintana, E., Shackleton, M., Sabel, M.S., Fullen, D.R., Johnson, T.M., and Morrison, S.J. (2008) Efficient tumour formation by single human melanoma cells. *Nature*, **456** (7222), 593–598.

30 Santisteban, M., Reiman, J.M., Asiedu, M.K., Behrens, M.D., Nassar, A., Kalli, K.R., Haluska, P., Ingle, J.N., Hartmann, L.C., Manjili, M.H., Radisky, D.C., Ferrone, S., and Knutson, K.L. (2009) Immune-induced epithelial to mesenchymal transition *in vivo* generates breast cancer stem cells. *Cancer Res.*, **69** (7), 2887–2895.

31 Gravdal, K., Halvorsen, O.J., Haukaas, S.A., and Akslen, L.A. (2007) A switch from E-cadherin to N-cadherin expression indicates epithelial to mesenchymal transition and is of strong and independent importance for the progress of prostate cancer. *Clin. Cancer Res.*, **13** (23), 7003–7011.

32 Finger, E.C. and Giaccia, A.J. (2010) Hypoxia, inflammation, and the tumor microenvironment in metastatic disease. *Cancer Metastasis Rev.*, **29** (2), 285–293.

33 Wu, Y. and Zhou, B.P. (2010) TNF-alpha/NF-kappaB/Snail pathway in cancer cell migration and invasion. *Br. J. Cancer*, **102** (4), 639–644.

34 Andreu, P., Johansson, M., Affara, N.I., Pucci, F., Tan, T., Junankar, S., Korets, L., Lam, J., Tawfik, D., DeNardo, D.G., Naldini, L., de Visser, K.E., De Palma, M., and Coussens, L.M. (2010) FcRgamma activation regulates inflammation-associated squamous carcinogenesis. *Cancer Cell*, **17** (2), 121–134.

35 Gieseg, M.A., Man, M.Z., Gorski, N.A., Madore, S.J., Kaldjian, E.P., and Leopold, W.R. (2004) The influence of tumor size and environment on gene expression in commonly used human tumor lines. *BMC Cancer*, **4**, 35.

36 Holbeck, S., Chang, J., Best, A.M., Bookout, A.L., Mangelsdorf, D.J., and Martinez, E.D. (2010) Expression profiling of nuclear receptors in the NCI60 cancer cell panel reveals receptor–drug and receptor–gene interactions. *Mol. Endocrinol.*, **24** (6), 1287–1296.

37 Weinberg, R.A. (2008) Mechanisms of malignant progression. *Carcinogenesis*, **29** (6), 1092–1095.

38 Quigley, D. and Balmain, A. (2009) Systems genetics analysis of cancer susceptibility: from mouse models to humans. *Nat. Rev. Genet.*, **10** (9), 651–657.

39 Schoenfelder, S., Clay, I., and Fraser, P. (2010) The transcriptional interactome: gene expression in 3D. *Curr. Opin. Genet. Dev.*, **20** (2), 127–133.

40 Kolch, W. and Pitt, A. (2010) Functional proteomics to dissect tyrosine kinase signalling pathways in cancer. *Nat. Rev. Cancer*, **10** (9), 618–629.

41 Tan, C.S. and Linding, R. (2009) Experimental and computational tools useful for (re)construction of dynamic kinase–substrate networks. *Proteomics*, **9** (23), 5233–5242.

3
Endogenous DNA Damage and Its Relevance for the Initiation of Carcinogenesis

Bernd Epe and Markus Fußer

3.1
Introduction

From the chemical point of view, DNA is only a moderately stable molecule. It is prone to spontaneous hydrolysis and can easily react with intermediates generated in the regular cellular metabolism, in particular electrophiles and reactive oxygen species (ROS). Therefore, various types of DNA modifications ("DNA damage") are continuously formed in all types of cells even under normal (physiological) conditions [1]. These lesions give rise to "spontaneous" mutations during DNA replication and thus contribute to the initiation of both carcinogenesis and other age-related diseases and the aging process itself [2]. The actual relevance of the endogenous DNA damage in comparison to the exogenous damage caused by various environmental agents (ionizing and UV radiation, chemical carcinogens, etc.) for cancer formation in humans is not yet known (Figure 3.1). However, there appears to be a discrepancy between the high cancer incidence in the human population on the one hand and the relatively low average concentrations and estimated mutagenic potencies of the known environmental carcinogens (except tobacco smoke) on the other hand [3, 4].

The generation of endogenous DNA modifications in the cells is attenuated by the presence of various types of protective molecules (radical scavengers, nucleophiles, and antioxidative enzymes). In addition, the adverse consequences of endogenous damage are counteracted by specific DNA repair mechanisms, which have evolved for apparently all important and sufficiently frequent types of endogenous DNA modification and which supplement the more general and unspecific repair mechanisms such as nucleotide excision repair, homologous recombination, and mismatch repair. The equilibrium between the continuous DNA damage generation and repair results in steady-state levels of the endogenous DNA modifications, which have been detected in many cell types and organisms [5–9]. The situation is illustrated in Figure 3.2. It is evident that the spontaneous mutation rates (and all their consequences, in particular the initiation of carcinogenesis) should depend on the ratio of the rates of DNA damage generation and repair.

Cancer Risk Evaluation: Methods and Trends,
Edited by Günter Obe, Burkhard Jandrig, Gary E. Marchant, Holger Schütz, and Peter M. Wiedemann.

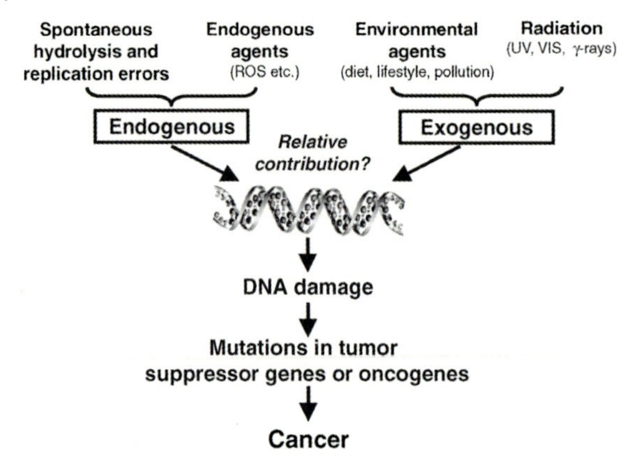

Figure 3.1 Sources of DNA damage that contribute to the initiation of carcinogenesis.

In this chapter, our knowledge about the various determinants of the model outlined in Figure 3.2 (generation, repair, and steady-state levels of endogenous DNA modifications) will be summarized and some experimental approaches to assess the relevance of the spontaneous DNA damage for the cancer risk will be described.

3.2
Types and Generation of Oxidative DNA Modifications

Endogenous DNA modifications result from three different mechanisms, namely, the spontaneous hydrolysis of DNA, the reaction of DNA with endogenous metabolites, and the misincorporation by DNA polymerases during replication.

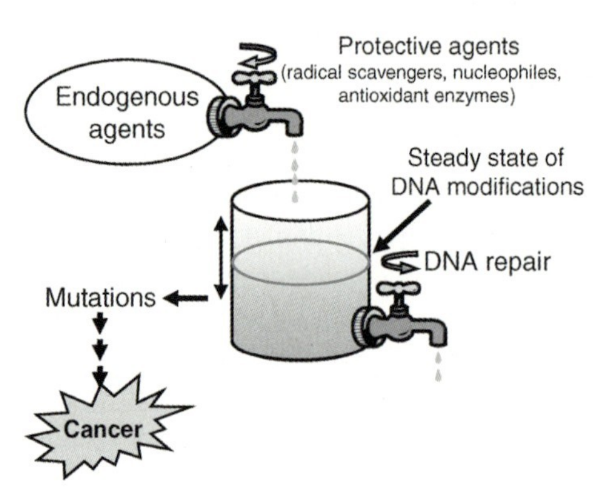

Figure 3.2 Basal (steady-state) levels of endogenous DNA modifications resulting from the balance between generation and repair determine the spontaneous mutation rates.

3.2.1
Spontaneous Hydrolysis Products in DNA

An important example of spontaneous hydrolysis is the formation of uracil from cytosine residues. It has been estimated that in a mammalian cell 100–500 cytosine residues are deaminated per day [1, 10]. Deamination of 5-methylcytosine, which is an epigenetic marker of silenced genes, is twofold more rapid [11] and probably more dangerous since the product in this case is the natural DNA base thymine. Saturation of the C5=C6 double bond as observed in UV-induced pyrimidine dimers also facilitates deamination [12–14]. Even more frequent than cytosine deamination is the hydrolysis of the glycosidic bonds in DNA, which gives rise to AP sites (apurinic and apyrimidinic sites). It has been calculated that more than 10 000 bases are lost from the genome of a mammalian cell per day [1]. Most likely, the spontaneous formation of AP sites is drastically enhanced when the cellular pH is decreased, for example, under hypoxic conditions in tumors. In contrast, the spontaneous hydrolysis of the sugar–phosphate backbone of the DNA, which would cause single-strand breaks (SSBs), is very slow and does not play a role under physiologically relevant conditions.

3.2.2
Oxidation Products in DNA

Among the reactive endogenous metabolites able to modify DNA, reactive oxygen species are often regarded most important. The chemical nature and source of the species that ultimately react with the DNA and contribute mostly to the basal levels of oxidative DNA modifications observed in mammalian cells (see below) are not known. Although the vast majority of the cellular ROS are generated in the mitochondria (approximately 0.2% of the total oxygen consumed is estimated to be converted into superoxide [15]), it is not clear whether ROS produced in the mitochondria can reach the nuclear DNA to a significant extent [16]. The participation in nuclear damage formation of cytosolic or nuclear oxidases, peroxides, and transition metals at the DNA (which potentially could generate ROS in a Fenton-type reaction) is also not clear.

The generation of ROS in cells and tissues is assumed to be highly variable. It is (per definition) increased under conditions of "oxidative stress" such as inflammation and reperfusion. Since the activity of the antioxidant defense system can also be modulated by environmental parameters such as the composition of the diet, the basal (steady-state) levels of oxidative DNA modifications are expected to strongly depend on exogenous factors, despite their endogenous nature.

There is increasing evidence that ROS have physiological functions as signaling molecules in various cellular pathways, for example, in the regulation of tyrosine kinase activities and apoptosis [17]. This means that DNA damage, mutagenesis, and tumor initiation by ROS is intimately linked to effects relevant for tumor promotion. It also explains why the cellular antioxidant defense system, which protects from oxidative DNA damage, is flexible and controlled by transcription factors such as Nrf2, which are associated with sensors of the cellular redox state [18].

Guanine is the DNA base most susceptible to oxidation by ROS [19], and 8-hydroxyguanine, which under physiological conditions exists nearly exclusively in its tautomeric form 8-oxo-7,8-dihydroguanine (8-oxoG) [20], appears to be the most frequent oxidative modification at least when DNA is exposed to mild (and therefore selective) oxidants [21]. Importantly, 8-oxoG can form stable base pairs with both adenine and cytosine, which explains its rather high mutagenic potential. For instance, when a single-stranded vector containing a single 8-oxoG residue was transfected into mammalian cells, G-to-T transversions were observed at a frequency of 0.3–1.8% [22, 23].

Besides 8-oxoG, many other oxidative DNA modifications have been characterized [24]. These include other base modifications, sites of base loss, and strand breaks of different types. An endogenous generation under physiological growth conditions has been demonstrated for only few of them, in part, because of analytical limitations (see below). Of particular relevance to the biological consequences might be clustered lesions, the yield of which has recently been shown to be unexpectedly high for the reaction of DNA with hydroxyl radicals [25].

3.2.3 Alkylation Products in DNA

One of the most important endogenous metabolites that can alkylate DNA is *S*-adenosylmethionine [26]. It not only serves as a cofactor for the physiological 5-methylation of cytosine residues but could also contribute to a detrimental generation of 7-methylguanine, O^6-methylguanine, 3-methyladenine, and other lesions [27]. Alkylation products resulting from unsaturated aldehydes such as malondialdehyde, acrolein, and 4-hydroxynonenal, which are generated during lipid peroxidation, constitute another class of DNA lesions [28]. They can be regarded as indirect products emanating from ROS formation. Characteristic endogenous modifications in DNA are etheno and propano adducts at guanine and adenine residues, which are known to be highly mutagenic [29].

3.2.4 DNA Modifications Resulting from DNA Replication and Repair

DNA polymerases easily incorporate dUTP instead of dTTP. The frequency of this misincorporation appears to be high, as concluded from the accumulation of uridine in the DNA of replicating – but not in that of arrested – fibroblasts, in which the repair by uracil glycosylase is knocked out [30, 31]. The relevance for genetic instability and carcinogenesis might be limited, however, since uracil is paired perfectly with adenine.

Misincorporation of 8-oxodGTP, which is generated in the cellular nucleotide pool from dGTP most probably by ROS, also takes place and is an endogenous source of A:T-to-C:G transversion mutations [32].

DNA double-strand breaks (DSBs) may easily arise from SSBs and possibly other lesions during replication (collapse of the replication fork) and then cause chromosomal aberrations and micronuclei formation [33].

During excision repair, sites of base loss and SSBs are regular intermediates (see below), which under certain conditions ("imbalanced repair") might cause mutations and other processing problems [34].

3.3 Repair of Endogenous DNA Modifications

3.3.1 Basic Mechanisms

To cope with the continuous generation of DNA modifications described above, evolution has developed efficient repair mechanisms, which, in contrast to the mechanisms evolved for exogenous DNA damage, are often specifically adapted to the substrate modifications. The most important pathway is the base excision repair (BER) [35–38]. It is initiated by substrate-specific or -selective repair glycosylases, which recognize the endogenous base modifications and catalyze the cleavage of the glycosidic bond, thereby generating sites of base loss (AP sites). In human cells, 11 different repair glycosylases have been detected by now, which are structural and functional homologues of similar enzymes found in apparently all types of organisms [37]. The AP site then is incised on its 5′ site by an AP endonuclease (APE1 in human cells). Subsequently, DNA polymerase beta adds the correct nucleotide to the free 3′ end generated by the AP endonuclease and at the same time removes the 5′-terminal deoxyribophosphate that is left on the other site of the nick. The final resealing is carried out by ligase III. The steps described are coordinated by a platform protein called XRCC1. The repair of endogenously generated AP sites proceeds by the same mechanism, except that the initial glycosylase activity is not needed. In some cases, variations of the pathway [39, 40] are possible. Thus, a displacement synthesis takes place if the 5′-deoxyribophosphate cannot be removed immediately (long-patch repair); it requires additional proteins such as FEN1. In addition, some repair glycosylases have an inherent AP lyase activity, which generates a nick on the 3′site of the AP site and therefore requires slightly different processing.

The known human repair glycosylases recognize and remove uracil residues (UNG, SMUG1, TDG, and MBD4), several oxidized pyrimidines and ring-opened purines (NTHL1), certain oxidized and ring-opened guanines (OGG1), and some alkylation products (MPG) [37]. The "endonuclease VIII-like glycosylases" NEIL1-3 appear to recognize several oxidized pyrimidines and purines preferentially in single-stranded DNA, that is, in the context of transcription and/or replication [41, 42]. Another glycosylase, MUTYH, specifically removes adenine residues from its mismatch with 8-oxoG, allowing to prevent the impending G:C-to-T:A transversion mutation even after the misincorporation of adenine has already occurred [35, 43]. (OGG1 does not remove 8-oxoG from a mismatch with adenine since that would directly lead to the fixation of a mutation.)

Another repair pathway dealing with endogenous modifications is direct damage reversion. Although this might be regarded as the most straightforward way to eliminate a DNA modification, it takes place only in a few cases. The demethylation of O^6-methylguanine by methylguanine methyltransferase (MGMT) is a well-known example [27]. Surprisingly, MGMT is not an enzyme, but a suicide protein because after reaction with the DNA modification it is not recycled but degrades. Only a couple of years ago, the damage reversal by another class of repair enzymes was characterized, which are homologues of the bacterial AlkB protein [44]. These enzymes are dioxygenases that carry out an oxidative dealkylation in a Fe(II)-dependent reaction with 2-oxoglutarate as a cofactor to revert lesions such as N^1-methyladenine, N^3-methylcytosine, and etheno adducts [45]. In human cells, ABH2 deals with these alkylation products in double-stranded DNA, while ABH3 prefers single-stranded nucleic acids [46]. Several other mammalian homologues of AlkB have been cloned and still await further characterization of their biological function.

In addition to the basic steps described, the repair of cellular lesions most probably requires various accompanying processes, such as chromatin remodeling, relocalization of repair proteins, and activation of signaling factors. In particular for BER, the underlying mechanisms are poorly understood, although their relevance has been demonstrated in several studies [38, 47, 48].

3.3.2
Analysis of Repair Activities

It is generally assumed that the rate-limiting step of BER is the recognition and incision by the repair glycosylase. The cellular repair capacity of one of the substrate modifications therefore can often be estimated in simple assays, in which crude cell extracts are incubated with a limiting amount of a synthetic oligonucleotide, which carries a single defined lesion (base modification, AP sites) in a central position and a label, mostly ^{32}P, at a terminus. The cleavage activity in the cell extract can then be determined from the amount of the oligonucleotide fragment of the correct size generated in a given time. The assay can be carried out with extracts from human lymphocytes and has been used in population studies to examine associations between repair capacities and cancer risk ([49] and see below).

3.4
Basal Steady-State Levels

According to the model shown in Figure 3.2, the induction of spontaneous mutations and the initiation of carcinogenesis by endogenous DNA damage should depend on (i) the steady-state levels of the lesions in the cellular DNA that result from the continuous generation and repair and (ii) the intrinsic mutagenic potencies of the lesions. The latter can be estimated, for example, from transfection experiments with DNA carrying defined lesions and, as already mentioned, appears to be quite high in many cases such as O^6-methylguanine [50], etheno adducts [51], and 8-oxoG [23].

Regrettably, the quantitation of the steady-state levels of endogenous DNA modifications is experimentally rather difficult, and many conflicting data have been reported in the literature. In addition to the analytical challenge to detect a few DNA base modifications in 10^7–10^9 normal bases, artifactual oxidation during DNA isolation and its subsequent hydrolysis have been recognized as a major source of error in the case of oxidative base modifications such as 8-oxoG, resulting in overestimations. Thus, oxidation of normal bases by radiolysis products was identified as a major problem in the application of the ^{32}P-postlabeling technique [52], and the derivatization step required for an analysis by GC–MS can also be a major source of error [53]. The quantitation of 8-oxoG by HPLC with electrochemical detection (HPLC-ECD) introduced by Floyd [54] is widely used as a robust and sensitive technique. Again, however, spurious oxidation turned out to be a problem, which was addressed by modified isolation protocols, resulting in lower levels of the detected lesions [55, 56]. Another powerful quantitation method for 8-oxoG and various other lesions is HPLC–MS/MS [57, 58]. Again, extraction and isolation procedures were shown to deserve special attention [58].

Problems with spurious oxidation are much less severe in the so-called enzymatic assays, in which repair glycosylases (see above) are used to convert the substrate modifications into SSBs, which then can be quantified in cellular (nuclear) DNA by techniques such as alkaline elution [59], comet assay [60], or DNA unwinding [61]. The sensitivity of these methods is often higher than that of the nonenzymatic methods [55], and lesions such as AP sites and SSBs can be quantified in addition (and in parallel) to DNA base modifications such as cyclobutane pyrimidine dimers, uracil, or oxidative purine modifications. Major constrains result from the limited substrate specificity of some of the repair enzymes that are used as probes (see Ref. [59] for a review) and the need for calibration by other techniques. Clustered lesions, which were recently shown to be generated by ionizing radiation in relatively high yields [62], could be counted with the enzymatic methods as single lesions.

DSBs can be very sensitively quantified in mammalian cells by means of antibodies against γH2AX and 53BP1. The number of foci detected microscopically probably represents DSBs that not only are directly generated by a damaging agent but could also arise as secondary lesions at (collapsed) replication forks [63].

In Table 3.1, steady-state levels of various types of endogenously generated lesion in mammalian cells are listed for comparison. These and similar data listed in previous reviews [6, 7] indicate that the steady-state levels of SSBs and AP sites in mammalian cells are close to the limit of detection by alkaline elution (<0.05 sites per 10^6 bp; equivalent to 300 lesions per diploid genome) [31]. Uracil levels are similarly low in particular in proliferating cells, but appear to be slightly higher in nonproliferating tissues [30, 31]. Etheno adducts and other DNA modifications that result from lipid peroxides were detected at highly variable levels, which might be explained by dietary differences (intake of unsaturated fatty acids) and other exogenous factors [64]. Relatively high endogenous damage levels were reported for some alkylation products such as 7-methylguanine and oxidative modifications, in particular 8-oxoG [6, 65], which, however, might be partly caused by artifactual oxidation, as discussed above.

Table 3.1 Basal levels of DNA modifications in mammalian cells.

DNA modification	Method	Cell type (species)	Lesions per 10^6 bp	References
8-OxoG/8-oxodG/Fpg	Comet/Fpg	Lymphocytes (human)	0.15	[78]
	GC–MS	Lymphocytes (human)	120	[79]
	HPLC–EC	Lymphocytes (human)	4.3	[78]
	HPLC–EC	Leukocytes (human)	5.4 ± 3	[80]
	HPLC–EC	PBMC (human)	2.5 ± 3	[81]
	HPLC–MS/MS	Monocyte cell line (human)	1.2 ± 0.8	[82]
	LC–MRM/MS	Bronchoalveolar cells (human)	0.4	[58]
	UPLC—MS/MS	Lymphocytes (human)	1.6	[57]
	AE/Fpg	Lymphocytes (human)	0.24 ± 0.03	[5]
	Comet/Fpg	HeLa cells (human)	0.05–2	[83, 84]
	AE/Fpg	HeLa cells (human)	0.25–0.4	[83, 84]
	AU/Fpg	HeLa cells (human)	0.45–0.6	[83, 84]
	HPLC–EC	HeLa cells (human)	0.4–4.5	[83, 84]
	HPLC–MS	HeLa cells (human)	2.5–45	[84]
	GC–MS	HeLa cells (human)	1.5–150	[83, 84]
8-OxoA/8-oxodA	HPLC–MS/MS	Monocyte cell line (human)	0.4 ± 0.4	[82]
FapyGua	GC–MS	Leukocytes (human)	1.2–2.2	[85]
	HPLC–MS/MS	Monocyte cell line (human)	12.8 ± 11.6	[82]
Thymine glycol	HPLC–MS/MS	Monocyte cell line (human)	1.6 ± 2.4	[82]
Ethenoadenine	LC–MS	Placental DNA (human)	0.011	[86]
	IPPA	Liver (human)	0.003–0.015	[87]
	IPPA	Leukocytes (human)	0.002–4.5	[64]
Ethenocytosine	IPPA	Liver (human)	0.001–0.12	[87]
	IPPA	Leukocytes (human)	0.005–3.5	[64]
AP sites	ARP	Calf thymus DNA	200–1000	[88]
	ARP	Liver	18–20	[89]
	AE	CHO cell line	<0.05	[5]
Double-strand breaks	γH2AX foci	Melanocytes (human)	0.4/cell	[90]

bp: base pair; 8-oxoG: 8-oxo-7,8-dihydroguanine; Fpg: Fpg-sensitive lesions; HPLC: high-performance liquid chromatography; UPLC: ultrahigh-performance liquid chromatography; IPPA: immunoaffinity/^{32}P-postlabeling; EC: electrochemical detection; GC: gas chromatography; MS: mass spectrometry; MRM: multiple reaction monitoring; AE: alkaline elution; ARP: aldehyde reactive probe; PBMC: peripheral blood mononuclear cells.

It is interesting to note that oxidative DNA base modifications (both free bases and nucleosides) are also present in urine [66]. The origin of these modifications is still a matter of debate; not only DNA repair but also oxidation within the cellular nucleotide pool and (oxidative) degradation of DNA or nucleosides from dying or dead cells in the organism have been taken into consideration (see Ref. [67] for a review). It appears clear, however, that the levels of these modifications in urine do not reflect the repair capacity or the steady-state levels in the DNA, but rather damage generation, according to the principle that under steady-state conditions "what comes in, comes out."

3.5 Contribution of Endogenous DNA Modifications to Cancer Risk

The knowledge of the contribution of endogenous DNA modifications of different types relative to the contribution of exogenous (environmental) DNA modifications is of obvious interest for cancer prevention. Conclusive experimental approaches to address this question, however, are rare. An important point is that DNA damage (and the subsequent generation of mutations) only represents the initiation step of carcinogenesis, and it appears possible that the later steps (referred to as tumor promotion) may be rate limiting under many conditions.

According to the model shown in Figure 3.2, it would make sense to modulate the steady-state level of a given type of endogenous DNA modification – by influencing either the generation or the repair of the lesion – and correlate the increase or decrease of the steady-state level with the resulting effect on the overall spontaneous mutation frequency or the spontaneous cancer incidence. Interventions with the damage generation, however, appear less promising since influences on other parameters relevant for mutagenesis and tumor development (cell proliferation, cytotoxicity, and signal transduction) are difficult to avoid.

So what can we learn from repair deficiencies? To which extent does a given increase in the basal steady-state levels (Table 3.1) translate into a higher cancer incidence? In *Ogg1*$^{-/-}$ mice, for example, the repair of oxidative purine modifications including 8-oxoG is significantly retarded and the steady-state levels of these lesions are elevated in various organs [68], in accordance with the model shown in Figure 3.2. The overall spontaneous mutation frequencies are also increased several fold, and most of the accumulated mutations are G:C-to-T:A transversions, in agreement with the expectation. The quantitative data suggest that an additional burden of less than one 8-oxoG residue per 10^6 bp in the genome already doubles the spontaneous mutation rate in a nontranscribed (transgenic) locus [68, 69]. Interestingly, the spontaneous cancer incidence in the *Ogg1*$^{-/-}$ mice is not significantly increased, although the absolute number of animals that has been studied is probably too small to detect small differences. This "negative" observation might be an indication of the rate-limiting character of tumor promotion, in particular because the knockout animals develop a higher number of preneoplastic foci than wild-type animals when treated with a peroxysome proliferator, which stimulates cell proliferation in liver and thus acts as a tumor promoter [70]. Increased spontaneous tumor development is also observed in *Ogg1*$^{-/-}$*Mutyh*$^{-/-}$ double-knockout mice, in which the repair deficiency of 8-oxoG is more pronounced [71].

While the results described clearly demonstrate the mutagenic and carcinogenic potency of elevated levels of endogenously generated 8-oxoG, a comparison with the levels of DNA modifications generated by exogenous carcinogens and the resulting cancer frequencies remains difficult since the exposure times and times for tumor development generally strongly differ.

Results with mice deficient in other repair glycosylases support the notion that a moderate increase in the levels of endogenous base damage does not directly result into a high spontaneous cancer incidence [72, 73], possibly because tumor promotion is absent. In addition, the actual increase in the steady-state levels of the substrate modifications has not always been analyzed and may often be low because of the existence of efficient backup repair mechanisms. Interestingly, mice deficient in proteins involved in later (and common) steps of BER, for example, APE1, XRCC1, or ligase III, are not viable [74]. It may be either concluded that the total absence of BER is lethal or that an accumulation of repair intermediates might represent a bigger problem for the cells than elevated levels of endogenous base damage.

Observations regarding repair deficiencies in human cells appear to support the conclusions from the animal data. Thus, mutations in repair glycosylases have rarely been detected in human tumors. One of the exceptions is MUTYH, the mismatch repair glycosylase for 8-oxoG, which was found to have mutated in some cases of hereditary colon cancer [75]. Various polymorphisms have been observed in human genes involved in base excision repair. Some of the variants have been linked to an increased cancer risk in population studies [76]. For example, a Ser326Cys variant of human OGG1, the allelic frequency of which is in the range of 20–40%, was reported to be associated with a higher incidence of lung, gastric, prostate, and orolaryngeal cancer [77]. However, the effects so far cannot be correlated with a corresponding increase in the steady-state levels of DNA modifications or spontaneous mutation rates.

In another approach to assess the relevance of endogenous oxidative base damage for cancer, the repair capacity in extracts of human lymphocytes was determined by means of the cleavage assay described above and compared between cancer patients and matched controls. In lung cancer patients, the average repair activity for 8-oxoG was found to be lower than that in nonaffected persons [49], and it was hypothesized that the reduced repair capacity (and the anticipated – but not demonstrated – elevated levels of 8-oxoG) is a causal factor for the development of the tumors.

In summary, a quantitative assessment of the contribution of endogenous DNA modifications to the cancer risk (and a comparison with the risk resulting from DNA damage caused by exogenous agents) cannot be given. Not only does this have partly experimental reasons (the lack of exact data that could link basal DNA damage levels and cancer incidence) but this is also a consequence of the fact that some types of endogenous DNA damage, in particular those resulting from ROS, can be strongly influenced by exogenous factors such as inflammatory diseases and dietary constituents. Qualitatively, however, the increases in spontaneous mutation rates and cancer incidences observed in repair-deficient animals demonstrate a high hazard imposed by endogenously generated DNA modifications.

References

1 Lindahl, T. (1993) Instability and decay of the primary structure of DNA. *Nature*, **362**, 709–715.

2 Maynard, S., Schurman, S.H., Harboe, C., de Souza-Pinto, N.C., and Bohr, V.A. (2009) Base excision repair of oxidative DNA damage and association with cancer and aging. *Carcinogenesis*, **30**, 2–10.

3 Lichtenstein, P., Holm, N.V., Verkasalo, P.K., Iliadou, A., Kaprio, J., Koskenvuo, M., Pukkala, E., Skytthe, A., and Hemminki, K. (2000) Environmental and heritable factors in the causation of cancer: analyses of cohorts of twins from Sweden, Denmark, and Finland. *N. Engl. J. Med.*, **343**, 78–85.

4 Ames, B.N. and Gold, L.S. (2000) Paracelsus to parascience: the environmental cancer distraction. *Mutat. Res.*, **447**, 3–13.

5 Pflaum, M., Will, O., and Epe, B. (1997) Determination of steady-state levels of oxidative DNA base modifications in mammalian cells by means of repair endonucleases. *Carcinogenesis*, **18**, 2225–2231.

6 De Bont, R. and van Larebeke, N. (2004) Endogenous DNA damage in humans: a review of quantitative data. *Mutagenesis*, **19**, 169–185.

7 Azqueta, A., Shaposhnikov, S., and Collins, A.R. (2009) DNA oxidation: investigating its key role in environmental mutagenesis with the comet assay. *Mutat. Res./Genet. Toxicol. Environ. Mutat.*, **674**, 101–108.

8 Collins, A.R. (2009) Investigating oxidative DNA damage and its repair using the comet assay. *Mutat. Res./Rev. Mutat. Res.*, **681**, 24–32.

9 van Loon, B., Markkanen, E., and Hübscher, U. (2010) Oxygen as a friend and enemy: how to combat the mutational potential of 8-oxo-guanine. *DNA Repair*, **9**, 604–616.

10 Frederico, L.A., Kunkel, T.A., and Shaw, B.R. (1990) A sensitive genetic assay for the detection of cytosine deamination: determination of rate constants and the activation energy. *Biochemistry*, **29**, 2532–2537.

11 Shen, J.C., Rideout, W.M., and Jones, P.A. (1994) The rate of hydrolytic deamination of 5-methylcytosine in double-stranded DNA. *Nucleic Acids Res.*, **22**, 972–976.

12 Barak, Y., Cohenfix, O., and Livneh, Z. (1995) Deamination of cytosine-containing pyrimidine photodimers in UV-irradiated DNA: significance for UV light mutagenesis. *J. Biol. Chem.*, **270**, 24174–24179.

13 Tu, Y.Q., Dammann, R., and Pfeifer, G.P. (1998) Sequence and time-dependent deamination of cytosine bases in UVB-induced cyclobutane pyrimidine dimers *in vivo*. *J. Mol. Biol.*, **284**, 297–311.

14 Cannistraro, V.J. and Taylor, J.S. (2009) Acceleration of 5-methylcytosine deamination in cyclobutane dimers by G and its implications for UV-induced C-to-T mutation hotspots. *J. Mol. Biol.*, **392**, 1145–1157.

15 St-Pierre, J., Buckingham, J.A., Roebuck, S.J., and Brand, M.D. (2002) Topology of superoxide production from different sites in the mitochondrial electron transport chain. *J. Biol. Chem.*, **277**, 44784–44790.

16 Hoffmann, S., Spitkovsky, D., Radicella, J.P., Epe, B., and Wiesner, R.J. (2004) Reactive oxygen species derived from the mitochondrial respiratory chain are not responsible for the basal levels of oxidative base modifications observed in nuclear DNA of mammalian cells. *Free Radic. Biol. Med.*, **36**, 765–773.

17 Poli, G., Leonarduzzi, G., Biasi, F., and Chiarpotto, E. (2004) Oxidative stress and cell signalling. *Curr. Med. Chem.*, **11**, 1163–1182.

18 Kensler, T.W., Wakabayashi, N., and Biswal, S. (2007) Cell survival responses to environmental stresses via the Keap1-Nrf2-ARE pathway. *Annu. Rev. Pharmacol. Toxicol.*, **47**, 89–116.

19 Steenken, S. and Jovanovic, S.V. (1997) How easily oxidizable is DNA? One-electron reduction potentials of adenosine and guanosine radicals in aqueous solution. *J. Am. Chem. Soc.*, **119**, 617–618.

20 Culp, S.J., Cho, B.P., Kadlubar, F.F., and Evans, F.E. (1989) Structural and

conformational-analyses of 8-hydroxy-2′-deoxyguanosine. *Chem. Res. Toxicol.*, **2**, 416–422.

21 Epe, B. (1995) DNA damage profiles induced by oxidising agents. *Rev. Physiol. Biochem. Pharmacol.*, **127**, 223–249.

22 Wood, M.L., Esteve, A., Morningstar, M.L., Kuziemko, G.M., and Essigmann, J.M. (1992) Genetic effects of oxidative DNA damage: comparative mutagenesis of 7,8-dihydro-8-oxoguanine and 7,8-dihydro-8-oxoadenine in *Escherichia coli*. *Nucleic Acids Res.*, **20**, 6023–6032.

23 Moriya, M. (1993) Single-stranded shuttle phagemid for mutagenesis studies in mammalian cells: 8-oxoguanine in DNA induces targeted G.C → T.A transversions in simian kidney cells. *Proc. Natl. Acad. Sci. USA*, **90**, 1122–1126.

24 Cadet, J., Douki, T., and Ravanat, J.L. (2008) Oxidatively generated damage to the guanine moiety of DNA: mechanistic aspects and formation in cells. *Acc. Chem. Res.*, **41**, 1075–1083.

25 Bergeron, F., Auvre, F., Radicella, J.P., and Ravanat, J.L. (2010) HO center dot radicals induce an unexpected high proportion of tandem base lesions refractory to repair by DNA glycosylases. *Proc. Natl. Acad. Sci. USA*, **107**, 5528–5533.

26 Rydberg, B. and Lindahl, T. (1982) Non-enzymatic methylation of DNA by the intracellular methyl-group donor S-adenosyl-L-methionine is a potentially mutagenic reaction. *EMBO J.*, **1**, 211–216.

27 Sedgwick, B., Bates, P.A., Paik, J., Jacobs, S.C., and Lindahl, T. (2007) Repair of alkylated DNA: recent advances. *DNA Repair*, **6**, 429–442.

28 Møller, P. and Wallin, H. (1998) Adduct formation, mutagenesis and nucleotide excision repair of DNA damage produced by reactive oxygen species and lipid peroxidation product. *Mutat. Res./Rev. Mutat. Res.*, **410**, 271–290.

29 Levine, R.L., Yang, I.-Y., Hossain, M., Pandya, G.A., Grollman, A.P., and Moriya, M. (2000) Mutagenesis induced by a single 1, N^6-ethenodeoxyadenosine adduct in human cells. *Cancer Res.*, **60**, 4098–4104.

30 Nilsen, H., Rosewell, I., Robins, P., Skjelbred, C.F., Andersen, S., Slupphaug, G., Daly, G., Krokan, H.E., Lindahl, T., and Barnes, D.E. (2000) Uracil-DNA glycosylase (UNG)-deficient mice reveal a primary role of the enzyme during DNA replication. *Mol. Cell*, **5**, 1059–1065.

31 Andersen, S., Heine, T., Sneve, R., Konig, I., Krokan, H.E., Epe, B., and Nilsen, H. (2005) Incorporation of dUMP into DNA is a major source of spontaneous DNA damage, while excision of uracil is not required for cytotoxicity of fluoropyrimidines in mouse embryonic fibroblasts. *Carcinogenesis*, **26**, 547–555.

32 Sakumi, K., Furuichi, M., Tsuzuki, T., Kakuma, T., Kawabata, S., Maki, H., and Sekiguchi, M. (1993) Cloning and expression of cDNA for a human enzyme that hydrolyzes 8-oxo-dGTP, a mutagenic substrate for DNA synthesis. *J. Biol. Chem.*, **268**, 23524–23530.

33 Vilenchik, M.M. and Knudson, A.G. (2003) Endogenous DNA double-strand breaks: production, fidelity of repair, and induction of cancer. *Proc. Natl. Acad. Sci. USA*, **100**, 12871–12876.

34 Clauson, C.L., Oestreich, K.J., Austin, J.W., and Doetsch, P.W. (2010) Abasic sites and strand breaks in DNA cause transcriptional mutagenesis in *Escherichia coli*. *Proc. Natl. Acad. Sci. USA*, **107**, 3657–3662.

35 David, S.S., O'Shea, V.L., and Kundu, S. (2007) Base-excision repair of oxidative DNA damage. *Nature*, **447**, 941–950.

36 Fortini, P. and Dogliotti, E. (2007) Base damage and single-strand break repair: mechanisms and functional significance of short- and long-patch repair subpathways. *DNA Repair*, **6**, 398–409.

37 Robertson, A.B., Klungland, A., Rognes, T., Leiros, I. (2009) DNA repair in mammalian cells: base excision repair: the long and short of it. *Cell Mol. Life Sci.*, **66**, 981–993.

38 Bjelland, S. and Seeberg, E. (2003) Mutagenicity, toxicity and repair of DNA base damage induced by oxidation. *Mutat.*

Res./Fund. Mol. Mech. Mutagen., **531**, 37–80.

39 Frosina, G., Fortini, P., Rossi, O., Carrozzino, F., Raspaglio, G., Cox, L.S., Lane, D.P., Abbondandolo, A., and Dogliotti, E. (1996) Two pathways for base excision repair in mammalian cells. *J. Biol. Chem.*, **271**, 9573–9578.

40 Klungland, A. and Lindahl, T. (1997) Second pathway for completion of human DNA base excision-repair: reconstitution with purified proteins and requirement for DNase IV (FEN1). *EMBO J.*, **16**, 3341–3348.

41 Hazra, T.K., Izumi, T., Boldogh, I., Imhoff, B., Kow, Y.W., Jaruga, P., Dizdaroglu, M., and Mitra, S. (2002) Identification and characterization of a human DNA glycosylase for repair of modified bases in oxidatively damaged DNA. *Proc. Natl. Acad. Sci. USA*, **99**, 3523–3528.

42 Hazra, T.K. and Mitra, S. (2006) Purification and characterization of NEIL1 and NEIL2, members of a distinct family of mammalian DNA glycosylases for repair of oxidized bases. *DNA Repair*, **408**, 33–48.

43 Michaels, M.L., Tchou, J., Grollman, A.P., and Miller, J.H. (1992) A repair system for 8-oxo-7,8-dihydrodeoxyguanine. *Biochemistry*, **31**, 10964–10968.

44 Falnes, P.O., Johansen, R.F., and Seeberg, E. (2002) AlkB-mediated oxidative demethylation reverses DNA damage in *Escherichia coli*. *Nature*, **419**, 178–182.

45 Ringvoll, J., Nordstrand, L.M., Vagbo, C.B., Talstad, V., Reite, K., Aas, P.A., Lauritzen, K.H., Liabakk, N.B., Bjork, A., Doughty, R.W., Falnes, P.O., Krokan, H.E., and Klungland, A. (2006) Repair deficient mice reveal mABH2 as the primary oxidative demethylase for repairing 1meA and 3meC lesions in DNA. *EMBO J.*, **25**, 2189–2198.

46 Sundheim, O., Vagbo, C.B., Bjoras, M., Sousa, M.M.L., Talstad, V., Aas, P.A., Drablos, F., Krokan, H.E., Tainer, J.A., and Slupphaug, G. (2006) Human ABH3 structure and key residues for oxidative demethylation to reverse DNA/RNA damage. *EMBO J.*, **25**, 3389–3397.

47 Campalans, A., Amouroux, R., Bravard, A., Epe, B., and Radicella, J.P. (2007) UVA irradiation induces relocalisation of the DNA repair protein hOGG1 to nuclear speckles. *J. Cell Sci.*, **120**, 23–32.

48 Amouroux, R., Campalans, A., Epe, B., and Radicella, J.P. (2010) Oxidative stress triggers the preferential assembly of base excision repair complexes on open chromatin regions. *Nucl. Acids Res.*, **38**, 2878–2890.

49 Paz-Elizur, T., Krupsky, M., Blumenstein, S., Elinger, D., Schechtman, E., and Livneh, Z. (2003) DNA repair activity for oxidative damage and risk of lung cancer. *J. Natl. Cancer Inst.*, **95**, 1312–1319.

50 Ellison, K.S., Dogliotti, E., and Essigmann, J.M. (1989) Construction of a shuttle vector containing a single O^6-methylguanine: a probe for mutagenesis in mammalian cells. *Mutat. Res./Rev. Genet. Toxicol.*, **220**, 93–100.

51 Barbin, A. (2000) Etheno-adduct-forming chemicals: from mutagenicity testing to tumor mutation spectra. *Mutat. Res./Rev. Mutat. Res.*, **462**, 55–69.

52 Phillips, D.H. (1997) Detection of DNA modifications by the P-32-postlabelling assay. *Mutat. Res./Fund. Mol. Mech. Mutagen.*, **378**, 1–12.

53 Cadet, J., Douki, T., and Ravanat, J.L. (1997) Artifacts associated with the measurement of oxidized DNA bases. *Environ. Health Perspect.*, **105**, 1034–1039.

54 Floyd, T.R., Cicero, S.E., Fazio, S.D., Raglione, T.V., Hsu, S.H., Winkle, S.A., and Hartwick, R.A. (1986) Mixed-mode hydrophobic ion-exchange for the separation of oligonucleotides and DNA fragments using HPLC. *Anal. Biochem.*, **154**, 570–577.

55 ESCODD (2002) Comparative analysis of baseline 8-oxo-7,8-dihydroguanine in mammalian cell DNA, by different methods in different laboratories: an approach to consensus. *Carcinogenesis*, **23**, 2129–2133.

56 Helbock, H.J., Beckman, K.B., Shigenaga, M.K., Walter, P.B., Woodall, A.A., Yeo, H.C., and Ames, B.N. (1998) DNA oxidation matters: the HPLC-electrochemical detection assay of 8-oxo-deoxyguanosine and 8-oxo-guanine. *Proc. Natl. Acad. Sci. USA*, **95**, 288–293.

57 Boysen, G., Collins, L.B., Liao, S., Luke, A.M., Pachkowski, B.F., Watters, J.L., and Swenberg, J.A. (2010) Analysis of 8-oxo-7,8-dihydro-2′-deoxyguanosine by ultra high pressure liquid chromatography-heat assisted electrospray ionization-tandem mass spectrometry. *J. Chromatogr. B*, **878**, 375–380.

58 Mangal, D., Vudathala, D., Park, J.H., Lee, S.H., Penning, T.M., and Blair, I.A. (2009) Analysis of 7,8-dihydro-8-oxo-2′-deoxyguanosine in cellular DNA during oxidative stress. *Chem. Res. Toxicol.*, **22**, 788–797.

59 Epe, B. and Hegler, J. (1994) Oxidative DNA damage: endonuclease fingerprinting, in *Oxygen Radicals in Biological Systems, Part D*, Academic Press, San Diego, CA, pp. 122–131.

60 Collins, A.R. (2004) The comet assay for DNA damage and repair: principles, applications, and limitations. *Mol. Biotechnol.*, **26**, 249–261.

61 Hartwig, A., Dally, H., and Schlepegrell, R. (1996) Sensitive analysis of oxidative DNA damage in mammalian cells: use of the bacterial Fpg protein in combination with alkaline unwinding. *Toxicol. Lett.*, **88**, 85–90.

62 Sutherland, B.M., Bennett, P.V., Sidorkina, O., and Laval, J. (2000) Clustered damages and total lesions induced in DNA by ionizing radiation: oxidized bases and strand breaks. *Biochemistry*, **39**, 8026–8031.

63 Nakamura, A.J., Rao, V.A., Pommier, Y., and Bonner, W.M. (2010) The complexity of phosphorylated H2AX foci formation and DNA repair assembly at DNA double-strand breaks. *Cell Cycle*, **9**, 389–397.

64 Nair, J., Vaca, C.E., Velic, I., Mutanen, M., Valsta, L.M., and Bartsch, H. (1997) High dietary omega-6 polyunsaturated fatty acids drastically increase the formation of etheno-DNA base adducts in white blood cells of female subjects. *Cancer Epidemiol. Biomarkers Prev.*, **6**, 597–601.

65 Neeley, W.L. and Essigmann, J.M. (2006) Mechanisms of formation, genotoxicity, and mutation of guanine oxidation products. *Chem. Res. Toxicol.*, **19**, 491–505.

66 Cadet, J. and Poulsen, H. (2010) Measurement of oxidatively generated base damage in cellular DNA and urine. *Free Radic. Biol. Med.*, **48**, 1457–1459.

67 Cooke, M.S., Olinski, R., Loft, S., and ESCULA (2008) Measurement and meaning of oxidatively modified DNA lesions in urine. *Cancer Epidemiol. Biomarkers Prev.*, **17**, 3–14.

68 Klungland, A., Rosewell, I., Hollenbach, S., Larsen, E., Daly, G., Epe, B., Seeberg, E., Lindahl, T., and Barnes, D.E. (1999) Accumulation of premutagenic DNA lesions in mice defective in removal of oxidative base damage. *Proc. Natl. Acad. Sci. USA*, **96**, 13300–13305.

69 Trapp, C., Reite, K., Klungland, A., and Epe, B. (2007) Deficiency of the Cockayne syndrome B (CSB) gene aggravates the genomic instability caused by endogenous oxidative DNA base damage in mice. *Oncogene*, **26**, 4044–4048.

70 Trapp, C., Schwarz, M., and Epe, B. (2007) The peroxisome proliferator WY-14,643 promotes hepatocarcinogenesis caused by endogenously generated oxidative DNA base modifications in repair-deficient $Csb^{m/m}/Ogg1^{-/-}$ mice. *Cancer Res.*, **67**, 5156–5161.

71 Russo, M.T., De Luca, G., Degan, P., Parlanti, E., Dogliotti, E., Barnes, D.E., Lindahl, T., Yang, H., Miller, J.H., and Bignami, M. (2004) Accumulation of the oxidative base lesion 8-hydroxyguanine in DNA of tumor-prone mice defective in both the Myh and Ogg1 DNA glycosylases. *Cancer Res.*, **64**, 4411–4414.

72 Xu, G., Herzig, M., Rotrekl, V., and Walter, C.A. (2008) Base excision repair, aging and health span. *Mech. Ageing Dev.*, **129**, 366–382.

73 Larsen, E., Meza, T.J., Kleppa, L., and Klungland, A. (2007) Organ and cell specificity of base excision repair mutants in mice. *Mutat. Res./Fund. Mol. Mech. Mutagen.*, **614**, 56–68.

74 Friedberg, E.C. and Meira, L.B. (2006) Database of mouse strains carrying targeted mutations in genes affecting biological responses to DNA damage. Version 7. *DNA Repair*, **5**, 189–209.

75 Al-Tassan, N., Chmiel, N.H., Maynard, J., Fleming, N., Livingston, A.L., Williams, G.T., Hodges, A.K., Davies, D.R., David, S.S., Sampson, J.R., and Cheadle, J.R. (2002) Inherited variants of MYH associated with somatic G:C → T:A mutations in colorectal tumors. *Nat. Genet.*, **30**, 227–232.

76 Goode, E.L., Ulrich, C.M., and Potter, J.D. (2002) Polymorphisms in DNA repair genes and associations with cancer risk. *Cancer Epidemiol. Biomarkers Prev.*, **11**, 1513–1530.

77 Weiss, J.M., Goode, E.L., Ladiges, W.C., and Ulrich, C.M. (2005) Polymorphic variation in hOGG1 and risk of cancer: a review of the functional and epidemiologic literature. *Mol. Carcinog.*, **42**, 127–141.

78 Collins, A.R., Duthie, S.J., Fillion, L., Gedik, C.M., Vaughan, N., and Wood, S.G. (1997) Oxidative DNA damage in human cells: the influence of antioxidants and DNA repair. *Biochem. Soc. Trans.*, **25**, 326–331.

79 Podmore, I.D., Griffiths, H.R., Herbert, K.E., Mistry, N., Mistry, P., and Lunec, J. (1998) Vitamin C exhibits pro-oxidant properties. *Nature*, **392**, 559–1559.

80 Foksinski, M., Gackowski, D., Rozalski, R., Siomek, A., Guz, J., Szpila, A., Dziaman, T., and Olinski, R. (2007) Effects of basal level of antioxidants on oxidative DNA damage in humans. *Eur. J. Nutr.*, **46**, 174–180.

81 Breton, J., Sichel, F., Pottier, D., and Prevost, V. (2005) Measurement of 8-oxo-7,8-dihydro-2′-deoxyguanosine in peripheral blood mononuclear cells: optimisation and application to samples from a case-control study on cancers of the oesophagus and cardia. *Free Radic. Res.*, **39**, 21–30.

82 Pouget, J.P., Frelon, S., Ravanat, J.L., Testard, I., Odin, F., and Cadet, J. (2009) Formation of modified DNA bases in cells exposed either to gamma radiation or to high-LET particles 1. *Radiat. Res.*, **157**, 589–595.

83 ESCODD (2003) Measurement of DNA oxidation in human cells by chromatographic and enzymic methods. *Free Radic. Biol. Med.*, **34**, 1089–1099.

84 Collins, A., Gedik, C., Vaughan, N., Wood, S., White, A., Dubois, J., Duez, P., Dehon, G., Rees, J.F., Loft, S., Moller, P., Poulsen, H., Riis, B., Weimann, A., Cadet, J., Douki, T., Ravanat, J.L., Sauvaigo, S., Faure, H., Morel, I., Morin, B., Epe, B., Phoa, N., Hartwig, A., Pelzer, A., Dolara, P., Casalini, C., Giovannelli, L., Lodovici, M., Olinski, R., Bialkowski, K., Foksinski, M., Gackowski, D., Durackova, Z., Hlincikova, L., Korytar, P., Sivonova, M., Dusinska, M., Mislanova, C., Vina, J., Lloret, A., Moller, L., Hofer, T., Nygren, J., Gremaud, E., Herbert, K., Chauhan, D., Kelly, F., Dunster, C., Lunec, J., Cooke, M., Evans, M., Patel, P., Podmore, I., Wild, C., Hardie, L., Olliver, J., and Smith, E. (2002) Comparative analysis of baseline 8-oxo-7,8-dihydroguanine in mammalian cell DNA, by different methods in different laboratories: an approach to consensus. *Carcinogenesis*, **23**, 2129–2133.

85 Kirkali, G., Tunca, M., Genc, S., Jaruga, P., and Dizdaroglu, M. (2008) Oxidative DNA damage in polymorphonuclear leukocytes of patients with familial Mediterranean fever. *Free Radic. Biol. Med.*, **44**, 386–393.

86 Doerge, D.R., Churchwell, M.I., Fang, J.L., and Beland, F.A. (2000) Quantification of etheno-DNA adducts using liquid chromatography, on-line sample processing, and electrospray tandem mass spectrometry. *Chem. Res. Toxicol.*, **13**, 1259–1264.

87 Nair, J., Barbin, A., Velic, I., and Bartsch, H. (1999) Etheno DNA-base adducts from endogenous reactive species. *Mutat. Res./Fund. Mol. Mech. Mutagen.*, **424**, 59–69.

88 Yanagisawa, H., Hirano, A., and Sugawara, M. (2004) A dot-blot method for quantification of apurinic/apyrimidinic sites in DNA using an avidin plate and liposomes encapsulating a fluorescence dye. *Anal. Biochem.*, **332**, 358–367.

89 Nakamura, J. and Swenberg, J.A. (1999) Endogenous apurinic apyrimidinic sites in genomic DNA of mammalian tissues. *Cancer Res.*, **59**, 2522–2526.

90 Warters, R.L., Adamson, P.J., Pond, C.D., and Leachman, S.A. (2005) Melanoma cells express elevated levels of phosphorylated histone H2AX foci. *J. Invest. Dermatol.*, **124**, 807–817.

4
The IARC Monographs Programme: Cancer Hazard Identification as a First Step in Cancer Risk Assessment and Cancer Prevention

Robert A. Baan and Vincent J. Cogliano

4.1
Introduction

4.1.1
The Origin of the IARC Monographs Programme

The International Agency for Research on Cancer (IARC) was established in 1965 at the initiative of a group of 12 French intellectuals – among them writer François Mauriac, architect Charles Le Corbusier, and oncologist Antoine Lacassagne – with the support of President Charles de Gaulle of France.

In its early years, the IARC received questions about the carcinogenic risks of chemicals, including frequent requests for lists of known and suspected human carcinogens. To adequately summarize the complexity of the available information would clearly be a difficult task, and the IARC began to consider means of obtaining international expert opinion on this topic. In 1970, the IARC Advisory Committee on Environmental Carcinogenesis recommended that "a compendium on carcinogenic chemicals be prepared by experts and that the biological activity and evaluation of practical importance to public health should be referenced and documented." The *IARC Monographs Programme* was formally established in 1971 when the Governing Council recommended that the IARC should prepare "Monographs on the evaluation of carcinogenic risks of chemicals to man," which became the initial title of the book series. In the years that followed, the scope of the Programme broadened as monographs were developed for groups of related chemicals, complex mixtures, occupational exposures, physical and biological agents, and lifestyle factors. For this reason, the title of the series was changed in the late 1980s to what it is today: *IARC Monographs on the Evaluation of Carcinogenic Risks to Humans.*

Cancer Risk Evaluation: Methods and Trends,
Edited by Günter Obe, Burkhard Jandrig, Gary E. Marchant, Holger Schütz, and Peter M. Wiedemann.

4.1.2
The IARC Monographs Programme: Objective and Scope

Through the Monographs Programme, the IARC seeks to identify the causes of human cancer. This is the first step toward cancer prevention, which is needed as much today as when the IARC was established. The global burden of cancer is high and continues to increase: the annual number of new cancer cases was estimated at 12 million in 2008 and is expected to reach 27 million by 2030 [1]. With current trends in demographics and exposure conditions, the cancer burden is shifting from high-resource countries to low- and medium-resource countries. Guided by the evaluations in the *IARC Monographs*, national health agencies have been able to take measures, based on scientific arguments, to reduce human exposure to carcinogens in the workplace and in the environment.

The objective of the IARC Monographs Programme is to prepare, with the help of international Working Groups of experts, critical reviews and assessments of the weight of the evidence that certain exposures could alter the incidence of cancer in humans. The monographs may also indicate where additional research efforts are needed. It should be noted that the monographs neither do extrapolate beyond the range of data available nor do they extrapolate from experimental data to the human situation. The *IARC Monographs* evaluations may assist national and international authorities in making risk assessments and in taking preventive action, but they do not recommend legislation or regulation, for example, with respect to exposure limits. Since its inception, the IARC Monographs Programme has convened more than 100 working groups and published evaluations on nearly 950 chemicals, chemical mixtures, physical and biological agents, occupational exposures, and lifestyle factors. A complete list of evaluations is available at the IARC Monographs web site http://monographs.iarc.fr.

4.1.3
Selection of Agents for Review

Agents are selected for review on the basis of two main criteria: (a) there is evidence of human exposure and (b) there is some evidence or suspicion of carcinogenicity. Ad hoc advisory groups convened by the IARC in 1984, 1989, 1991, 1993, 1998, 2003, and 2008 made recommendations as to which agents should be evaluated with priority in the *Monographs* series. Recent recommendations are available on the Monographs Programme web site. The IARC may schedule other agents for review when it becomes aware of new scientific information or when national health agencies identify an urgent public health need related to cancer.

4.1.4
The Evaluation Process

Once an agent has been scheduled for evaluation, the openly available scientific literature is surveyed for published data relevant to an assessment of its carcinogenicity.

An IARC Monographs Working Group then develops its evaluations through a series of distinct steps. The process begins with separate evaluations of the evidence of cancer in humans and of cancer in experimental animals, each choosing one of the descriptors "sufficient evidence," "limited evidence," "inadequate evidence," or "evidence suggesting lack of carcinogenicity." The criteria established in 1971 to evaluate carcinogenic risks to humans were adopted by the working groups whose deliberations resulted in the first 16 volumes of the *Monographs* series and have been regularly updated and refined in the past three decades by different ad hoc advisory groups, notably those on the use of mechanistic information in cancer hazard identification [2, 3]. The descriptors mentioned above are defined in detail in the "Preamble" to the *IARC Monographs*, which was revised in 2006 and is published in each volume. The "Preamble" is also available on the IARC Monographs website.

The two evaluations of the evidence of cancer in humans and in experimental animals are combined (see Scheme 4.1 and accompanying text) into an evaluation indicating that the agent is "carcinogenic to humans" (Group 1), "probably carcinogenic to humans" (Group 2A), "possibly carcinogenic to humans" (Group 2B), "not classifiable as to its carcinogenicity to humans" (Group 3), or "probably not carcinogenic to humans" (Group 4). It should be noted that these "categories of concern" were formally introduced only in 1987 (Supplement 7, see below). Mechanistic and other relevant data are also considered to determine whether the "default" evaluation should be modified. The working group then makes an overall evaluation that reflects the weight of the evidence derived both from studies in humans and experimental animals and from mechanistic and other relevant data.

The first of the two examples given below illustrates the dynamics of IARC's evaluation process, its scientific principles and standards, and its consistent use of well-defined evaluation criteria by different working groups: it follows the historical development of cancer hazard evaluations of a single agent – formaldehyde – over a span of nearly 30 years.

The second example highlights the use of mechanistic information that helped a recent working group to arrive at a Group 1 classification for an agent for which specific epidemiological evidence in humans will probably never become available: the identification of aristolochic acids as the carcinogenic agent in herbal mixtures or crops that contain plants of the genus *Aristolochia*.

4.2 Formaldehyde, Nasopharyngeal Cancer, and Leukemia: Evolution in Evaluation

The history of *IARC Monographs* evaluations of the evidence with respect to an association between the exposure to formaldehyde and the risks for cancer in humans spans nearly three decades. It started with discussions by a working group in 1981, the outcome of which was published in volume 29 of the *IARC Monographs* [4]. Subsequently, formaldehyde was reviewed in Supplement 7 [5], volume 62 [6], volume 88 [7], and most recently in volume 100-F [8]. The epidemiological data that were considered by the different working groups and the conclusions drawn

IARC Monographs classification ("default") of agents on the basis of the strength of evidence of carcinogenicity to humans and to experimental animals.

ANIMAL / HUMAN	*Sufficient*	*Limited*	*Inadequate* (or lack of data)
Sufficient	**1**	**1**	**1**
Limited	**2A**	**2B**	**2B**
Inadequate (or lack of data)	**2B**	**3**	**3**

Criteria for *sufficient*, *limited,* and *inadequate evidence* of carcinogenicity to humans and to experimental animals are detailed in the Monographs' *Preamble*, which is published in every Monograph volume and available at http://monographs.iarc.fr.

An overall evaluation of the carcinogenicity of the agent to humans is reached by considering the body of evidence as a whole, including mechanistic and other relevant data. In this process, Working Groups are guided by the following paragraphs in the *Preamble*:

Group 1: The agent is *carcinogenic to humans*. This category is used when there is *sufficient evidence* of carcinogenicity in humans. Exceptionally, an agent may be placed in this category when evidence of carcinogenicity in humans is less than sufficient, but there is *sufficient evidence* of carcinogenicity in experimental animals and strong evidence in exposed humans that the agent acts through a relevant mechanism of carcinogenicity.

Group 2: This category includes agents for which, at one extreme, the degree of evidence of carcinogenicity in humans is almost sufficient, as well as those for which, at the other extreme, there are no human data but for which there is evidence of carcinogenicity in experimental animals. Agents are assigned to either Group 2A (*probably carcinogenic to humans*) or Group 2B (*possibly carcinogenic to humans*) on the basis of epidemiological and experimental evidence of carcinogenicity and mechanistic and other relevant data. The terms *probably carcinogenic* and *possibly carcinogenic* have no quantitative significance and are used simply as descriptors of different levels of evidence of human carcinogenicity, with *probably carcinogenic* signifying a higher level of evidence than *possibly carcinogenic*.

Group 2A: The agent is *probably carcinogenic to humans*. This category is used when there is *limited evidence* of carcinogenicity in humans and *sufficient evidence* of carcinogenicity in experimental animals. In some cases, an agent may be classified in this category when there is *inadequate evidence* of carcinogenicity in humans and *sufficient evidence* of carcinogenicity in experimental animals and strong evidence that the carcinogenesis is mediated by a mechanism that also operates in humans. Exceptionally, an agent may be classified in this category solely on the basis of *limited evidence* of carcinogenicity in humans. An agent may be assigned to this category if it clearly belongs, based on mechanistic considerations, to a class of agents for which one or more members have been classified in Group 1 or Group 2A.

Group 2B: The agent is *possibly carcinogenic to humans*. This category is used for agents for which there is *limited evidence* of carcinogenicity in humans and less than sufficient evidence of carcinogenicity in experimental animals. It may also be used when there is *inadequate evidence* of carcinogenicity in humans but there is *sufficient evidence* of carcinogenicity in experimental animals. In some instances, an agent for which there is *inadequate evidence* of carcinogenicity in humans and less than sufficient evidence of carcinogenicity in experimental animals together with supporting evidence from mechanistic and other relevant data may be placed in this group. An agent may be classified in this category solely on the basis of strong evidence from mechanistic and other relevant data.

Scheme 4.1

Group 3: The agent is *not classifiable as to its carcinogenicity to humans*. This category is used most commonly for agents for which the evidence of carcinogenicity is inadequate in humans and inadequate or limited in experimental animals. Exceptionally, agents for which the evidence of carcinogenicity is inadequate in humans but sufficient in experimental animals may be placed in this category when there is strong evidence that the mechanism of carcinogenicity in experimental animals does not operate in humans. Agents that do not fall into any other group are also placed in this category. An evaluation in Group 3 is not a determination of noncarcinogenicity or overall safety. It often means that further research is needed, especially when exposures are widespread or the cancer data are consistent with differing interpretations.

Criteria for Group 4 (the agent is *probably not carcinogenic to humans*) are also given in the *Preamble*, and not further discussed here.

Scheme 4.1 *(Continued)*

from these five evaluations are summarized below, with special emphasis on the growing evidence of an increased risks for lymphohematopoietic cancers in workers exposed to formaldehyde.

4.2.1 Summary of Epidemiological Data Reviewed in Volume 29 [4]

Three mortality studies of workers engaged in manufacturing formaldehyde and other chemicals or using formaldehyde were inconclusive with respect to increased cancer risks. The first study of embalmers who used fluids containing formaldehyde showed a proportional excess of deaths from skin cancer, which increased with both duration of employment in embalming and intensity of exposure. Men involved in embalming also had increased mortality from cancers of the brain and kidney [9]. The second study showed a significant excess of deaths from prostatic cancer in the period after 20 years from first employment in a formaldehyde factory [10]. The third study showed an excess mortality from digestive tract cancer in the youngest age group among those exposed to formaldehyde, but in men exposed for less than 5 years to formaldehyde and after more than 20 years of onset of exposure, there was no overall excess of mortality [11]. In each of these three studies, the numbers of deaths or person-years observed after a suitable latent period were small and would be insufficient to show an increased risks of an uncommon cancer.

One epidemiological study on formaldehyde and lymphohematopoietic cancers was mentioned in volume 29. In this mortality study of pathologists and medical laboratory technicians in the United Kingdom, a total of 2709 pathologists alive and active between January 1, 1955 and December 31, 1973 and 12 944 medical laboratory technicians registered between August 1963 and December 1973 were followed up to the end of 1973. Mortality data were compared with those of the population of England and Wales or Scotland. The standardized mortality ratios (SMRs) for all causes of death were 0.60 for pathologists and 0.67 for medical laboratory technicians. In male pathologists, a statistically significant increase in lymphoma and hematoma was observed (8 observed, 3.3 expected; SMR, 2.42; 95% CI, 1.05–4.78; $p < 0.01$). Similar findings were not seen in laboratory technicians. In this study, no information was provided on the actual exposure to formaldehyde [12].

Overall, the epidemiological studies available to the working group in 1981 provided *inadequate evidence* to assess the carcinogenicity to humans from exposure to formaldehyde. The evidence of the carcinogenicity of formaldehyde in experimental animals (squamous cell carcinoma in the nasal cavity of exposed rats) was considered *sufficient* (details not shown here). According to the criteria discussed above, this overall evidence would have resulted in a classification of formaldehyde in Group 2B, *possibly carcinogenic to humans.*

4.2.2
Summary of Epidemiological Data Reviewed in Supplement 7 [5]

In 1987, an IARC Working Group reviewed all the agents that had been evaluated in volumes 1–42 of the *IARC Monographs.* Each agent was placed in one of the five categories of concern, that is, Group 1, 2A, 2B, 3, or 4, on the basis of the criteria that had been developed during the preceding 15 years (see Scheme 4.1). The outcome of this exercise was published as Supplement 7 of the *IARC Monographs.* For some of the agents, the data from previous reviews formed the sole basis of the classification. In the case of formaldehyde, a number of relevant new studies had been published since the review in 1982 (volume 29, Ref. [4]), which warranted a more detailed reevaluation.

The risks for nasopharyngeal cancer – a rare malignancy in Western countries – was elevated in a cohort study of industrial workers [13] and in three case–control studies [14–16]. Among industrial workers exposed to formaldehyde-containing particulates, SMRs for nasopharyngeal cancer increased with cumulative exposure to formaldehyde: 1.92 (one death) for <0.5 ppm-years, 4.03 (two deaths) for 0.5–5.5 ppm-years, and 7.46 (two deaths) for >5.5 ppm-years. There was a similar trend with duration of exposure to formaldehyde, and all five cases held jobs in which hourly exposure concentrations exceeded 4.0 ppm [17].

Mortality from leukemia, predominantly of the myeloid subtype, was found to be elevated among different professional groups: embalmers [18, 19], anatomists [20], and undertakers [21]. In the study by Stroup *et al.* [20], 5 of the 10 observed deaths were due to myeloid leukemia and the SMR for chronic myeloid leukemia was statistically significantly elevated (3 deaths; SMR, 8.8; 95% CI, 1.8–25.5) in the period 1969–1979 for which cell type-specific mortality rates were available. Increased risks for leukemia were not found among industrial workers exposed to formaldehyde, except for a small, nonsignificant excess reported in one study [22].

In summary, the evidence of a possible involvement of formaldehyde in cancer was considered the strongest for the nasal cavity and the nasopharynx. The occurrence of these cancers showed an exposure–response gradient in more than one study, but the numbers of exposed cases were often small and some studies did not show an excess. The nose and nasopharynx could come into direct contact with formaldehyde through inhalation, which would lend mechanistic plausibility to the association. The excess mortality from leukemia was observed in four studies of professional groups, but generally not seen among industrial workers, suggesting that the excess for these cancers among professionals is due to factors other than formaldehyde.

Overall, this epidemiological evidence was considered by the working group to be *limited*. Taking into account the *sufficient evidence* of carcinogenicity in experimental animals, the 1987 Working Group placed formaldehyde in Group 2A, *probably carcinogenic to humans*.

4.2.3
Summary of Epidemiological Data Reviewed in Volume 62 [6]

Excess cases of nasopharyngeal cancers were associated with occupational exposure to formaldehyde in two of the six cohort studies on industrial or professional groups, in three of the four case–control studies, and in meta-analyses (see volume 62 for references). In one cohort study performed in 10 plants in the United States [13, 23], the risk increased with category of increasing cumulative exposure. In three of the case–control studies, the risk was highest in people in the highest category of exposure and among people exposed 20–25 years before death. The meta-analyses found significantly higher risks for lung and nasopharyngeal cancers among people estimated to have had substantial exposure than among those with low/medium or no exposure [24, 25]. The observed associations between exposure to formaldehyde and risks for cancer could not be reasonably attributed to other occupational agents, including wood dust, or to tobacco smoking. Taken together, the working group found the epidemiological studies suggestive of a causal relationship between exposure to formaldehyde and nasopharyngeal cancer, but the numbers of observed and expected cases in the cohort studies were small. The studies of the industrial cohorts showed low or no risks for lymphohematopoietic cancers.

The cohort studies of embalmers, anatomists, and other professionals who use formaldehyde did show excess risks for lymphatic or hematopoietic cancers [26, 27] (see also references cited in Supplement 7), although they were often based on small numbers. Hayes *et al.* [26] analyzed mortality records of 4046 embalmers and funeral directors from across the United States (3649 whites and 397 nonwhites). Proportionate mortality ratios (PMRs) were calculated on the basis of expected numbers from race- and sex-specific groups of the general population, adjusted for 5-year age and calendar time categories. Significantly elevated PMRs were those for lymphatic and hematopoietic tumors for whites (1.3; 95% CI, 1.1–1.6; 100 observed) and nonwhites (2.4; 95% CI, 1.4–4.0; 15 observed). The PMR was 1.6 for myeloid leukemia (95% CI, 1.0–2.3; 24 observed) and 2.3 for other and unspecified leukemia (95% CI, 1.4–3.5; 20 observed).

Overall, the 1995 Working Group considered this epidemiological evidence to be *limited* and maintained the Group 2A classification for formaldehyde.

4.2.4
Summary of Epidemiological Data Reviewed in Volume 88 [7]

This reevaluation took into account the extended follow-up of three major cohort studies and three new case–control studies (see volume 88 for references). In the largest and most informative cohort study of industrial workers exposed to formaldehyde [28], a statistically significant excess of deaths from nasopharyngeal cancer was observed in comparison with the US national population, with statistically significant trends in exposure–response relationships for peak and cumulative exposure. An excess of deaths from nasopharyngeal cancer was also observed in a proportionate mortality analysis of the largest US cohort of embalmers [26] and in a Danish study of proportionate cancer incidence among workers at companies that used or manufactured formaldehyde [29]. In three other cohort studies of US garment manufacturers, British chemical workers, and US embalmers, cases of nasopharyngeal cancer were fewer than expected.

The relationship between the risks for nasopharyngeal cancer and exposure to formaldehyde was also investigated in seven case–control studies, five of which found elevated risks for overall exposure to formaldehyde or in higher exposure categories. In one of these studies, the increase in risk was statistically significant [30]. A meta-analysis published in 1997 found an increased overall meta-relative risk for nasopharyngeal cancer of 1.3 (95% CI, 1.2–1.5) based on 12 studies with a total of 455 cases [31].

The working group considered that it was "improbable that all of the positive findings could be explained by bias or by unrecognized confounding effects" and concluded that there is *sufficient evidence* that formaldehyde causes nasopharyngeal cancer in humans.

Increased mortality from leukemia, especially myeloid leukemia, was found in six of the seven cohort studies of embalmers, funeral parlor workers, pathologists, and anatomists. These findings had previously been discounted because an increased incidence of leukemia had not been seen in industrial workers. Recent updates, however, reported a greater incidence than before of leukemia in US industrial workers [32] and garment workers [33], but not in chemical workers in the United Kingdom [34]. The working group concluded that there is a "strong but not sufficient evidence" of a causal association between leukemia and exposure to formaldehyde, mainly because a mechanism for induction of leukemia by formaldehyde could not be identified.

On the basis of the *sufficient evidence* from epidemiological studies with regard to nasopharyngeal cancer and the *sufficient evidence* of carcinogenicity in experimental animals, formaldehyde was classified in Group 1.

4.2.5
Summary of Epidemiological Data Reviewed in Volume 100-F [8]

A major project of the IARC Monographs Programme is the compilation of a special review of known human carcinogens, to be published in the form of volume 100 of the *IARC Monographs*. This volume will present an update of IARC's assessments of the more than 100 agents that have been classified as *carcinogenic to humans* (Group 1) in *Monographs* 1–99. This landmark volume is being developed in six parts (A–F) with the help of six working groups that convened between October 2008 and October

2009. Because of its recent classification as a Group 1 carcinogen (volume 88, see above) formaldehyde was reconsidered in volume 100-F.

Since the previous evaluation, several new studies had been published on the association between exposure to formaldehyde and increased cancer risk to humans. In particular, the evidence on lymphohematopoietic cancers had become stronger by a recent nested case–control study of workers in the funeral industry. In addition, several studies – possibly triggered by the previous evaluation – focused on identifying a plausible mechanism by which formaldehyde could act as a leukemogen.

The working group confirmed that there is *sufficient evidence* that occupational exposure to formaldehyde causes nasopharyngeal cancer in humans. One industrial cohort study showed both a strong overall association and highest risks in the highest exposure category [28]. Based on eight cases, a significant excess mortality from nasopharyngeal cancer was observed among formaldehyde-exposed workers in comparison with the national population (SMR, 2.10; 95% CI, 1.05–4.21). A highly significant ($p_{\text{trend}} < 0.001$) exposure–response relationship was seen between peak exposure to formaldehyde and risks for nasopharyngeal cancer. No association was observed in the two other large industrial cohort studies. Positive associations were also found in many of the case–control studies, in particular those of larger size and with higher-quality exposure assessment. The working group noted that it was unlikely that confounding or bias could explain the observed association.

With regard to leukemia, an update to the NCI cohort and a nested case–control study of workers in the funeral industry had been published since the previous evaluation [35, 36]. Positive associations for leukemia were observed in two of the three largest industrial cohort studies, which were somewhat stronger for myeloid leukemia. Although no such association was seen in the third cohort study, there is no strong evidence that confounding or bias explains the positive associations seen in multiple settings.

Excess mortality from leukemia has been observed relatively consistently in studies of professional workers, that is, among embalmers, funeral parlor workers, pathologists, and anatomists. Six mortality studies showed positive associations [18–21, 26, 27] and one did not [37]. It may be noted that four of these studies had already been mentioned more than two decades earlier, in Supplement 7 [5].

A drawback of the mortality studies among professionals has been the lack of adequate exposure assessment. A recently published case–control study conducted among workers in the funeral industry examined lifetime work practices and exposures in this industry to develop exposure metrics among this group, which included duration of jobs held while embalming, number of embalmings, average intensity of embalming, and peak exposure to formaldehyde [36]. Positive associations were seen at many levels of exposure and for multiple exposure metrics for hematopoietic malignancies of nonlymphoid origin, notably myeloid leukemia. Embalming was significantly associated with an increased risks for this leukemia type, with significant trends for cumulative years of embalming ($p_{\text{trend}} = 0.020$) and for increasing peak exposure to formaldehyde ($p_{\text{trend}} = 0.036$).

In support of the epidemiological evidence described above, a recent study [38] of a small group of workers exposed to formaldehyde showed numerical chromosomal aberrations in myeloid progenitor cells (chromosome-7 monosomy, chromosome-8 trisomy), consistent with myeloid leukemia, and hematological changes in peripheral blood that are indicative of effects on bone marrow. Although they provide no rigid proof, these data strengthen the plausibility that formaldehyde can act as a leukemogen in humans. On balance, the working group concluded that there was *sufficient evidence* that formaldehyde causes leukemia and arrived at the following overall evaluation [39]:

- Formaldehyde is *carcinogenic to humans* (Group 1).
- Formaldehyde causes nasopharyngeal cancer and leukemia.

4.3
Herbal Medicines, *Aristolochia* Plant Species, and Aristolochic Acid Nephropathy

Traditional herbal medicines encompass an extremely diverse group of preparations and originate from many different cultures. Many herbal medicines have emerged from healing traditions around the world. Digitalis and quinine are well-known examples of valuable therapeutic products of botanical origin. Some herbal products in current use in many parts of the world, such as ginseng and valerian, have long been known for their modest efficacy and few side effects. Some, however, such as ephedra have been imported from traditional healing systems to be used for indications – weight loss, enhancement of athletic performance – that were never contemplated in the traditions from which they emerged.

Until recently, rather few data on possible carcinogenic hazards of any of these substances had been collected. In the *IARC Monograph* on "Some traditional herbal medicines," volume 82 [40], some medicinal plants and other natural products were evaluated for the first time. This section summarizes the review of the carcinogenic hazards from *Aristolochia* plant species and some of their chemical constituents.

Aristolochia, the most diverse genus of the family Aristolochiaceae, comprises about 120 plant species distributed throughout the tropics and subtropics [41, 42]. Several of these species, for example, *Aristolochia debilis*, *A. contorta*, *A. manshuriensis*, and *A. fangchi*, are found in traditional Chinese medicinal preparations. Roots of some of these plants have recently been imported from China and sold in Europe in powdered form, to be taken orally in capsules as an aid to bodyweight reduction. The active herbal ingredients are often traded under their common Chinese Pin Yin name, which can lead to confusion. For example, the name "Fang Ji" can be used to describe the roots of *A. fangchi*, *Stephania tetrandra*, or *Cocculus* species [43] and adulteration of herbal products with *Aristolochia* has indeed occurred – as illustrated below – through inadvertent substitution of *S. tetrandra* (han fang ji) with *Aristolochia* species (e.g., guang fang ji).

4.3.1
Summary of Epidemiological Data in Volume 82 [40]

An outbreak of rapidly progressive renal fibrosis in Belgium involved at least 100 patients, mostly middle-aged women undergoing a weight loss regimen that included the use of a mixture of Chinese herbs containing *Aristolochia* species incorrectly labeled as *S. tetrandra* [44–47]. This syndrome was initially called "Chinese herb nephropathy" (CHN). Additional cases of rapidly progressive renal disease involving Chinese herbs were reported from France, Germany, Japan, Spain, Taiwan, the United Kingdom, and the United States. Because a number of early cases of urothelial cancer were recorded among Belgian patients suffering from CHN, individuals with end-stage renal disease were offered prophylactic excision of the native kidneys and ureters, prior to transplantation or dialysis. This led to the discovery of a high prevalence of preinvasive and invasive neoplastic lesions of the renal pelvis, the ureter, and the urinary bladder in these patients. In one study, a total of 18 urothelial carcinomas were detected in 39 women undergoing bilateral removal of the kidneys and ureters. This greatly exceeds the expected number of these uncommon tumors [48]. The mean cumulative dose ($\pm$ SD) of herbs labeled as *S. tetrandra* (which upon analysis proved to contain various amounts of *A. fangchi*) taken by 18 patients with urothelial cancer was 226 ± 23 g, which was significantly higher than the dose ingested by the remaining 21 patients without cancer (167 ± 17 g; $p = 0.035$). Among the 24 patients with a cumulative dose of 200 g or less, 8 cases (33%) of urothelial cancer were recorded, while 10 cases of this cancer were seen among the remaining 15 patients (67%) who had ingested more than 200 g of these herbs ($p = 0.05$). The inadvertent substitution of *Stephania* by *Aristolochia* was confirmed by phytochemical analysis of the herbal mixture consumed by these patients, which led to the identification of aristolochic acids I and II, the active chemicals characteristic for *Aristolochia* [49].

This mixture of acids has been reported to function as a phospholipase-A2 inhibitor and as antineoplastic, antiseptic, anti-inflammatory, and bactericidal agent [50]. When metabolically activated by nitroreduction, aristolochic acids may form DNA adducts, as indicated in Figure 4.1. These specific aristolochic acid–DNA adducts were found in urothelial tissue specimens from all the urothelial cancer patients studied [51], providing conclusive evidence of exposure to plants of the genus *Aristolochia* as the underlying cause of the urothelial cancers. The term Chinese herb nephropathy has since been replaced with "aristolochic acid nephropathy" (AAN).

These data led the 2002 Working Group to conclude that herbal remedies containing plant species of the genus *Aristolochia* are *carcinogenic to humans* (Group 1). On the basis of *sufficient evidence* in experimental animals of the carcinogenicity of aristolochic acids (forestomach carcinomas in mice and rats, kidney tumors in rabbits; details not shown) the working group placed naturally occurring mixtures of aristolochic acids in Group 2A, taking into account the lack of epidemiological data on human cancer induced by these specific chemicals alone.

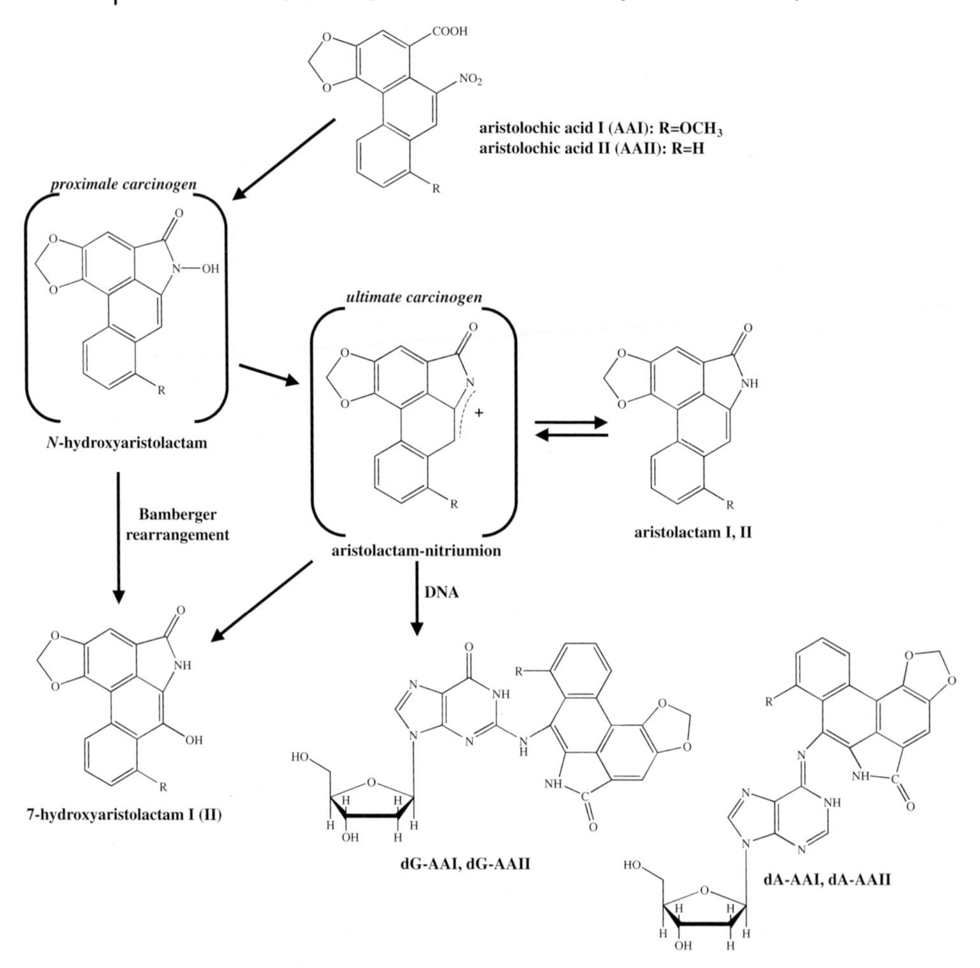

Figure 4.1 Metabolic activation and DNA-adduct formation of aristolochic acid I (AAI, R = OCH_3) and II (AAII, R = H), 7-(deoxyadenosin-N^6-yl)aristolactam I or II (dA–AAI or dA–AAII), and 7-(deoxyguanosin-N^2-yl)aristolactam I or II (dG–AAI or dG–AAII) (from Ref. [51], with permission).

4.3.2
Summary of Mechanistic Data on *Aristolochia*-Induced Urothelial Cancer (Volume 100-*A*; Ref. [52]): Identification of Aristolochic Acid as the Key Carcinogen

The review of *Aristolochia*-containing herbal remedies as a Group 1 agent, as part of *IARC Monograph* volume 100, was mainly focused on the assessment of recently published mechanistic studies on the specific role of aristolochic acids in the development of AAN-associated urothelial tumors.

The toxic effects of aristolochic acids I and II have been inferred from effects seen in patients suffering from kidney nephropathy as a result of ingestion of herbal mixtures containing *Aristolochia* species, and in experimental animals, where high doses of aristolochic acids administered orally or intravenously caused severe necrosis of the renal tubules, atrophy of the spleen and thymus, and ulceration of the forestomach, followed by hyperplasia and hyperkeratosis of the squamous epithelium [53]. The aristolochic acid-specific DNA adducts mentioned above were also detected in experimental animals exposed to aristolochic acid or botanical products containing aristolochic acid [51].

On the basis of clinical and histopathological similarities, it has been suggested [53] that AAN may have an etiological agent in common with another renal disease, known for over 50 years, namely, Balkan endemic nephropathy (BEN). This is a form of interstitial nephritis that slowly progresses to end-stage renal disease and urothelial malignancy. It was first identified in the 1920s in several small, discrete communities along the Danube River and its major tributaries, in the modern countries of Croatia, Bosnia/Herzegovina, Serbia, Romania, and Bulgaria. The first official publication describing the disease appeared in 1956 in Bulgaria [54]. The etiology was initially thought to be related to a possible intoxication with heavy metals because the villages affected were supplied with metal-contaminated water. It has also been suggested that contamination of foodstuffs with ochratoxin A, a fungal mycotoxin, was the underlying cause of the disease [55]. Recently, chronic exposure to dietary aristolochic acid has been mentioned as a major risk factor for BEN [56]. In this case, the aristolochic acid comes from *A. clematitis* (birthwort), a plant native to the endemic regions. More than 40 years ago, it was hypothesized that seeds of *A. clematitis*, which is prevalent in crop fields in Croatia, may become admixed with wheat grain during harvest and that a toxic constituent of the plant may thus end up in the bread [57]. This hypothesis went unnoticed until the results of a case–control epidemiological study among inhabitants of endemic and disease-free villages in Croatia excluded the possibility that the exposure to aristolochic acid was due to the herbal medicine use and suggested that dietary exposure to the seeds of *A. clematitis*, which contain 0.65% aristolochic acid, is an important risk factor for BEN [58]. In a subsequent study, DNA adducts derived from aristolochic acids were detected in renal tissues from four patients with BEN and in tumor tissue from three long-term residents of endemic villages who had upper urinary tract cancer. No such adducts were found in five control patients with common forms of chronic renal disease or in five control patients with upper urinary tract transitional cell cancers who resided in a nonendemic region in Croatia. In addition, analysis of *TP53* mutations in this urothelial cancer tissue revealed a high predominance of A:T-to-T:A transversions [59]. In rodent tumors, the major DNA adduct formed by aristolochic acid (see Figure 4.1) has been associated with the activation of *Ras* oncogenes through an A:T-to-T:A transversion mutation at position 2 in codon 61 [60]. Such A:T-to-T:A transversions were also the predominant mutation type in studies using human *TP53* knock-in mouse fibroblasts treated with aristolochic acid [61, 62].

In summary, key steps in the mechanism by which aristolochic acid causes tumors in experimental animals have been identified and are consistent with events occurring in patients with urothelial cancers associated with AAN and BEN. The same DNA adducts identified in humans were also found in experimental animals exposed to the natural mixture or the pure major components [51]. A:T-to-T:A transversions in the *TP53* gene in urothelial tumors of AAN and BEN patients correspond to the predominant mutations found in human *TP53* knock-in mouse fibroblasts treated with aristolochic acid. Collectively, these data support the conclusion that aristolochic acid – a mixture of aristolochic acids I and II – is a human carcinogen.

The working group for volume 100-A recognized that a final evaluation should no longer be limited to herbal medicine use and arrived at the following overall evaluations [63]:

- Plants containing aristolochic acid are *carcinogenic to humans* (Group 1).
- Aristolochic acid is *carcinogenic to humans* (Group 1).

4.4 Concluding Remarks

4.4.1 Formaldehyde

The overview of the evaluations of cancer hazards to humans from exposure to formaldehyde, conducted by five different IARC working groups over a period of almost 30 years, demonstrates the scientific rigor and consistency in IARC's evaluation process. Nonetheless, some observations can be made.

Nasopharyngeal Cancer

With regard to the association with nasopharyngeal carcinoma, it is noted that the Group 2A classification of formaldehyde remained unchanged for almost 20 years, between the evaluations in Supplement 7 [5] and volume 88 [7]. In hindsight, the question can be raised whether the "suggestive epidemiological evidence of a causal relationship between exposure to formaldehyde and nasopharyngeal cancer," as indicated by the working group for volume 62 [6], could have been strengthened by mechanistic information and other relevant data. In this context, key information could have come from, for example, the occurrence of (cyto)genetic or genotoxic effects in exposed workers or from effects on genes involved in cell cycle control.

Despite the fact that formaldehyde is consistently genotoxic in a large variety of experimental systems ranging from bacteria to rodents *in vivo*, adequate data on genetic and cytogenetic effects of formaldehyde in humans were not available in 1995. The large majority of studies showing micronucleus formation in the nasal mucosa of formaldehyde-exposed workers were published since then (and are cited in volumes 88 and 100-F).

With respect to effects on cell regulatory genes, one study showed that about 50% of formaldehyde-induced nasal tumors in rats carried a point mutation in the *p53* tumor suppressor gene [64]. Similar evidence in humans, that is, from data on *TP53* mutations in formaldehyde-induced human nasal tumors was and still is, however, not available. In fact, mutation in *TP53* does not seem to be a necessary component in nasopharyngeal carcinogenesis in humans [65–67].

A number of studies mentioned in volume 62 [6] reported histopathological changes – for example, dysplasia, significantly elevated histological scores – in the nasal mucosa of workers exposed to formaldehyde. This evidence, along with the experimental finding of nasal cancer in formaldehyde-exposed rats and the *limited evidence* from epidemiological data, could have supported the notion that formaldehyde is a carcinogen for the nose and the nasopharynx in humans. This information might have counterbalanced the strong reliance on solid, positive epidemiological findings, which often means that preventive measures are not taken until cancers are seen in exposed humans.

Lymphohematopoietic Cancer

The association between exposure to formaldehyde and an increased risks for lymphohematopoietic malignancies has taken a long time to become clear. The first epidemiological study hinting at such an increased risk among British pathologists was published in 1975, but adequate exposure data were lacking. Subsequent studies among other professional groups reported similar results, but were discounted because an increased incidence of this type of tumor had not been seen in the cohort studies among industrial workers. When subsequent follow-up analyses of two of the three large industrial cohorts in 2004 did show increased incidences of leukemia, the epidemiological evidence fell just short of *sufficient*, mainly because a plausible mechanism for induction of leukemia by formaldehyde could not be identified. Since then, the epidemiological evidence has again become stronger and a number of other studies have given insight into possible mechanisms by which formaldehyde can affect bone marrow and act as a human leukemogen. This highlights the value of these mechanistic studies, which in only 5 years have replaced previous assertions of biological implausibility with new evidence that formaldehyde can cause blood cell abnormalities that are characteristic of leukemia development.

4.4.2
Aristolochia spp and Aristolochic Acid

Herbal remedies containing plant species of the genus *Aristolochia* were evaluated as showing *sufficient evidence* of carcinogenicity to humans, based on four case reports and two ecological studies [40]. In these studies, there were no control groups for nephropathy or cancer, that is, no comparison was made with patients who did not use the herbs. In making the evaluation of a causal association between the use of the herbs and the urothelial cancers, the working group considered the following: (a) all the patients had used Chinese herbs contaminated with *Aristolochia*, (b) there was no other common exposure, (c) the prevalence of this type of cancer in these patients was remarkably high compared to the prevalence of malignant disease in this age group, and (d) there was a strong temporal association between the use of the herbs and the nephropathy.

Alerted by clinical and histopathological similarities, researchers started looking with renewed interest at another renal disorder, namely, the Balkan endemic nephropathy. The etiology of this disease had remained a mystery for decades, until it was demonstrated that in this case the carcinogenic exposure comes from *A. clematitis* (birthwort), a plant native to the endemic regions. Remarkably, the hypothesis that seeds of this plant may contaminate the wheat grain during harvest had been published 40 years ago in the Croatian journal *Liječnički vjesnik* (*Doctor's Herald*) [57].

In the 6 years since the initial evaluation of Chinese herbal medicines [40], a host of mechanistic studies convincingly showed that aristolochic acid was responsible for the high risks for nephropathy and cancer in individuals who ingested material from *Aristolochia* plants – in the form of weight loss pills in Belgium and from cereal fields in the Balkans where *Aristolochia* plants grew as weeds.

As illustrated above, the most recent evaluations of formaldehyde and aristolochic acid as Group 1 human carcinogens were supported by mechanistic and other relevant data from recent studies on the molecular biology of these two agents: formaldehyde as a leukemogen and aristolochic acid as a mutagen for genes involved in cell cycle control. In addition, careful exposure assessment contributed to the identification of the plant species *Aristolochia* as the key factor in the nephropathies and kidney cancers associated with the intake of traditional herbal medicines and with the use of contaminated wheat flour in the Balkans.

By highlighting knowledge gaps, the IARC Monographs Programme may stimulate research in areas relevant to cancer hazard identification [68]. Whether or not the recent studies mentioned above were triggered by the earlier *IARC Monographs* evaluations, the two examples given in this chapter demonstrate the great value of including mechanistic information in the hazard identification and evaluation process. In the future, cancer assessments will increasingly rely on molecular epidemiology and on information about mechanisms of carcinogenesis, two areas that are rapidly expanding and progressing. In cases where current epidemiological evidence falls short, a better understanding of the carcinogenic mechanism can provide invaluable support to conclusions on the carcinogenic properties of agents before the cancer outcome is actually confirmed in exposed humans. This could lead to preventive measures at an earlier stage than before, to the benefit of public health.

Acknowledgment

We thank Dr. Neela Guha and Dr. Christopher P. Wild for useful comments.

References

1 Boyle, P. and Levin, B. (eds) (2008) *World Cancer Report*, World Health Organization, Geneva.
2 IARC (1983) Approaches to classifying chemical carcinogens according to mechanism of action. IARC Internal Technical Report No. 83/001.
3 IARC (1991) A consensus report of an IARC Monographs Working Group on the use of mechanisms of carcinogenesis in

risk identification. IARC Internal Technical Report No. 91/002.

4 IARC (1982) *IARC Monographs on the Evaluation of the Carcinogenic Risk of Chemicals to Humans: Some Industrial Chemicals and Dyestuffs*, vol. 29, International Agency for Research on Cancer, Lyon, pp. 345–389.

5 IARC (1987) *IARC Monographs on the Evaluation of Carcinogenic Risks to Humans. Suppl. 7. Overall Evaluations of Carcinogenicity: An Updating of IARC Monographs*, vols. 1–42, International Agency for Research on Cancer Lyon, pp. 211–216.

6 IARC (1995) *IARC Monographs on the Evaluation of Carcinogenic Risks to Humans: Wood Dust and Formaldehyde*, vol. **62**, International Agency for Research on Cancer, Lyon, pp. 217–362.

7 IARC (2006) *IARC Monographs on the Evaluation of Carcinogenic Risks to Humans: Formaldehyde, 2-Butoxyethanol and 1-tert-Butoxypropan-2-ol*, vol. 88, International Agency for Research on Cancer, Lyon, pp. 39–325.

8 IARC (2009) *IARC Monographs on the Evaluation of Carcinogenic Risks to Humans: A Review of Human Carcinogens. Part F, Chemical Agents and Related Occupations*, vol. 100, International Agency for Research on Cancer, Lyon, in preparation.

9 Walrath, J. and Fraumeni, J.R., Jr. (1982) Proportionate mortality among New York embalmers, in *Proceedings of the Third Annual CIIT Conference: Formaldehyde Toxicity*, Hemisphere Publishing Corp., New York.

10 Wong, O. (1982) An epidemiologic mortality study of a cohort of chemical workers potentially exposed to formaldehyde, with a discussion on SMR and PMR, in *Proceedings of the Third Annual CIIT Conference: Formaldehyde Toxicity*, Hemisphere Publishing Corp., New York.

11 Marsh, G.M. (1983) Proportional mortality among chemical workers exposed to formaldehyde, in *Proceedings of the Third CIIT Annual Conference: Formaldehyde Toxicity*, Hemisphere Publishing Corp., New York.

12 Harrington, J.M. and Shannon, H.S. (1975) Mortality study of pathologists and medical laboratory technicians. *Br. Med. J.*, **4**, 329–332.

13 Blair, A., Stewart, P., O'Berg, M., Gaffey, W., Walrath, J., Ward, J., Baies, R., Kaplan, S., and Cubit, D. (1986) Mortality among industrial workers exposed to formaldehyde. *J. Natl. Cancer Inst.*, **76**, 1071–1084.

14 Olsen, J.H., Jensen, S.P., Hink, M., Faurbo, K., Breum, N.O., and Jensen, O.M. (1984) Occupational formaldehyde exposure and increased nasal cancer risk in man. *Int. J. Cancer*, **34**, 639–664.

15 Vaughan, T.L., Strader, C., Davis, S., and Daling, J.R. (1986) Formaldehyde and cancers of the pharynx, sinus and nasal cavity. I. Occupational exposures. *Int. J. Cancer*, **38**, 677–683.

16 Vaughan, T.L., Strader, C., Davis, S., and Daling, J.R. (1986) Formaldehyde and cancers of the pharynx, sinus and nasal cavity. II. Residential exposures. *Int. J. Cancer*, **38**, 685–688.

17 Blair, A., Stewart, P.A., Hoover, R.N., Fraumeni, J.F., Jr., Walrath, J., O'Berg, M., and Gaffey, W. (1987) Cancers of the nasopharynx and oropharynx and formaldehyde exposure. *J. Natl. Cancer Inst.*, **78**, 191–192.

18 Walrath, J. and Fraumeni, J.F., Jr. (1983) Mortality patterns among embalmers. *Int. J. Cancer*, **31**, 407–411.

19 Walrath, J. and Fraumeni, J.F., Jr. (1984) Cancer and other causes of death among embalmers. *Cancer Res.*, **44**, 4638–4641.

20 Stroup, N.E., Blair, A., and Erikson, G.E. (1986) Brain cancer and other causes of death in anatomists. *J. Natl. Cancer Inst.*, **77**, 1217–1224.

21 Levine, R.J., Andjelkovich, D.A., and Shaw, L.K. (1984) The mortality of Ontario undertakers and a review of formaldehyde-related mortality studies. *J. Occup. Med.*, **26**, 740–746.

22 Fayerweather, W.E., Pell, S., and Bender, J.R. (1983) Case–control study of cancer deaths in DuPont workers with potential exposure to formaldehyde, in

Formaldehyde: Toxicology, Epidemiology, Mechanisms (eds J.J. Clary, J.E. Gibson, and R.S. Waritz), Marcel Dekker, New York, pp. 47–125.

23 Blair F A., Stewart, P.A., and Hoover, R.N. (1990) Mortality from lung cancer among workers employed in formaldehyde industries. *Am. J. Ind. Med.*, **17**, 683–699.

24 Blair, A., Saracci, R., Stewart, P.A., Hayes, R.B., and Shy, C. (1990) Epidemiologic evidence on the relationship between formaldehyde exposure and cancer. *Scand. J. Work Environ. Health*, **16**, 381–393.

25 Partanen, T. (1993) Formaldehyde exposure and respiratory cancer: a meta-analysis of the epidemiologic evidence. *Scand. J. Work Environ. Health*, **19**, 8–15.

26 Hayes, R.B., Blair, A., Stewart, P.A., Herrick, R.F., and Mahar, H. (1990) Mortality of US embalmers and funeral directors. *Am. J. Ind. Med.*, **18**, 641–652.

27 Hall, A., Harrington, J.M., and Aw, T.-C. (1991) Mortality study of British pathologists. *Am. J. Ind. Med.*, **20**, 83–89.

28 Hauptmann, M., Lubin, J.H., Stewart, P.A., Hayes, R.B., and Blair, A. (2004) Mortality from solid cancers among workers in formaldehyde industries. *Am. J. Epidemiol.*, **159**, 1117–1130.

29 Hansen, J. and Olsen, J.H. (1995) Formaldehyde and cancer morbidity among male employees in Denmark. *Cancer Causes Control*, **6**, 354–360.

30 Vaughan, T.L., Stewart, P.A., Teschke, K., Lynch, C.F., Swanson, G.M., Lyon, J.L., and Berwick, M. (2000) Occupational exposure to formaldehyde and wood dust and nasopharyngeal carcinoma. *Occup. Environ. Med.*, **57**, 376–384.

31 Collins, J.J., Acquavella, J.F., and Esmen, N.A. (1997) An updated meta-analysis of formaldehyde exposure and upper respiratory tract cancers. *J. Occup. Environ. Med.*, **39**, 639–651.

32 Hauptmann, M., Lubin, J.H., Stewart, P.A., Hayes, R.B., and Blair, A. (2003) Mortality from lymphohematopoietic malignancies among workers in formaldehyde industries. *J. Natl. Cancer Inst.*, **95**, 1615–1623.

33 Pinkerton, L., Hein, M., and Stayner, L. (2004) Mortality among a cohort of garment workers exposed to formaldehyde: an update. *Occup. Environ. Med.*, **61**, 193–200.

34 Coggon, D., Harris, E.C., Poole, J., and Palmer, K.T. (2003) Extended follow-up of a cohort of British chemical workers exposed to formaldehyde. *J. Natl. Cancer Inst.*, **21**, 1608–1614.

35 Beane Freeman, L.E., Blair, A., Lubin, J.H., Stewart, P.A., Hayes, R.B., Hoover, R.N., and Hauptmann, M. (2009) Mortality from lymphohematopoietic malignancies among workers in formaldehyde industries: the National Cancer Institute Cohort. *J. Natl. Cancer Inst.*, **101**, 751–761.

36 Hauptmann, M., Stewart, P.A., Lubin, J.H., Beane Freeman, L.E., Hornung, R.W., Herrick, R.F., Hoover, R.N., Fraumeni, J.F., Jr., Blair, A., and Hayes, R.B. (2009) Mortality from lymphohematopoietic malignancies and brain cancer among embalmers exposed to formaldehyde. *J. Natl. Cancer Inst.*, **101**, 1696–1708.

37 Logue, J.N., Barrick, M.K., and Jessup, G.L., Jr. (1986) Mortality of radiologists and pathologists in the Radiation Registry of Physicians. *J. Occup. Med.*, **28**, 91–99.

38 Zhang, L., Tang, X., Rothman, N., Vermeulen, R., Ji, Z., Shen, M., Qiu, C., Guo, W., Liu, S., Reiss, B., Beane Freeman, L., Ge, Y., Hubbard, A.E., Hua, M., Blair, A., Galvan, N., Ruan, X., Alter, B.P., Xin, K.X., Li, S., Moore, L.E., Kim, S., Xie, Y., Hayes, R.B., Azuma, M., Hauptmann, M., Xiong, J., Stewart, P., Li, L., Rappaport, S.M., Huang, H., Fraumeni, J.F., Jr., Smith, M.T., and Lan, Q. (2010) Occupational exposure to formaldehyde, hematotoxicity and leukemia-specific chromosome changes in cultured myeloid progenitor cells. *Cancer Epidemiol. Biomarkers Prev.*, **19** 80–88.

39 Baan, R., Grosse, Y., Straif, K., Secretan, B., El Ghissassi, F., Bouvard, V., Benbrahim-Tallaa, L., Guha, N., Freeman, C., Galichet, L., Cogliano, V., and WHO International Agency for Research on Cancer Monograph Working Group (2009) A review of human carcinogens – Part F: chemical agents and related occupations. *Lancet Oncol.*, **10**, 1143–1144.

40 IARC (2002) *IARC Monographs on the Evaluation of Carcinogenic Risks to Humans: Some Traditional Herbal Medicines, Some Mycotoxins, Naphthalene and Styrene*, vol. 82, International Agency for Research on Cancer, Lyon, pp. 43–168.

41 Mabberley, D.J. (1995) *The Plant Book*, 2nd edn, Cambridge University Press, Cambridge, UK.

42 WHO (1997) Medicinal plants in China, in *A Selection of 150 Commonly Used Species, Western Pacific Series No. 2*, WHO Regional Publications, World Health Organization, Manila.

43 EMEA (European Agency for the Evaluation of Medicinal Products) (2000) Working Party on Herbal Medicinal Products: Position paper on the risks associated with the use of herbal products containing *Aristolochia* species (EMEA/HMPWP/23/00), London.

44 Vanherweghem, J.-L., Depierreux, M., Tielemans, C., Abramowicz, D., Dratwa, M., Jadoul, M., Richard, C., Vandervelde, D., Verbeelen, D., Vanhaelen-Fastre, R., and Vanhaelen, M. (1993) Rapidly progressive interstitial renal fibrosis in young women: association with slimming regimen including Chinese herbs. *Lancet*, **341**, 387–391.

45 Cosyns, J.P., Jadoul, M., Squifflet, J.P., Van Cangh, P.J., and van Ypersele de Strihou, C. (1994) Urothelial malignancy in nephropathy due to Chinese herbs [letter]. *Lancet*, **344** (8916), 188.

46 Depierreux, M., Van Damme, B., Vanden Houte, K., and Vanherweghem, J.-L. (1994) Pathologic aspects of a newly described nephropathy related to the prolonged use of Chinese herbs. *Am. J. Kidney Dis.*, **24**, 172–180.

47 Vanherweghem, J.-L. (1998) Misuse of herbal remedies: the case of an outbreak of terminal renal failure in Belgium (Chinese herbs nephropathy). *J. Altern. Complement. Med.*, **4**, 9–13.

48 Nortier, J.L., Martinez, M.C., Schmeiser, H.H., Arlt, V.M., Bieler, C.A., Petein, M., Depierreux, M.F., De Pauw, L., Abramowicz, D., Vereerstraeten, P., and Vanherweghem, J.-L. (2000) Urothelial carcinoma associated with the use of a Chinese herb (*Aristolochia fangchi*). *N. Engl. J. Med.*, **342**, 1686–1692.

49 Vanhaelen, M., Vanhaelen-Fastre, R., But, P., and Vanherweghem, J.-L. (1994) Identification of aristolochic acid in Chinese herbs (Letter to the Editor). *Lancet*, **343**, 174.

50 Buckingham, J. (ed.) (2001) *Dictionary of Natural Products on CD-ROM*, CRC Press/Chapman & Hall, Boca Raton, FL.

51 Arlt, V.M., Stiborova, M., and Schmeiser, H.H. (2002) Aristolochic acid as a probable human cancer hazard in herbal remedies: a review. *Mutagenesis*, **17**, 265–277.

52 IARC (2009) *IARC Monographs on the Evaluation of Carcinogenic Risks to Humans: A Review of Human Carcinogens. Part A, Pharmaceuticals*, vol. 100, International Agency for Research on Cancer, Lyon, in preparation.

53 Cosyns, J.P. (2003) Aristolochic acid and 'Chinese herbs nephropathy': a review of the evidence to date. *Drug Safety*, **26**, 33–48.

54 Tanchev, I., Evstatiev, Ts., Dorosiev, D., Pencheva, Zh., and Tsvetkov, G. (1956) Study of nephritis in Vrattsa district. *Suvr. Med. (Sofia)*, **7**, 14–29 (in Bulgarian).

55 O'Brien, E. and Dietrich, D.R. (2005) Ochratoxin A: the continuing enigma. *Crit. Rev. Toxicol.*, **35**, 33–60.

56 Grollman, A.P. and Jelaković, B. (2007) Role of environmental toxins in endemic (Balkan) nephropathy. *J. Am. Soc. Nephrol.*, **18**, 2817–2823.

57 Ivić, M. (1969) The problem of etiology of endemic nephropathy. *Liječ Vjes*, **91**, 1273–1281.

58 Hranjec, T., Kovač, A., Kos, J., Mao, W., Chen, J.J., Grollman, A.P., and Jelaković, B. (2005) Endemic nephropathy: the case for chronic poisoning by *Aristolochia*. *Croat. Med. J.*, **46**, 116–125.

59 Grollman, A.P., Shibutani, S., Moriya, M., Miller, F., Wu, L., Moll, U., Suzuki, N., Fernandes, A., Rosenquist, T., Medverec, Z., Jakovina, K., Brdar, B., Slade, N., Turesky, R.J., Goodenough, A.K., Rieger, R., Vukelić, M., and Jelaković, B. (2007) Aristolochic acid and the etiology of endemic (Balkan)

nephropathy. *Proc. Natl. Acad. Sci. USA*, **104**, 12129–12134.

60 Schmeiser, H.H., Janssen, J.W., Lyons, J., Scherf, H.R., Pfau, W., Buchmann, A., Bartram, C.R., and Wiessler, M. (1990) Aristolochic acid activates *Ras* genes in rat tumors at deoxyadenosine residues. *Cancer Res.*, **50**, 5464–5469.

61 Liu, Z., Hergenhahn, M., Schmeiser, H.H., Wogan, G.N., Hong, A., and Hollstein, M. (2004) Human tumor p53 mutations are selected for in mouse embryonic fibroblasts harboring a humanized p53 gene. *Proc. Natl. Acad. Sci. USA*, **101**, 2963–2968.

62 Feldmeyer, N., Schmeiser, H.H., Muehlbauer, K.R., Belharazem, D., Knyazev, Y., Nedelko, T., and Hollstein, M. (2006) Further studies with a cell immortalization assay to investigate the mutation signature of aristolochic acid in human p53 sequences. *Mutat. Res.*, **608**, 163–168.

63 Grosse, Y., Baan, R., Straif, K., Secretan, B., El Ghissassi, F., Bouvard, V., Benbrahim-Tallaa, L., Guha, N., Galichet, L., Cogliano, V., and WHO International Agency for Research on Cancer Monograph Working Group (2009) A review of human carcinogens – Part A: pharmaceuticals. *Lancet Oncol.*, **10**, 13–14.

64 Recio, L., Sisk, S., Pluta, L., Bermudez, E., Gross, E.A., Chen, Z., Morgan, K., and Walker, C. (1992) p53 Mutations in formaldehyde-induced nasal squamous cell carcinomas in rats. *Cancer Res.*, **52**, 6113–6116.

65 Effert, P., McCoy, R., Abdel-Hamid, M., Flynn, K., Zhang, Q., Busson, P., Tursz, T., Liu, E., and Raab-Traub, N. (1992) Alterations of the *p53* gene in nasopharyngeal carcinoma. *J. Virol.*, **66**, 3768–3775.

66 Sun, Y., Hegamyer, G., Cheng, Y.J., Hildesheim, A., Chen, J.Y., Chen, I.H., Cao, Y., Yao, K.T., and Colburn, N.H. (1992) An infrequent point mutation of the *p53* gene in human nasopharyngeal carcinoma. *Proc. Natl. Acad. Sci. USA*, **89**, 6516–6520.

67 Spruck, C.H., 3rd, Tsai, Y.C., Huang, D.P., Yang, A.S., Rideout, W.M., 3rd, Gonzalez-Zulueta, M., Choi, P., Lo, K.W., Yu, M.C., and Jones, P.A. (1992) Absence of *p53* gene mutations in primary nasopharyngeal carcinomas. *Cancer Res.*, **52**, 4787–4790.

68 Ward, E., Schulte, P.A., Straif, K., Hopf, N.B., Caldwell, J.C., Carreón, T., DeMarini, D.M., Fowler, B.A., Goldstein, B.D., Hemminki, K., Husgafvel-Pursiainen, K., Kuempel, E., Lewtas, J., Lunn, R.M., Lynge, E., McElvenny, D.M., Muhle, H., Nasu-Nakajima, T., Robertson, L.W., Rothman, N., Ruder, A.M., Schubauer-Berigan, M.K., Siemiatycki, J., Silverman, D., Smith, M.T., Sorahan, T., Steenland, K., Stevens, R.G., Vineis, P., Zahm, S.H., Zeise, L., and Cogliano, V.J. (2010) Research recommendations for selected IARC-classified agents. *Env. Health Perspect.*, **118**, 1355–1362.

Part Two
Epidemiological Research

Cancer Risk Evaluation: Methods and Trends,
Edited by Günter Obe, Burkhard Jandrig, Gary E. Marchant, Holger Schütz, and Peter M. Wiedemann.

5
The Role of Epidemiology in Cancer Risk Assessment of Nonionizing Radiation

Joachim Schüz, Gabriele Berg-Beckhoff, Brigitte Schlehofer, and Maria Blettner

5.1
Introduction

Epidemiology is the study of the occurrence and distribution of diseases in populations [1, 2]. Its ultimate goal is to determine prevalence and causes of disease, thereby leading to effective preventive measures. In the field of disease etiology, epidemiological studies are usually observational and are therefore vulnerable to bias and confounding.

Different types of observational studies have different weights in risk assessment: more confidence can be placed on results derived from well-conducted prospective cohort studies than on those from case–control studies, whereas cross-sectional or ecological studies do not allow firm conclusions. Because the different study types have different advantages and disadvantages, and because of the caveats associated with observational studies, conflicting epidemiological data are often difficult to reconcile [3]. Conclusions can be drawn from epidemiological studies only when their strengths and limitations have been assessed critically.

Epidemiological studies on possible adverse health effects from exposure to electromagnetic fields (EMFs) are particularly challenging. First, at exposure levels below those of international guidelines to protect the public from EMF-related health effects [4], no mechanism of interaction between the fields and the human body has been established; hence, it is a difficult question how to measure exposure: Is the relevant exposure best measured by lifetime cumulative exposure, exposure during certain time windows, peak exposure, average exposure, or change in exposure over time? Second, exposure to both extremely low-frequency (ELF) and radio frequency (RF) EMFs is ubiquitous; hence, rather than investigating disease risk in nonexposed versus exposed, one has to examine disease risk at various exposure levels, often with little variability of exposure in the population or an intraindividual variability of exposure over time that is much larger than interindividual variability. Third, assessment of peoples' exposure to EMF is difficult and expensive to conduct. Therefore, many observational studies use proxy measures of exposure, for example, amount of use of mobile phones as a proxy for RF exposure to the head. These proxy measures often rely on self-reported information collected from study participants

Cancer Risk Evaluation: Methods and Trends,
Edited by Günter Obe, Burkhard Jandrig, Gary E. Marchant, Holger Schütz, and Peter M. Wiedemann.

with uncertain accuracy of precision. Fourth, many of the outcomes investigated in relation to EMF are rare, for example, childhood leukemia that affects about 5 per 100 000 children per year [5] or malignant brain tumors that affect about 4–6 per 100 000 persons per year [6]. Thus, studies have to be large enough to achieve sufficient statistical power to detect even modest associations, if there are any.

In this chapter, two examples from EMF research are used to illustrate the difficulties in risk assessment based on epidemiological studies. The first example is the relationship between ELF EMF and the risk of childhood leukemia. Based on epidemiological studies, there appears to be a broad consensus that ELF magnetic fields are possibly carcinogenic to humans, reflecting a high consistency among epidemiological studies showing an approximately twofold increased risk with average exposure above 0.3–0.4 μT but a lack of support from experimental research [7–10]. The second example is the relationship between RF fields, specifically those emitted from mobile phones, and the risk of brain tumors. Most scientific reviews come to similar conclusions, namely, that there is no evidence for an increased risk from up to 10 years of mobile phone use and little evidence for an increased risk from longer term use, although data of longer term users are sparse and a risk increase cannot be ruled out [10–12]. However, there is ongoing public concern and some stakeholders demand stricter exposure limits (e.g., the BioInitiative group). The focus of this chapter is to describe difficulties in risk assessment when there is large uncertainty in the data.

5.2 Brief Outline of Common Epidemiological Study Designs

The observational study types range from rather simple descriptive studies (e.g., ecological studies) to analytical studies such as cross-sectional, case–control, and cohort studies. In ecological studies, exposure information is available only at an aggregate level, that is, disease rates across populations are compared, while one particular exposure level is assigned to each population irrespective of the exposure distribution within the respective population. In cross-sectional studies, information on risk factors and on the effects of interest is obtained simultaneously; hence, such an approach is often not appropriate for studying chronic diseases. In case–control studies, one group consists of people having the disease of interest (cases) while the other group consists of people free of this disease (controls). The two groups are then compared with regard to differences in their past exposures. The nature of a case–control study is therefore retrospective, as the direction of inquiry is from disease to exposure. Advantages of case–control studies are the shorter time frame for getting results and thereby lower cost compared to cohort studies and that they allow the investigation of a wide range of possible risk factors for the disease of interest. The challenge in a case–control study is to find a suitable control group representative of the total population from which the cases arose and to obtain an accurate measurement of past exposures. Exposure is often assessed by conducting interviews, the accuracy of which depends on the individual's ability to recall the past events.

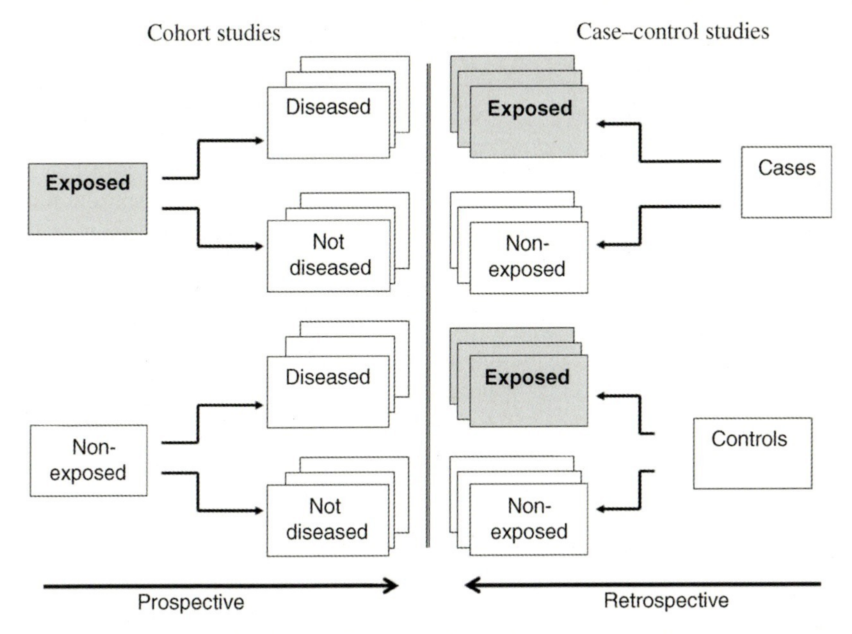

Figure 5.1 Overview of the design of cohort studies and case–control studies.

Cohort studies reflect the time course of a disease, from cause to effect. In cohort studies, groups are selected on the basis of their exposure status and then followed up in time with respect to occurrence of the diseases of interest. Hence, multiple outcomes can be studied for any one exposure. Cohort studies, however, are usually expensive and time consuming, especially for rare diseases with long induction periods. Challenges are how to update exposure status over time and how to avoid losses of participants during follow up. Many cohort studies are retrospective, based on the use of preexisting records for exposure assessment. The direction of inquiry in retrospective cohort studies is still prospective, but the date of entry into the cohort is in the past (Figure 5.1 illustrates the design of case–control and cohort studies).

A large majority of studies on ELF magnetic fields and childhood leukemia fall into two types of case–control studies. Type 1 studies are register-based case–control studies, which use the distance between a power line and the child's residence and the power load of the respective power line to estimate the residential ELF magnetic field exposure. Cases are obtained from population-based cancer registries and controls are drawn randomly from population registers [13]. These studies were conducted only in the Nordic countries due to the availability of high-quality registers for cancer, population statistics, and power distribution. Their main advantage is the coverage of a representative population, as there are virtually no losses of subjects due to missing data or nonparticipation (as no personal contact is required). The disadvantages are that no other sources of ELF magnetic fields than power lines are captured, that little information is available on covariates, and that the Nordic countries have rather small populations.

Type 2 studies are case–control studies using measurements of ELF EMF [14]; whereas like in the register-based studies the sources of cases and controls are complete disease and population registers, this type of study requires active participation of the study families, and participation rates in these studies were usually low, that is, between 50 and 60% [7]. An advantage of type 2 studies, however, is that they collect relatively detailed information on exposure and capture all relevant sources, especially from the indoor electrical wiring and low-voltage installations that produce a larger proportion of ELF exposure in Western European homes than the nearby power lines [15].

RF EMF studies mostly focus on mobile phone use. Almost all studies on mobile phone use and the risk of brain tumors are interview-based case–control studies [12]. Due to the poor prognosis of patients with malignant brain tumors, cases have to be ascertained actively in hospitals [16], otherwise some of them die before interview. Although population registers can be used for sampling controls, interview-based studies require consent and active participation; however, participation rates are low, especially among controls, for example, 54% in the largest case–control study on this subject [16, 17]. Information on past mobile phone use is assessed by questionnaire or personal interview; however, it is challenging to obtain accurate information on phone use many years earlier [11, 12]. In a retrospective cohort study conducted in Denmark [18], 420 000 mobile phone subscribers between 1982 and 1995 were followed for brain tumors until the end of 2002 [19]. Although the entire adult population of Denmark was included in this study, neither in the cohort nor in the comparison population did the rather crude exposure proxy allow categorizing the cohort by phone use and other important exposure variables such as the side of the head the mobile phone were not available. Prospective cohort studies are under way, but they are expensive and need a long time before they can provide first results (Cosmos study).

5.3
Criteria for Evaluating the Plausibility of Epidemiological Findings

When epidemiological data are used in risk assessment, criteria are needed to evaluate the overall evidence. The most widely accepted criteria are "Hill's" criteria, described in 1965 [20]. There are plenty of examples with suggestions of slightly modified versions [3, 21], but the nature of the assessment suggested by Hill remains unchanged. Since the criteria list is a mixture of quantitative and qualitative criteria and lacks clear-cut definitions in many criteria and weights for each criterion, it is prone to subjective judgment. Poole suggests that the "criteria" are rather values or viewpoints [22], which are in line with their original proposal by Hill [20].

With regard to the first criterion, namely, *temporal relation consistent with cause and effect*, it can be a difficult question of how much time has to elapse between cause and effect before an excess risk is observable in an epidemiological study. Temporality is closely linked to biological plausibility as the biological pathway suggests the time period needed for the development of the disease. In studies on mobile phones and the risk of brain tumors, it was argued that both short-and long-term exposures may be of interest; if microwaves act as a promoter or progressor, an effect may already become apparent after a few years of exposure [18]. Cohort studies follow the natural cause–effect path and are therefore particularly suitable for addressing temporality. In case–control studies relying on the reconstruction of past exposures, it may sometimes be difficult to define the exact timing of the onset of exposure, for example, because interviewees have difficulties in recalling the time period of first exposure or, like in the studies on ELF EMF, the measurement of exposure is conducted after the diagnosis of disease but used to predict the exposure years before the date of diagnosis.

The second criterion, *strength of the association,* is a quantitative criterion suggesting that a strong relationship between exposure and effect is more likely to be a causal effect. However, if there is only a weak association, lack of a strong empirical association cannot be used as a counterargument against causality. The point estimates should also be interpreted in the context of the level of their confidence; a strong association with a high level of uncertainty may be less convincing than a more moderate association with little statistical uncertainty. In line with this, it was suggested to evaluate both strength of the association and level of significance [23]. A combined relative risk estimate from pooling the ELF EMF studies on childhood leukemia yielded a doubling in risk at average exposures of $\geq$0.4 μT [24], indicating a rather moderate association with a lower confidence boundary of 1.27. Most studies on mobile phones and brain tumors did not show an increased risk estimate, with the exception of a series of studies conducted in Sweden [25], but most of their elevated risk estimates were still below 2.

Further evidence in favor of causality is provided if increasing levels of exposure are associated with an increasing strength of the effect (*dose–response relationship,* the third criterion). Unfortunately, many researchers test only for linearity, although statistical approaches are available and well described for allowing a variety of possible dose–response relationships [26]. The issue of dose–response relationships has to be interpreted in the context of biological plausibility. Possible shapes are a linear relationship, a threshold effect, a U-shaped relationship, or other even more complex scenarios. The meta-analysis of ELF EMF and childhood leukemia studies of Ahlbom *et al.* [24] showed a threshold effect with a doubling in risk at field levels $\geq$0.4 μT but no risk increase at lower field levels. A recent update of this meta-analysis [27] focusing on ELF EMF exposures at night showed a dose–response relationship that was actually indicative of a monotonous increase. As the same trend was seen using the exposure metric applied in the first meta-analysis, it is likely that this trend is not due to the restriction of the exposure period to nighttime but due to the exclusion of the studies using calculated ELF EMF as an exposure metric (in which no distinction between day and night was made) and the inclusion of one more recent study from Germany. Most studies on mobile phones and brain tumors do not show any convincing dose–response relationship [12].

Consistency within and across studies (the fourth criterion) provides further causal evidence. If consistent results have been found across different studies, the association is more likely to be causal. There is, however, no clear-cut definition of how consistent study results have to be called "consistent." From a statistical point of view, tests for heterogeneity are available, but it may be more appropriate to approach consistency from a qualitative perspective. Consistency of results could also be due to studies suffering from the same type and magnitude of bias. Lack of consistency does not necessarily indicate absence of causality but may be due to biased studies showing no effect, while higher quality studies show an effect. Examples of lack of internal consistency are risk estimates driven by implausible outliers or strong dependency on the choice of the categorization of the exposure.

Meta-analysis is a useful tool to numerically summarize the evidence, but if substantial heterogeneity is identified, a structured approach trying to clarify the sources of such heterogeneity is more important than any calculation of pooled estimates. A good meta-analysis or review can be seen as a study of studies; hence, like original studies, they vary considerably in quality. If the individual studies suffer from bias, the pooled effect estimates are biased too. A number of meta-analyses have been published on mobile phone use and the risk of brain tumors, but all fail to provide more insight than the original studies, as they simply provide summary estimates without attempting to comment on the heterogeneity of results they detect across studies (e.g., Ref. [28] and letter by Samkange-Zeeb *et al.* [29]). If the original data can be achieved for conducting a pooled analysis rather than a literature-based meta-analysis, this has the advantage that the analytical approaches can be harmonized, for example, the choice of cutoff points [30]. With regard to the ELF EMF and childhood leukemia evidence, the pooled analyses by Ahlbom *et al.* [24] (Table 5.1) and Greenland *et al.* [31] are regarded as milestones, as none of the individual studies had enough statistical power to detect the modest association seen in the pooled effect estimates. Further, it was shown that after harmonizing the analytical approaches, the heterogeneity across studies was rather small.

Table 5.1 Studies on residential extremely low-frequency magnetic fields and the risk of childhood leukemia (studies pooled in a meta-analysis by Ahlbom *et al.* [24]).

		Leukemia cases	
	Relative risk[a)] **$\geq$0.4 μT versus $<$0.1 μT**	**Observed** **$\geq$0.4 μT**	**Expected[b)]** **$\geq$0.4 μT**
Canada	1.55 (0.65–3.68)	13	10
United States	3.44 (1.24–9.54)	17	5
United Kingdom	1.00 (0.30–3.37)	4	4
Norway	0 cases, 10 controls	0	3
Germany	2.00 (0.26–15.17)	5	2
Sweden	3.74 (1.23–11.4)	5	2
Denmark	2 cases, 0 controls	2	0
Finland	6.21 (0.68–56.9)	1	0
New Zealand	0 cases, 0 controls	0	0
Total	2.00 (1.27–3.13)	47	26

a) Relative risk estimated by the odds ratio and 95% confidence interval.
b) Rounded to whole numbers.

The fifth criterion, *specificity of an association*, occurs when a particular exposure increases the risk of one disease but not others. This argument has been used to strengthen the case in favor of causality, particularly in the evaluation of potential reporting bias in interview-based case–control studies. While a similar effect seen across all case groups would indicate reporting bias, an association specific to one case group provides some evidence in favor of causality, if appropriately balanced against the possibility of chance due to multiple statistical comparisons. With regard to ELF EMF, the association with childhood leukemia appears to be stronger than that for other childhood cancers, but data for other cancers are also sparse [32]. For mobile phones and brain tumors, there appear to be suggestions of an increased risk at higher exposure levels for glioma but not meningioma [17].

Noncausal explanations for observed exposure–effect associations include *bias and confounding* (described below). Absence of substantial bias and confounding would strengthen the evidence in favor of a true effect, but some bias operates in every observational study. Therefore, in practice, it is important to discuss or even make attempts to quantify the impact of bias by applying sensitivity analyses, so this assessment can be regarded as a sixth criterion. It should be noted that bias and confounding may also mask a true effect and hence their impact has to be thoroughly discussed for studies showing no associations.

Biological plausibility of the observed effects is very important (the seventh criterion). Epidemiological studies provide an estimate of the empirical association between exposure and effect. Hence, knowledge of a plausible biological mechanism or evidence from experimental results that are coherent with the epidemiological findings strongly supports positive empirical associations. As a minimum requirement, the study report should clearly state the study hypothesis and how it was addressed. Lack of biological plausibility weakens the epidemiological evidence especially when a reasonable number of sound experimental studies have been conducted; nevertheless epidemiologists like to point out that they observed the association between smoking and lung cancer years before it was accepted to be biologically plausible.

Lack of support from experimental studies is also hampering the interpretation of the epidemiological finding of a doubling of childhood leukemia risk with average ELF EMF exposures $\geq 0.3/0.4\ \mu T$ [7]. However, the development of childhood leukemia is a unique process [33] and it is not certain whether the animal models used in genotoxicity or cell proliferation studies are adequate to test the complex origin of childhood leukemia. There is strong evidence that the majority of cases of acute lymphoblastic leukemia in children are a result of two genetic "hits," and it has been proposed that the conversion rate of the preleukemic clone to overt leukemia is rather low, that is, about 1% [33]. It is unclear whether this process requires a second direct DNA damage or whether some activation of cell proliferation of otherwise resting cells leads to further damage. In the latter case, factors qualifying for this activation are not necessarily mutagenic. For RF fields, there is little evidence from experimental studies supportive of a carcinogenic effect at exposure levels typical for mobile phones [10].

A summary of the criteria applied to the ELF–childhood leukemia and the RF–brain tumor evidence is shown in Table 5.2.

Table 5.2 Discussion of the evidence from epidemiological studies on extremely low-frequency electromagnetic fields (ELF EMF) and the risk of childhood leukemia and studies on radio frequency electromagnetic fields (RF EMF) from the use of mobile phones and the risk of brain tumors.

Criterion	ELF EMF and childhood leukemia	RF EMF and brain tumors
Temporality	Fulfilled	One study[a] shows association already at low exposure levels; among them after short periods of use that appears to be biologically unlikely; most studies show no association, but an increased risk among longer term users cannot be excluded
Strength of the association	Small to modest association	Most studies show no association, but suggest weak associations in subgroups; only one study[a] shows larger effects
Dose–response relationship	Sparse data at higher exposure levels, both compatible with a threshold effect at 0.3–0.4 μT or monotonous increase from 0.1 μT	Interphone shows no clear dose–response relationship, as an increased risk is seen in the highest decile of cumulative use but not in the nine preceding deciles
Consistency of studies	Studies of higher quality show relatively high consistency of results, but numbers of exposed children are small in each individual study	Most studies are consistent as they show no association, but cannot exclude a risk increase among longer term users; only one study[a] shows elevated risks at most exposure levels; these results are in contradiction with each other
Specificity	Too few data to investigate risk by subtype of childhood leukemia	Odds ratios in the highest category appear to be slightly higher for glioma than meningioma, but the J-shape is present for all investigated tumor types (Interphone)
Absence of bias and confounding	Selection bias found to lead to an overestimation of the association, but it is unclear whether it can explain the association in its entirety; other biases and confounding not shown to play any major role	Selection bias may explain the reduced risk in the Interphone study found for ever regular users of mobile phones; recall bias appears to play a major role in all mobile phone studies and hampers a straightforward interpretation of the results
Biological plausibility	Weak biological plausibility, but natural history of childhood leukemia is not well described	Weak biological plausibility

a) The series of case–control studies by Hardell and colleagues [25] is referred to as "one study" throughout the table; the term "most other studies" applies to the international Interphone study [17] and the Danish cohort study [18, 19].

5.4
Bias and Errors in Epidemiological Studies

All observational studies are susceptible to bias and confounding, and studies on the relationship between EMF and cancer are no exception. Most important in epidemiological studies are the following three types of bias: (i) selection and participation bias, (ii) information bias in general and recall bias in particular, and (iii) confounding.

Selection bias is due to different selection criteria during the data collection. In case–control studies, it is crucial that controls are derived from the same population as cases, which can be difficult to achieve if complete population registers are lacking. In studies requiring active participation, selection bias is often due to participation patterns even if the sampling frame is excellent. The reasons for participation are mostly nonrandom. Often, case participation is higher than control participation and, among control subjects, the ones with better education or more healthy lifestyles are more willing to participate. If participation is related to the exposure of interest, bias in the risk estimates will occur, and it is important to keep this bias to a minimum. In cohort studies, one often has to deal with a healthy cohort effect; that is, cohorts constructed of persons being employed are healthier than the general population.

Information bias is due to incorrect information given by the participants themselves or results from errors in exposure assessment. In studies using interview data, recall bias is of concern, because it is generally difficult to remember past events with high accuracy. Often there is a reporting difference between cases and controls, as case reporting may be influenced by the disease itself. The physical measurements of exposure may also introduce error, as it often occurs that spot measurements are used to predict exposures that occurred long times ago.

Confounding is a problem of all studies; confounding arises if a third factor related to both exposure and disease produces a spurious effect or masks a true effect. If the confounder is known, statistical corrections can be applied, but confounding by an unknown or unmeasured factor is always a concern. Statistical tools are available to correct for confounding, under the prerequisite that the nature and magnitude of bias and error is, at least to some extent, known.

Validation studies and sensitivity analyses are helpful tools to get a more complete picture about selection bias, information bias, and confounding. Studies making no attempt in collecting additional information to evaluate the impact of bias and confounding should not be considered to be state of the art.

In the risk assessment by the International Agency for Research on Cancer [7], ELF magnetic fields were classified as possibly carcinogenic to humans (group 2B), but chance, bias, and confounding could not be ruled out with reasonable confidence. Chance alone seems unlikely, given the results of the meta-analyses [24, 27, 31]. With respect to confounding, many potential confounders have been proposed but none has been confirmed to date [9]. In statistical simulations, it has been shown that a plausible confounder would have to be in itself quite a strong risk factor for childhood leukemia and, at the same time, highly correlated with field strengths from all sources of elevated magnetic fields [34]. With regard to exposure assessment, there is huge potential for misclassification; however, it is unlikely that the extent of misclassification was associated with disease status, that is, the misclassification being different for cases and controls. In situations of a dichotomous exposure estimate affected by nondifferential exposure misclassification, the expected bias in the effect estimate would underestimate an association [1]. With regard to selection bias, the low participation rates in many of these studies were a concern, especially in studies in which exposure was measured [9]. The findings suggest that families of lower socioeconomic status were particularly underrepresented among controls. As residential ELF EMFs are inversely associated with social status, the deficit of controls with low social status presumably leads to a deficit of controls with higher ELF exposure; hence, the observed association might have been overestimated. A simulation study using data of the German study estimated that 66% of the observed association was attributable to selection bias [35]. Although selection bias alone is probably not sufficient to explain the association entirely, it clearly led to an overestimate of the relative risk estimates (Figure 5.2).

Table 5.3 shows the results for glioma from the largest case–control study on mobile phones [17]. The biologically implausible decrease in risk observed in relation to use of mobile phones for 1–9 years and the low response rate among controls suggest that mobile phone use may be associated with the likelihood of participation, either because of socioeconomic status or, more directly, as less communicative people may want to avoid both mobile phones and participation in scientific studies. The findings from a nonresponder survey carried out in the context of this study confirm this hypothesis to some extent, showing more mobile phone users among those participating than among nonparticipating subjects (both cases and controls), while participation rates were generally higher among cases [36]. The possibility that

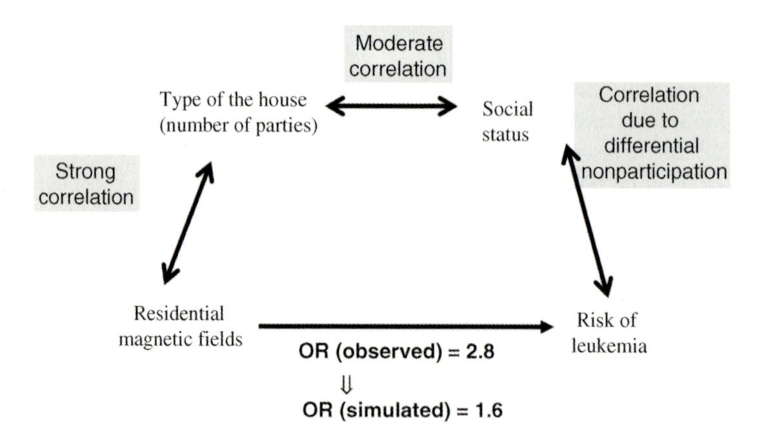

Figure 5.2 The route how selection bias has an impact on the observed association between residential extremely low-frequency electromagnetic fields and the risk of childhood leukemia; there is a correlation between field strength and type of the house and between type of the house and socioeconomic position of the family, while controls of lower socioeconomic position were underrepresented in the sample due to higher nonparticipation in this group.

Table 5.3 Mobile phone use and the risk of glioma: relative risk estimates for regular mobile phone use[a)] and for ipsilateral versus contralateral use[b)] (modified from Interphone [17]).

Exposure	Regular use relative risk[c)]	Ipsilateral relative risk[c)]	Contralateral relative risk[c)]	Ratio I/C[d)]
Time since first regular use				
Not exposed	1.00	1.00	1.00	
1–1.9 years	0.62 (0.46–0.81)	0.77 (0.49–1.20)	0.38 (0.20–0.71)	2.03
2–4 years	0.84 (0.70–1.00)	0.80 (0.62–1.04)	0.81 (0.57–1.14)	0.99
5–9 years	0.81 (0.67–0.97)	0.81 (0.62–1.05)	0.65 (0.44–0.95)	1.25
10+ years	0.98 (0.76–1.26)	1.21 (0.80–1.80)	0.70 (0.42–1.15)	1.73

a) One call per week or more over a period of 6 months or longer.
b) Ipsilateral use: preferred side of the head during mobile phone use corresponds to the side of the head of the tumor; contralateral use: preferred side of the head during mobile phone use corresponds to the side of the head opposite of the tumor.
c) Relative risk estimated by the odds ratio and 95% confidence interval.
d) Ratio of the relative risk estimates for ipsilateral use by contralateral use.

people with brain tumors might avoid using mobile phones because of prodromal symptoms of the disease may add to this spurious protective effect [17]. The finding that the effect among people who reported using the mobile phone mainly to the side of the head with the tumor (ipsilateral users) might be interesting due to the highly localized RF EMF exposure to the brain [37]. However, it is also an item susceptible to differential reporting error by case or control status [38]. In fact, the cases know which side of their head is affected by the tumor, while the controls do not know which side of their head will be relevant for the analyses (in a matched study, it is the side of the head where the tumor occurred in their corresponding matched case). Therefore, overreporting of the affected side of the head among cases may occur. Assuming causality, one would also expect that the effect of laterality becomes stronger with increasing exposure; that is, the ratio of the two effect estimates for ipsilateral and contralateral use would be close to one among short-term or occasional mobile phone users but would then grow with increasing exposure. However, this was not the case in the Interphone study [17], as shown in Table 5.3.

Reliance on self-reported information for assessing past mobile phone use is a major source of bias: first, because it is generally difficult to accurately recall past events (like the number of minutes of mobile phone use many years ago), most likely leading to random error. Second, because the disease itself may affect memory, this could lead to a systematic bias. A validation study within the Interphone study showed that the length of phone calls was overestimated [39]. When such reporting errors are equal in cases and controls, the strength of the association will be underestimated. Alternatively, a spurious increase in risk will be found when cases tend to overestimate more strongly than the controls. A small study based on past traffic records of Interphone participants suggested an overestimation by cases in more distant periods, which could result in a positive bias in the relative risk estimates [40]. Also, in the main study, more implausible values of mobile phone use were reported by cases compared to controls [17]. A summary of the possible impact of bias and error for the three major mobile phone studies is given in Table 5.4.

5.5
Compatibility between Study Findings and Time Trends in the Occurrence of Disease

It is important to compare risks derived from studies with time trends or regional distributions of age- and sex-specific incidence rates of the respective disease. Common risk factors, such as smoking, diet, and sun exposure, go along with visible changes in the time trends of tobacco- and diet-related cancers or skin cancer, respectively. But as many other factors have an impact on incidence rates, such as competing risk factors, improvements in diagnosis, changes in disease classification, and completeness in disease registration, the absence of concordance has to be assessed critically. If risk factors are rare or the effect is small, incidence rates are unlikely allow any insight.

The association between ELF magnetic fields and childhood leukemia is an example where incidence rates provide little useful information. The attributable risk is estimated to be around 1% of childhood leukemia cases in Western Europe and 3–4% in North America [41]. Even if this effect is assumed to be causal, such differences are too small to be seen in cancer registry data with their variable completeness by a few percent over time.

The situation for mobile phones is very different, because the widespread use of mobile phones started only in the mid-1990s and reached a very high penetration of the technology within very few years. Men in their 40s and 50s were the first to use mobile phones frequently; therefore, an effect would be expected to emerge first among middle-aged men. Deltour *et al.* [42] analyzed the age- and sex-specific incidence rates in Denmark, Sweden, Norway, and Finland up to 2003 and, based on 60 000 cases, found no change in incidence rates compatible with a strong risk related to mobile phone use (Figure 5.3). Among 20–39-year-old men, the risk for glioma has decreased in recent years, while the strongest increase was seen for meningioma among elderly women – opposite of what would be expected under a causal mobile phone effect. The authors conclude that the effect from mobile phones is either small, becomes apparent only after induction periods longer than 10 years, or there is no effect. Further follow-up of age- and sex-specific incidence rates, including the topographical distribution of tumors within the brain [37], is considered to be very informative.

5.6
Discussion

Risk assessment based on epidemiological studies on EMF is particularly difficult due to weak biological plausibility. For this reason, studies with positive associations need to be carefully evaluated to explore whether bias and error have produced spurious associations. However, it cannot be ruled out that EMF acts through a yet

Table 5.4 Discussion of bias and error in epidemiological studies on mobile phones and the risk of brain tumors.

	Interphone study [17]	**Swedish case–control study series [25]**	**Danish cohort study [18, 19]**
Selection bias	Some evidence that mobile phone users were overrepresented especially in the control sample but unclear whether this was the same for occasional and heavy users; some evidence that cases with early symptoms of the disease did not use mobile phones; both biases may lead to a spurious protective effect	Difficult to evaluate due to little information on methodology provided in the publications; concern about accurate reporting of response rates has been raised; nonconcurrent sampling between cases and controls may have lead to an overestimation of the observed association; cases with poor prognosis were excluded from the study	Whole Danish adult population included in the study
Healthy cohort effect	Randomly drawn sample	Randomly drawn sample	Lower risk of tobacco-related cancers confirms healthier lifestyle of early mobile phone users; however, there is little evidence that lifestyle factors are related to brain tumor risk
Confounding	None of the factors used for correction for confounding had any impact on the risk estimates	None of the factors used for correction for confounding had any impact on the risk estimates	No adjustment for confounding possible; however, as seen in the case–control studies, no confounding factor is known
Accuracy of exposure	Validation studies show discrepancy between self-reported mobile phone use and operator data, showing that it is generally difficult to recall past mobile phone use; this leads to a large random error that might attenuate a true association, if there is any	No attempt has been made to examine the accuracy of exposure assessment; it has not been well described in the methods of the study when and how telephone interviews were used to complement the paper questionnaires	Subscriber status predicts usage of a mobile phone only to some extent; moreover, no data on amount of use or side of the head during use were available, all leading to a possible underestimation of a true effect; therefore, only a substantial risk can be ruled out

(Continued)

Table 5.4 (*Continued*)

	Interphone study [17]	Swedish case–control study series [25]	Danish cohort study [18, 19]
Recall bias	There is some evidence that overreporting of amount of use was stronger among cases; implausible values of mobile phone use have been reported more often among cases; this might lead to a spurious effect in the highest exposure levels	No validation data	Subscriber status is objective information collected before the diagnosis of disease
Other biases	Reporting of the preferred side of the head during use of mobile phones is heavily affected by reporting bias and hampers the interpretation of these results	Reporting of the preferred side of the head during use of mobile phones is heavily affected by reporting bias and hampers the interpretation of these results	Data too sparse to investigate long-term effects by brain tumor type; therefore, results might have been diluted if there is only an association with one particular subgroup of tumors
Compatibility with brain tumor incidence	Compatible with no overall association; suggestions of small increased risks in subgroups would not affect incidence rates until now	Incompatible; increased risks seen for occasional users and after short-term use would be visible in the incidence rates	Compatible with no overall association
Summary of findings	No overall association, both for short- and long-term users; there are suggestions of an increased risk for glioma at the highest exposure levels, but bias and error prevent a causal interpretation	Elevated risks for all types of users, stronger for malignant than benign tumors; risk becomes stronger with increasing exposure; risk estimates are incompatible with trends in the brain tumor incidence rates, but too few details on methodology are available to better evaluate bias and error affecting this study	No overall association, small risk decrease in long-term users might be a chance finding; study provides evidence against a substantial brain tumor risk but leaves open the possibility of a small risk increase, especially among heavy users

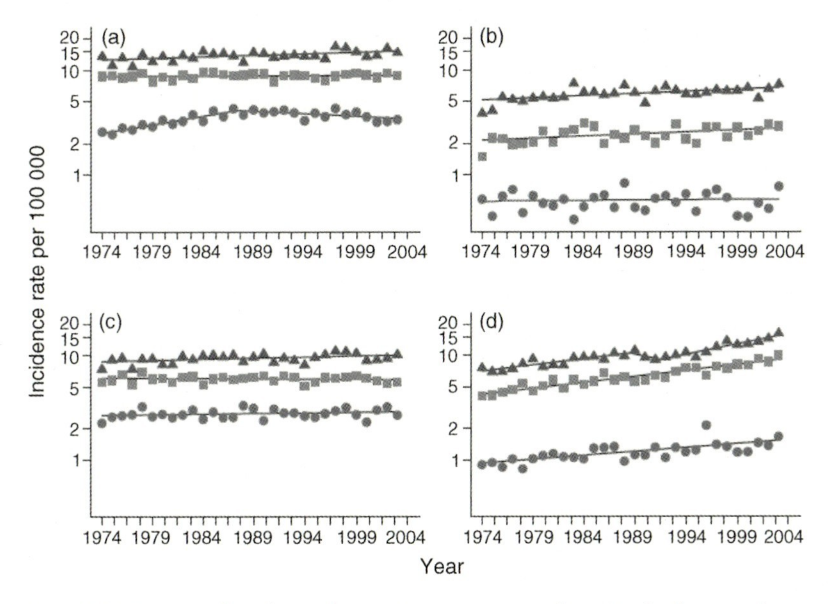

Figure 5.3 Time trends in the incidence rates of glioma and meningioma in Denmark, Finland, Norway, and Sweden, 1974–2003. (a) Glioma in men. (b) Meningioma in men. (c) Glioma in women. (d) Meningioma in women. Circles: rates for those aged 20–39 years; squares: rates for those aged 40–59 years; triangles: rates for those aged 60–79 years; solid line: regression curve. Published in Ref. [42].

unknown mechanism, which is why positive results that cannot be explained by bias need to be taken very seriously. This is especially so as exposure to EMF is ubiquitous and, particularly for widely used mobile phones, even small associations may lead to large numbers of affected persons. Epidemiological studies showing no effect also need to be carefully evaluated for bias. For example, exposure is very difficult to measure, and there is great potential that misclassification of exposure might attenuate or even completely mask a true effect.

Looking at ELF EMF and childhood leukemia, study methodology has improved considerably over time and it was therefore striking that improved studies confirmed the weak associations with a higher leukemia risk seen in the early studies [24, 27, 31]. Nevertheless, low response rates in studies with ELF magnetic field measurements over 24 h or longer raise concerns about selection bias producing a spurious positive association, as participation was associated with the parents' education and living conditions; that is, the more affluent they are, the more likely they participated [9]. It is very important to reconcile the data, not only to overcome the impasse between epidemiology and biology for this research question but also for the whole field of childhood leukemia etiology. If the ELF EMF finding is an artifact, reported associations with, for example, parental smoking, use of household insecticides, or day care attendance need to be revisited, as these positive associations are based on similarly obtained epidemiological data.

Looking at RF EMF and brain tumors, it is unlikely that further case–control studies of the same type will reveal further insights. As it appears to be very difficult for people to remember their past use of mobile phones, the inaccuracy in reporting is too large to allow detection of modest associations. At the same time, it looks very challenging to find an appropriate control group. Prospective studies may overcome the problems related to selection and reporting bias, but the studies have to be large and need a long time before firm conclusions can be drawn. Due to the widespread use of mobile phones, incidence studies are informative under the assumption that the effect is large enough to be detectable. More sophisticated study designs, making use of the fact that the RF EMF exposure from mobile phones is highly localized in the head, are needed to provide further insight. Further studies are also merited because the proportion of "heavy" long-term mobile phone users in studies to date has been rather small.

A number of methodological recommendations can be made. In order to learn from the difficulties of previous studies, all shortcomings must be appropriately documented, so that new studies can try to avoid them. Use of exposure metrics is an example: some simple metrics, such as distance from fixed EMF sources, have been shown to be poor predictors of the exposure of interest and should not be used in future studies. In general, the researchers should demonstrate how well their exposure proxies measure the actual exposure of interest, and, as described earlier, validation studies are a useful tool to compare, in a smaller sample, the comprehensive measurement data with the exposure proxy that will be applied to a larger sample. The aim of sensitivity analyses is to provide a more complete picture of an epidemiological study, and they are a valuable part of a publication; nevertheless, they must be clearly identified as such, so that less experienced readers do not misinterpret them as the main findings. The reporting of the results of epidemiological studies should follow good practice, as recently laid out by the STROBE consortium [43].

In conclusion, epidemiological studies conducted with appropriate diligence can play a key role in identifying real health risks.

References

1 Dos Santos Silva, I. (1999) *Cancer Epidemiology: Principles and Methods*, IARC Press, Lyon

2 Ahrens, W. and Pigeot, I. (2007) *Handbook of Epidemiology*, Springer, Berlin

3 Schüz, J. (2008) Epidemiology, in *The Role of Evidence in Risk Characterization: Making Sense of Conflicting Data* (eds P.M. Wiedemann and H. Schütz), Wiley-VCH Verlag GmbH, Weinheim

4 International Commission on Non-Ionizing Radiation Protection (1998) Guidelines for limiting exposure to time-varying electric, magnetic, and electromagnetic fields (up to 300 GHz). *Health Phys.*, **74**, 494–522.

5 Steliarova-Foucher, E., Stiller, C., Kaatsch, P., Berrino, F., Coebergh, J.W., Lacour, B., and Parkin, M. (2004) Geographical patterns and time trends of cancer incidence and survival among children and adolescents in Europe since the 1970s (the ACCIS project): an epidemiological study. *Lancet*, **364**, 2097–2105.

6 Bondy, M.L., Scheurer, M.E., Malmer, B., Barnholtz-Sloan, J.S., Davis, F.G., Il'yasova, D., Kruchko, C., McCarthy, B.J., Rajaraman, P., Schwartzbaum, J.A., Sadetzki, S., Schlehofer, B., Tihan, T., Wiemels, J.L., Wrensch, M., and Buffler, P.A. (2008) Brain tumor epidemiology: consensus from the Brain Tumor Epidemiology Consortium. *Cancer*, **113**, 1953–1968.

7 International Agency for Research on Cancer (2002) *IARC Monographs on the Evaluation of Carcinogenic Risks to Humans. Vol. 80. Non-Ionizing Radiation, Part 1: Static and Extremely Low-Frequency (ELF) Electric and Magnetic Fields*, IARC Press, Lyon

8 World Health Organization (2007) Extremely low frequency fields. Environmental Health Criteria Monograph No. 238, http://www.who.int/peh-emf/publications/elf_ehc/en/index.html (last accessed February 8, 2010).

9 Schüz, J. and Ahlbom, A. (2008) Exposure to electromagnetic fields and the risk of childhood leukaemia: a review. *Radiat. Prot. Dosimetry*, **132**, 202–211.

10 Scientific Committee on Emerging and Newly Identified Health Risks (2009) Health effects of exposure to EMF. European Commission, Directorate-General for Health & Consumers, http://ec.europa.eu/health/ph_risk/committees/04_scenihr/docs/scenihr_o_022.pdf (last accessed February 8, 2010).

11 Schüz, J., Lagorio, S., and Bersani, F. (2009) Electromagnetic fields and epidemiology: an overview inspired by the fourth course at the International School of Bioelectromagnetics. *Bioelectromagnetics*, **30**, 511–524.

12 Ahlbom, A., Feychting, M., Green, A., Kheifets, L., Savitz, D.A., Swerdlow, A.J., and ICNIRP Standing Committee on Epidemiology (2009) Epidemiologic evidence on mobile phones and tumor risk: a review. *Epidemiology*, **20**, 639–652.

13 Feychting, M. and Ahlbom, A. (1993) Magnetic fields and cancer in children residing near Swedish high-voltage power lines. *Am. J. Epidemiol.*, **138**, 467–481.

14 Linet, M.S., Hatch, E.E., Kleinerman, R.A., Robison, L.L., Kaune, W.T., Friedman, D.R., Severson, R.K., Haines, C.M., Hartsock, C.T., Niwa, S., Wacholder, S., and Tarone, R.E. (1997) Residential exposure to magnetic fields and acute lymphoblastic leukemia in children. *N. Engl. J. Med.*, **337**, 1–7.

15 Schüz, J., Grigat, J.P., Störmer, B., Rippin, G., Brinkmann, K., and Michaelis, J. (2000) Extremely low frequency magnetic fields in residences in Germany. Distribution of measurements, comparison of two methods for assessing exposure, and predictors for the occurrence of magnetic fields above background level. *Radiat. Environ. Biophys.*, **39**, 233–240.

16 Cardis, E., Richardson, L., Deltour, I., Armstrong, B., Feychting, M., Johansen, C., Kilkenny, M., McKinney, P., Modan, B., Sadetzki, S., Schüz, J., Swerdlow, A., Vrijheid, M., Auvinen, A., Berg, G., Blettner, M., Bowman, J., Brown, J., Chetrit, A., Christensen, H.C., Cook, A., Hepworth, S., Giles, G., Hours, M., Iavarone, I., Jarus-Hakak, A., Klaeboe, L., Krewski, D., Lagorio, S., Lönn, S., Mann, S., McBride, M., Muir, K., Nadon, L., Parent, M.E., Pearce, N., Salminen, T., Schoemaker, M., Schlehofer, B., Siemiatycki, J., Taki, M., Takebayashi, T., Tynes, T., van Tongeren, M., Vecchia, P., Wiart, J., Woodward, A., and Yamaguchi, N. (2007) The INTERPHONE study: design, epidemiological methods, and description of the study population. *Eur. J. Epidemiol.*, **22**, 647–664.

17 Interphone Study Group (2010) Brain tumour risk in relation to mobile telephone use: results of the INTERPHONE international case–control study. *Int. J. Epidemiol.*, **39**, 675–694.

18 Johansen, C., Boice, J.D., Jr., McLaughlin, J.K., and Olsen, J.H. (2001) Cellular telephones and cancer: a nationwide cohort study in Denmark. *J. Natl. Cancer Inst.*, **93**, 203–207.

19 Schüz, J., Jacobsen, R., Olsen, J.H., Boice, J.D., Jr., McLaughlin, J.K., and Johansen, C. (2006) Cellular telephone use and cancer risk: update of a nationwide

Danish cohort. *J. Natl. Cancer Inst.*, **98**, 1707–1713.

20 Hill, A.B. (1965) The environment and disease: association or causation? *J. R. Soc. Med.*, **58**, 295–300.

21 Weed, D.L. (2005) Weight of evidence: a review of concept and methods. *Risk Anal.*, **25**, 1545–1557.

22 Poole, C. (2001) Causal values. *Epidemiology*, **12**, 139–141.

23 Proctor, D.M., Otani, J.M., Finley, B.L., Paustenbach, D.J., Bland, J.A., Speizer, N., and Sargent, E.V. (2002) Is hexavalent chromium carcinogenic via ingestion? A weight-of-evidence review. *J. Toxicol. Environ. Health*, **65**, 701–746.

24 Ahlbom, A., Day, N., Feychting, M., Roman, E., Skinner, J., Dockerty, J., Linet, M., McBride, M., Michaelis, J., Olsen, J.H., Tynes, T., and Verkasalo, P.K. (2000) A pooled analysis of magnetic fields and childhood leukaemia. *Br. J. Cancer*, **83**, 692–698.

25 Hardell, L., Mild, K.H., Carlberg, M., and Söderqvist, F. (2006) Tumour risk associated with use of cellular telephones or cordless desktop telephones. *World J. Surg. Oncol.*, **4**, 74

26 Greenland, S. (1995) Dose–response and trend analysis in epidemiology: alternatives to categorical analysis. *Epidemiology*, **6**, 356–365.

27 Schüz, J., Svendsen, A.L., Linet, M., McBride, M.L., Roman, E., Feychting, M., Kheifets, L., Lightfoot, T., Mezei, G., Simpson, J., and Ahlbom, A. (2007) Night-time exposure to electromagnetic fields and childhood leukemia: an extended pooled analysis. *Am. J. Epidemiol.*, **166**, 263–269.

28 Myung, S.K., Ju, W., McDonnell, D.D., Lee, Y.J., Kazinets, G., Cheng, C.T., and Moskowitz, J.M. (2009) Mobile phone use and risk of tumors: a meta-analysis. *J. Clin. Oncol.*, **27**, 5565–5572.

29 Samkange-Zeeb, F., Schüz, J., Schlehofer, B., Berg-Beckhoff, G., and Blettner, M. (2010) Comparison of studies on mobile phone use and risk of tumors. *J. Clin. Oncol.*, **28**, e123

30 Blettner, M., Sauerbrei, W., Schlehofer, B., Scheuchenpflug, T., and Friedenreich, C. (1999) Traditional reviews, meta-analyses and pooled analyses in epidemiology. *Int. J. Epidemiol.*, **28**, 1–9.

31 Greenland, S., Sheppard, A.R., Kaune, W.T., Poole, C., Kelsh, M.A., and Childhood Leukemia-EMF Study Group (2000) A pooled analysis of magnetic fields, wire codes, and childhood leukemia. *Epidemiology*, **11**, 624–634.

32 Kheifets, L., Repacholi, M. Saunders, R., and van Deventer, E. (2005) The sensitivity of children to electromagnetic fields. *Pediatrics*, **116**, e303–e313.

33 Greaves, M. (2006) Infection, immune responses and the aetiology of childhood leukaemia. *Nat. Rev. Cancer*, **6**, 193–203.

34 Langholz, B. (2001) Factors that explain the power line configuration wiring code–childhood leukemia association: What would they look like? *Bioelectromagnetics*, **5**, S19–S31.

35 Schüz, J. (2007) Implications from epidemiologic studies on magnetic fields and the risk of childhood leukemia on protection guidelines. *Health Phys.*, **92**, 642–648.

36 Vrijheid, M., Richardson, L., Armstrong, B.K., Auvinen, A., Berg, G., Carroll, M., Chetrit, A., Deltour, I., Feychting, M., Giles, G.G., Hours, M., Iavarone, I., Lagorio, S., Lönn, S., McBride, M., Parent, M.E., Sadetzki, S., Salminen, T., Sanchez, M., Schlehofer, B., Schüz, J., Siemiatycki, J., Tynes, T., Woodward, A., Yamaguchi, N., and Cardis, E. (2009) Quantifying the impact of selection bias caused by nonparticipation in a case–control study of mobile phone use. *Ann. Epidemiol.*, **19**, 33–41.

37 Cardis, E., Deltour, I., Mann, S., Moissonnier, M., Taki, M., Varsier, N., Wake, K., and Wiart, J. (2008) Distribution of RF energy emitted by mobile phones in anatomical structures of the brain. *Phys. Med. Biol.*, **53**, 2771–2783.

38 Schüz, J. (2009) Lost in laterality: interpreting "preferred side of the head during mobile phone use and risk of brain tumour" associations. *Scand. J. Public Health*, **37**, 664–667.

39 Vrijheid, M., Deltour, I., Krewski, D., Sanchez, M., and Cardis, E. (2006) The effects of recall errors and of selection bias in epidemiologic studies of mobile phone use and cancer risk. *J. Expo. Sci. Environ. Epidemiol.*, **16**, 371–384.

40 Vrijheid, M., Armstrong, B.K., Bédard, D., Brown, J., Deltour, I., Iavarone, I., Krewski, D., Lagorio, S., Moore, S., Richardson, L., Giles, G.G., McBride, M., Parent, M.E., Siemiatycki, J., and Cardis, E. (2009) Recall bias in the assessment of exposure to mobile phones. *J. Expo. Sci. Environ. Epidemiol.*, **19**, 369–381.

41 Greenland, S. and Kheifets, L. (2006) Leukemia attributable to residential magnetic fields: results from analyses allowing for study biases. *Risk Anal.*, **26**, 471–482.

42 Deltour, I., Johansen, C., Auvinen, A., Feychting, M., Klaeboe, L., and Schüz, J. (2009) Time trends in brain tumor incidence rates in Denmark, Finland, Norway and Sweden, 1974–2003. *J. Natl. Cancer Inst.*, **101**, 1721–1724.

43 von Elm, E., Altman, D.G., Egger, M., Pocock, S.J., Gøtzsche, P.C., Vandenbroucke, J.P., and STROBE Initiative (2007) The Strengthening the Reporting of Observational Studies in Epidemiology (STROBE) statement: guidelines for reporting observational studies. *Lancet*, **370**, 1453–1457.

6
The Role of Epidemiology in Cancer Risk Assessment of Ionizing Radiation

Richard Wakeford

6.1
Introduction

Laboratory experiments have established that ionizing radiation can cause cancer [1, 2]. However, at present, it is not possible on the basis of experimental evidence alone to determine the degree of cancer risk in humans arising from a given level of exposure to radiation [3]. As a consequence, human cancer risk estimates are derived from the epidemiological study of suitably exposed populations [1–5].

Epidemiological studies of exposure to ionizing radiation, as with other epidemiological studies, need to be designed, conducted, and interpreted with considerable care if misleading findings are to be avoided – epidemiology is largely an observational (i.e., nonexperimental) science that makes use of data generated by the uncontrolled conditions of everyday life, so that in addition to the concern that a statistical association could have been generated by the play of chance alone (a concern that also affects experimental studies such as randomized controlled clinical trials), bias and confounding have to be seriously considered as potential explanations for epidemiological associations as well as a cause-and-effect interpretation. It is generally the case that the findings of epidemiological studies of exposure to radiation, especially small studies, need to be viewed against the backdrop of available scientific evidence to avoid the potentially misleading interpretation of the results of a single study. The United Nations Scientific Committee on the Effects of Atomic Radiation (UNSCEAR) in its 1994 Report [6] summarized the situation thus:

"Studies of disease in human populations must adhere strictly to epidemiological principles in order to achieve valid quantitative results. These include sound case ascertainment, an appropriate comparison group, sufficient follow-up, an accounting for confounding factors and well-characterised dosimetry."

Certain scientific units are used in the study of ionizing radiation. The *absorbed dose* of radiation is the energy deposited per unit mass of material and is measured in gray (Gy): 1 Gy = 1 J/kg. However, the relevant biological damage at the DNA level

Cancer Risk Evaluation: Methods and Trends,
Edited by Günter Obe, Burkhard Jandrig, Gary E. Marchant, Holger Schütz, and Peter M. Wiedemann.

produced by ionizing radiation depends on the ionization density of the particular radiation, and for the purposes of radiological protection, this is taken into account by the *equivalent dose*, which is the absorbed dose adjusted by the *radiation weighting factor* (w_R); for example, the w_R for sparsely ionizing γ-rays is 1, while the w_R for densely ionizing α-particles is 20. The equivalent dose is measured in sievert (Sv) [3].

Quantitative estimates of the risk of cancer following exposure to ionizing radiation derive from four categories of epidemiological studies: the Japanese atomic bomb survivors, those irradiated for medical purposes, occupational exposure, and irradiation from sources in the environment [3–5, 7].

6.2
Japanese Atomic Bomb Survivors

The Japanese survivors of the atomic bombings of Hiroshima and Nagasaki in August 1945 are treated as a separate study category because of the importance of this population as a source of quantitative information on the risk of cancer consequent to exposure to ionizing radiation [3–5]. Substantial effort has been expended in the epidemiological study of these survivors and the findings form the mainstay of radiation-related cancer risk estimates [3–5]. The survivors consist of a large group of people drawn from a general population of both sexes and all ages (although the proportion of men of an age to potentially serve in the military will have been lower than in peacetime) who were not selected for exposure for a particular reason (e.g., a known or suspected medical condition), but just happened to be in the wrong place at the wrong time. As such, the survivors represent a suitable basis for the estimation of the radiation-induced risk of cancer in a general population of people [3–5].

It was in the late 1940s that alert clinicians first noted an unusually large number of cases of leukemia among the Japanese survivors of the atomic bombings and this informal observation was one of the reasons behind the ambitious plan to conduct a large prospective study of the survivors [8]. The Japanese national census of October 1950 was taken as the opportunity to identify those Japanese citizens who had been present in Hiroshima or Nagasaki (or in a few unfortunate instances, both) during the atomic bombings of the cities. From this emerged the Life Span Study (LSS), a large cohort of survivors that is still the subject of epidemiological study today [9]. Established in 1958 and nested within the LSS is the Adult Health Study (AHS), consisting of $\sim$20 000 survivors, the members of which undergo biennial health examinations allowing disease morbidity to be investigated for a variety of conditions [10]. Other cohorts are also followed, such as the survivors who were *in utero* at the time of the explosions [11], and the offspring of the survivors (the F1 study) who were conceived after the bombings and who are investigated to determine any hereditary effects (including cancer) induced by irradiation [12].

One of the strengths of the LSS cohort is the individual organ dose estimates that have been reconstructed over the years. Each survivor was interviewed or responded to a questionnaire to determine his or her location during the explosions. Radiation source data from weapons tests in combination with sophisticated computational techniques for radiation transport and shielding, together with measurements of induced long-lived radioactivity in the environment, were used to determine the radiation dose to each organ of each survivor. The exposure was brief and mainly to high-energy γ-rays from the explosion itself and from the rising fireball, but a small component of the absorbed dose was due to neutrons, and this dose needs to be weighted to account for the higher ionization density of tracks produced by neutrons in tissue so that the organ doses of survivors are frequently expressed in sieverts. The most recent dose estimates are found in the Dosimetry System 2002 (DS02), which has superseded the previous DS86 database. Of the ~93 500 individuals that form the LSS (those within 10 km of the explosions and including almost all the survivors who were closest to the detonations), ~86 500 have been assigned DS02 doses; for ~7000 survivors, DS02 doses could not be computed. Of the survivors with DS02 doses, ~49 000 are estimated to have received nontrivial doses (doses ≥5 mSv) and ~31 500 (64%) of these survivors received doses <100 mSv, while nearly over 2000 (4%) received doses in excess of 1 Sv. So around two-thirds of the LSS members received doses that can be considered as "low," although much of the statistical power of the LSS derives from the survivors who received higher doses [13].

The Japanese family registration system (*koseki*) has permitted virtually complete knowledge of the vital status of LSS members, with underlying cause of death being determined from death certificates. Cancer incidence studies are also made possible by specialist cancer registries maintained in Hiroshima and Nagasaki for residents of the cities: cases of leukemia, lymphoma, and multiple myeloma have been recorded since 1950 and cases of other cancers from 1958. The task of collecting, collating, and conducting the initial analyses of the data pertaining to the Japanese atomic bomb survivors falls to the Radiation Effects Research Foundation (RERF), a joint Japanese–US organization based in Hiroshima and Nagasaki [13].

Leukemia was the first malignant neoplastic disease found to be in excess in the LSS, which is not surprising given the earlier informal observation of a notable excess of cases among the survivors in the late 1940s [8]. The latest analysis of leukemia mortality covering deaths among both sexes at all ages during 1950–2000 [13] found an excess relative risk (ERR), the proportional increase in risk over background, at an equivalent dose to the red bone marrow (RBM) of 1 Sv of 4.02 (90% confidence interval (CI): 3.02, 5.26). The slope of the dose response is highly significantly positive and shows upward curvature at moderate doses, so that it is linear–quadratic in shape. At high doses (>2 Sv), cell killing in the red bone marrow causes the slope of the dose response for leukemia to reduce and turn over. About half of the ~200 leukemia deaths among survivors who were nontrivially exposed are attributable to irradiation during the bombings. The ERR of leukemia at 1 Sv red bone marrow dose is markedly higher at younger ages at exposure, rising to ~70 around 7 years after exposure for a child exposed at 10 years of age; the ERR then attenuates so that at 25 years after exposure the ERR (at ~2) is at a level comparable to that experienced by those exposed as adults this long after exposure [14]. This "wave" of excess risk is much less notable at older ages at exposure, and for those aged 30 years or more at exposure, the ERR is essentially flat with time since exposure. Leukemia incidence, using data from the specialist registries in Hiroshima and Nagasaki, has been analyzed for 1950–1987 [15] and for both sexes and all ages during this period, the ERR at 1 Sv RBM dose was 4.84 (90% CI: 3.59, 6.44). The registration data allow an analysis by leukemia type, and acute lymphoblastic leukemia, acute myeloid leukemia, and chronic myeloid leukemia show an increasing incidence with dose, but this is not observed for chronic lymphatic leukemia (CLL), which is rare among Japanese.

For solid cancers (all types of cancer other than leukemia, lymphoma, and multiple myeloma) [13], just over 10 000 deaths occurred among the survivors during 1950–2000, of which almost 500 are attributable to exposure during the bombings; that is, ~8% of the just over 5800 deaths from a solid cancer that occurred among the survivors who received a nontrivial dose (average dose, 0.2 Sv) are attributable to radiation exposure from the bombings. Recently, the dose response for mortality from solid cancers has exhibited significant departure from linearity with upward curvature being detectable at moderate doses; a reduction in slope occurs above 2 Sv, although not as markedly as for leukemia. For both sexes and all ages over the entire period 1950–2000, the ERR of solid cancer mortality at 1 Sv equivalent dose to the colon (as a representative dose) is 0.48 (90% CI: 0.40, 0.57); the ERR for female survivors is around twice as great as that for males. The ERR at 1 Sv is greater for those exposed at a young age: the ERR at 1 Sv for those under 20 years of age at exposure being 0.80 (90% CI: 0.62, 1.00), while for those over 40 years of age at exposure, it is 0.28 (90% CI: 0.17, 0.41). The ERR attenuates with time since exposure, reducing by almost a factor of 2 in 30 years for those aged 10 years at exposure, but much less so for older ages at exposure.

Solid cancer incidence has been studied for the period 1958–1998 [16]; around 7850 cases of solid cancers occurred among those assessed to have received an organ dose in the range 0.005–4 Gy. About 850 cases (~11%) among the nontrivially exposed survivors are attributable to exposure during the bombings. The dose response does not exhibit a significant departure from linearity, although such a departure for the mortality data became apparent only after the inclusion of data from recent years of follow-up. The raised risk of solid cancers cannot be detected statistically below ~100 mGy, although, of course, this does not mean that such a risk is not present at low doses, it is only that the data do not permit any risk at low doses to be discerned; but the data are consistent with the absence of a threshold dose. For both sexes and all ages over the 41-year study period, the ERR at 1 Sv was 0.62 (90% CI: 0.55, 0.69), and as for mortality, the ERR for females was almost twice that for males. As with mortality, the ERR is initially higher and then falls away, more steeply at younger ages at exposure.

A number of studies of specific solid tumor types have been conducted. Thyroid cancer was the first solid tumor to exhibit a radiation-related excess, which was most marked among those exposed at a young age [17], while a raised risk of lung cancer was also recognized early in the follow-up. Of course, a significant challenge for RERF researchers is the unraveling of the influence of cigarette smoking upon the risk of lung cancer from that of radiation (and determining whether these two carcinogens interact and if so, how), and smoking data for a subset of survivors have permitted appropriate analysis that demonstrates that the pattern of radiation-induced lung cancer is not greatly different from that for most other solid tumors; the interaction of cigarette smoke and radiation appears to be closer to additive than multiplicative [18]. Female breast cancer is another cancer that has been the subject of particular study, and irradiation at a young age has been shown to have a particularly high risk coefficient [19]. Other cancers, such as of the pancreas, prostate, and rectum, do not demonstrate statistically significant relationships with dose, indicating a low sensitivity to radiation induction [16].

Although the epidemiological data relating to irradiation incurred during the atomic bombings of Hiroshima and Nagasaki are the product of intense study and are the principal source of information on the risk of cancer resulting from exposure to ionizing radiation, they are not perfect. The exposure was brief and at a high dose rate, so the data offer little direct evidence of the effect of protracted exposure to radiation as occurs, for example, in the workplace. Clearly, the doses received during the bombings had to be estimated retrospectively, which introduces uncertainty, and as noted above, it has not been possible to compute doses for $\sim$7000 survivors. The doses were largely from high-energy γ-rays (and a neutron component that is generally small) and do not provide direct information of the effect of, say, densely ionizing α-particle irradiation, as occurs on exposure to, for example, radon and its radioactive decay products. The study is of a Japanese population exposed at the end of a long war (and therefore to some extent malnourished), which raises the question of the transfer of risk to, say, a healthy European nuclear industry workforce. To enter the LSS, survivors had to be alive in 1950 and therefore to have lived through the difficult conditions that existed in the years immediately following the war, raising the possibility of biased risk estimates due to the "healthy survivor effect": it is the strongest and healthiest survivors who entered the study (particularly those who received the highest doses) and they may not be representative of the original survivor population in terms of their risk of radiation-induced cancer. Systematic collection of data only commenced in October 1950 and health event data prior to that date are largely missing, which has implications for those cancers (such as leukemia) with a short minimum latent period, and the minimum latent period for leukemia (excess cases of which certainly occurred in the late 1940s) cannot be determined from the survivor data. Finally, at the end of 1997, the end date of the period covered by the last update of mortality in the LSS, 48% of the study members were still alive [13], so the expression of risk many years after exposure, especially for those irradiated at a young age, is inevitably associated with a degree of uncertainty.

The deficiencies of the epidemiological data derived from the experience of the Japanese atomic bomb survivors in covering all aspects of the risk of cancer arising from exposure to ionizing radiation should not detract from the value of these data to the understanding of risk, which is why risk estimates remain largely based upon the results from the LSS [3–5]. However, these points do emphasize the importance in complementing the Japanese survivor data with studies of other suitably exposed groups, especially those with exposure circumstances not directly addressed by the Japanese data, such as Caucasian workers exposed protractedly to radiation or exposure to α-particle-emitting radon and its decay products. Such complementary data sets will now be examined under the broad classifications of medical irradiation, occupational exposure, and environmental exposure.

6.3
Medical Exposures

Patients are irradiated therapeutically to treat disease and diagnostically to detect or monitor disease. A number of groups treated with radiotherapy have been the subject of epidemiological study. Around 14 000 British ankylosing spondylitis patients treated with X-rays have been followed up, with doses being estimated for all the cases of leukemia and a sample of the cohort [20]. In the United States and the Nordic countries, nearly 105 000 women treated with radiation for cervical cancer have been followed, with cancer sites categorized by degree of irradiation received during radiotherapy [21]. These two large studies have been supplemented by other studies of cancer following therapeutic irradiation and the findings have been reviewed by Little [22]; relative risks tend to be lower in the medical studies due to cell killing.

An analysis of pooled data for leukemia in the Japanese LSS, the ankylosing spondylitis patients, and the US cervical cancer patients found that specific risk models were required for the different types of leukemia, but the data from the three studies were statistically compatible for these leukemia types [23]. Chronic lymphatic leukemia was not found to be associated with exposure to radiation, a finding that is generally confirmed by other studies.

A study of thyroid cancer following irradiation using the pooled data from the Japanese LSS and six studies of medical exposure found that childhood irradiation posed a particularly high ERR of cancer: 7.7 (95% CI: 2.1, 28.7) Sv^{-1} [24]. This is of some relevance to the large number of excess thyroid cancers that have occurred in the heavily contaminated regions of the former Soviet Union after the Chernobyl reactor accident. Pooled data analyses using data from the Japanese atomic bomb survivors and medically exposed groups have also been conducted for breast cancer [25].

Rather different exposure circumstances applied to those injected with the contrast medium Thorotrast, a thorium dioxide colloid that is radioactive [26]. The long-lived, α-particle-emitting material is retained within the body and has led to a large excess of liver cancers and also leukemia. The short-lived, α-emitter ^{224}Ra was used as a medical treatment in Germany in the mid-twentieth century; radium deposits on the bone surfaces and the ^{224}Ra led to a large excess of bone cancers, and possibly also leukemia [27]. The complex dosimetry of these exposures leads to difficulties in calculating reliable risk coefficients.

Of some interest with respect to the effect of low-level irradiation are the studies of diagnostic exposure. North American tuberculosis patients were routinely monitored with multiple fluoroscopic examinations of the chest, each giving a dose ~10 mGy. The subsequent risk of breast cancer was directly proportional to the number of exposures (i.e., the cumulative dose), indicating that each exposure contributed to the risk [28, 29]. However, there was no excess risk of lung cancer, although the influence of the severity of tuberculosis upon other risk factors for lung cancer must be considered. Patients with scoliosis have been monitored with multiple radiographic examinations; the average number of examinations was 25 and the average cumulative dose to the breast was 100 mGy. The risk of breast cancer was found to increase proportionally with the number of examinations, again indicating that low doses of radiation increase the risk of breast cancer [30].

A statistical association between childhood cancer (particularly leukemia) and exposure of the third trimester fetus to antenatal radiography has been consistently found by case–control studies carried out around the world [31, 32]. From the largest of these studies, the Oxford Survey of Childhood Cancers, an ERR coefficient for childhood cancer of 50 Gy^{-1} may be derived from the receipt *in utero* of a dose $\sim$10 mGy [33], which is consistent with the ERR coefficient for childhood leukemia obtained from the Japanese atomic bomb survivors irradiated as young children. Although a causal interpretation of this association is not universally accepted, these studies do provide additional evidence for the risk of cancer being increased by doses $\sim$10 mGy.

While studies of medical exposures to radiation offer a valuable complement to the evidence derived from the Japanese atomic bomb survivors, care is required in the interpretation of these studies. Exposure occurs because of a known or suspected disease and the presence of disease may influence the risk of radiation-induced cancer such that accurate generalization of risk estimates to a broadly healthy population may be in doubt. Further, radiotherapy involves doses that are designed to kill abnormal cells and these doses are frequently highly localized, leading to a heterogeneous distribution of dose to normal tissues – doses to tissues close to the target of radiotherapy can be sufficient to kill significant numbers of normal cells and modify the carcinogenic effect, while tissues distant from the target receive scattered radiation that is difficult to quantify. As a consequence, accurate dose estimates are often lacking in studies of medical irradiation, although modern dose reconstruction techniques are adding to the value of such studies in providing reliable risk estimates.

6.4 Occupational Exposures

The first persuasive evidence of the carcinogenic effect of ionizing radiation was reported before the atomic bomb explosions over Japan, when in 1944 findings were published of raised levels of leukemia among US radiologists, although earlier reports of elevated levels of leukemia among medical staff working with radiation had occurred, but their significance with respect to radiation exposure had not been recognized [34, 35]. Since then, studies of British radiologists, of radiographers in the United States and in China, and of other medical staff exposed to radiation have been conducted. The evidence has been reviewed by Yoshinaga *et al.* [36], who concluded that the firmest evidence was for an excess risk of leukemia (excluding CLL), although there are indications of raised risks of breast and skin cancers. Large numbers of medical staff have been exposed to radiation in the course of their work, but individual recorded doses from the earlier years of work with radiation, when exposures tended to be highest, are often lacking, which limits the quantitative information on risk that can be obtained from studies.

Aircrew are exposed to enhanced levels of cosmic radiation as a consequence of reduced atmospheric shielding, but the increase in annual dose is small (at a few millisieverts), although a significant proportion of this increase is due to neutron exposure. It is of interest that the average occupational dose to aircrew in the United States is now the highest of any employment sector. A number of epidemiological studies of aircrew have been conducted and a review of the results has found consistent evidence for an increase in melanoma and weaker evidence for an increase of nonmelanoma skin cancer and breast cancer [37]. Of course, the unconventional lifestyle of aircrew, including their exposure to solar UV radiation, must be taken into account when viewing these findings.

During the first half of the twentieth century, radium-based luminous paint was applied to instrument dials. Unfortunately, the dial painters tended to lick their brushes to achieve a fine point, leading to intakes of radium (principally, ^{226}Ra and ^{228}Ra) that were often substantial, particularly for young women in the United States [38]. Radium deposits on the bone surfaces and the US dial painters suffered a large excess of bone cancers, although the derived excess risks per unit dose are somewhat crude due to difficulties in determining accurate doses to the sensitive cells. The heavily exposed dial painters also experienced a large excess risk of cancers of the paranasal sinuses and mastoid air cells, which is attributed to the accumulation of radon gas that is generated on the decay of ^{226}Ra present in bone.

Underground hard rock miners (e.g., uranium, gold, and tin miners) are exposed to radon and its radioactive decay products in the course of their work. When mines had only crude ventilation, high levels of radon and its progeny could be inhaled, and studies of miners have found raised levels of lung cancer. Studies using the pooled data from groups of underground hard rock miners have been able to quantify the lung cancer risk in terms of the level of radon exposure and those factors that modify this risk, such as time since exposure and smoking habits. The BEIR VI Committee [39] developed a risk model based upon about 2700 lung cancer deaths among around 68 000 miners – the risk was found to be approximately directly proportional to the cumulative exposure to radon and to attenuate with increasing time since exposure and attained age, but did not depend upon age at exposure. Evidence was found for an inverse dose rate effect, although this may be confined to high cumulative exposures. Most of the miners were smokers and the BEIR VI Committee concluded that smoking and radon interact to produce a combined effect that is greater than additive, but which is also less than multiplicative. Little evidence was found for an excess risk of cancers other than lung cancer among the miners that could be attributed to radon [39]. The dose received by the sensitive cells in the bronchial epithelium from radon and its decay products is not easy to calculate, so that accurate risk coefficients for radon-induced lung cancer are problematic in their derivation and the risk tends to be expressed in terms of cumulative exposure. However, dosimetry models are continuously being improved, encouraging hope that risk coefficients of a reasonable reliability may be derived in the near future.

The reactor accident at Chernobyl in 1986 led to many workers being exposed in the initial emergency and subsequent recovery work – around 600 000 emergency and recovery workers (liquidators) were involved in total, of whom ~240 000 worked at the site during 1986–1987 when exposures would have been greatest (the mean effective dose for this group is assessed to have been 100 mSv). Studies are complicated by the attention that has been paid to the health of the Chernobyl liquidators making ascertainment bias a distinct possibility if comparisons are made with the general population, and accurate dosimetry is a problem for many of the workers. Nonetheless, the Chernobyl liquidators offer an opportunity to study the effects of protracted exposure to radiation on a large group of adults, although the limitations of the studies require recognition when interpreting results, especially the earlier studies that have made comparisons with health event rates for the general population. However, substantial effort has been expended on nested case–control studies of leukemia among cleanup workers from Ukraine [40] and other countries of the former Soviet Union (Russia, Belarus, and the Baltic states combined) [41] since leukemia would be predicted to be the first cancer to show an effect of exposure to radiation. Evidence for a raised risk of leukemia was found in these two nested case–control studies, although the ERR coefficient for chronic lymphocytic leukemia (CLL) was about the same as that for all other leukemias in both studies, and the study of the other countries of the former Soviet Union also found a raised risk of non-Hodgkin's lymphoma, findings that require caution to be exercised in their interpretation, and consequently of that of leukemia excluding CLL. The Chernobyl cleanup workers certainly justify further study, although the difficulties associated with such investigations should not be underestimated.

Participants in nuclear weapons tests would have been exposed, at some level, to radiation, and the participants have been the subject of a number of studies, the largest of these involving those in the US and UK tests. Overall, the evidence points to a possible increased risk of leukemia, although it is difficult to attribute this to exposure to radiation given the assessed levels of exposure [42].

Those who have worked in the nuclear weapons and energy industry perhaps offer the best opportunity of deriving reliable risk estimates for those exposed to protracted low levels of radiation [43]. At least in the West, dosimetry records are generally good, although their accuracy will have improved with time, and detailed personnel data usually allow unambiguous linkage to mortality and cancer incidence data. Nonetheless, the low cumulative doses received, on average, by most workforces imply that large numbers of workers must be followed up over long periods to provide reasonable statistical power and the opportunity to achieve acceptable precision; but workers in the early years of the nuclear industry tended to receive occupational doses that were higher than would be considered tolerable now, and these workers are presently reaching an age when they are likely to develop serious illnesses and die, so health event data for these workers are accumulating at an increased rate. Clearly, most power will be derived from national or international studies of nuclear industry workers, especially those that include mature workforces from the early years of the industry, and several such studies have been conducted.

The International Agency for Research on Cancer (IARC) has coordinated the study of nuclear workers from a number of countries, initially from 3 countries [44] but recently this has been increased to 15 countries [45]. The 3-country study involved workforces from the United States, United Kingdom, and Canada, and found nothing remarkable; but the 15-country study reported an ERR coefficient for all cancers other than leukemia of 0.97 (95% CI: 0.14, 1.97) Sv^{-1}, which is around sixfold greater than would be expected from the risk coefficient that provides the basis of radiological protection, and has generated considerable comment. The authors noted, however, that the ERR coefficient for lung cancer was particularly raised and that there was other evidence that tobacco smoke may have had a role in producing the increased risk estimate for all cancers other than leukemia, and they concluded that this estimate had been inflated by uncontrolled confounding by smoking to an extent that could not be quantified. Furthermore, the ERR coefficient for all cancers other than leukemia was notably influenced by data from Canada: although Canadian deaths accounted for only 4% of deaths from all cancers other than leukemia, if these deaths are removed from the analysis, the value of the coefficient falls by 40%, which is surprising. Indeed, the pattern of results from studies of Canadian workers suggests that there may be a problem with the database for workers from Atomic Energy of Canada Ltd (AECL) [43], and investigations have revealed deficiencies in the AECL data, although it has yet to be determined what effect this has on the result from the 15-country study.

The UK National Registry for Radiation Workers (NRRWs) is a large database containing the details of almost all radiation workers in the United Kingdom. The third analysis of the NRRW includes ~175 000 radiation workers with an average cumulative individual external recorded dose of 25 mSv [46]. The average follow-up is 22.3 years, and 7684 cancer deaths and 11 165 incident cancer cases are included in the latest analysis, making NRRW-3 the most powerful nuclear worker study that has been reported to date. The anticipated "healthy worker effect" is seen for mortality from all causes and from cancer when mortality rates are compared with the general population, so most emphasis is placed upon the variation of cancer mortality and incidence with cumulative dose. Unlike the 15-country study, there are no obvious problems with NRRW-3 as far as confounding or bias are concerned. The ERR coefficient for leukemia (excluding CLL) mortality is 1.71 (90% CI: 0.06, 4.29) Sv^{-1} and for mortality from cancers other than leukemia is 0.28 (90% CI: 0.02, 0.56) Sv^{-1}; for cancer incidence, the ERR coefficients are 1.78 (90% CI: 0.17, 4.36) and 0.27 (90% CI: 0.04, 0.51) Sv^{-1}, respectively. These statistically significant trends with dose are compatible with the findings of the Japanese atomic bomb survivors – for adult men, the ERR coefficients for mortality from leukemia (excluding CLL) and other cancers are 1.4 (90% CI: 0.1, 3.4) and 0.25 (90% CI: 0.26, 0.41) Sv^{-1}, respectively. It will be seen that the statistical power obtained by a study of radiation workers in just one country is now impressive, and careful international collaborative studies should be capable of producing power that can equal, if not exceed, that achieved by the studies of the Japanese atomic bomb survivors. Of course, NRRW-3 is just one study, and it would be naïve to assume that there are no unresolved questions relating to this study – for example, statistically significant ERR coefficients were found for mortality from cancers of the rectum and larynx and for incidence of cancers of the rectum and uterus, nonmelanoma skin cancer, and multiple myeloma, and this is not the pattern of excess risks that might be anticipated from other studies. Nonetheless, NRRW-3 encourages the view that the study of nuclear workers exposed protractedly to low levels of radiation can produce meaningful results.

Those working, or who have worked, in the nuclear industry also offer the opportunity to study the effects of radioactive material deposited within the body, usually via inhalation. A number of workforces have been exposed to a variety of radionuclides, although the calculation of organ-specific doses is substantially more complex for many radionuclides, especially those emitting short-range radiations that are of greatest interest in their health effects. For example, many workers at the Sellafield installation in the United Kingdom have been exposed to plutonium, and the plutonium workers there have been the subject of study, although risks expressed in terms of plutonium-specific doses have yet to be published. Of particular interest with respect to plutonium exposure (and also possibly to exposure to other radionuclides) are the workers at the Mayak nuclear complex in the Southern Urals of Russia who in the early years of operations in the former Soviet Union were exposed to high levels of plutonium (and external radiation). Again, the calculation of organ-specific doses is far from straightforward, although results with respect to plutonium exposure have been published [47].

6.5 Environmental Exposures

Certain populations are exposed to high levels of natural background radiation, for example, in Kerala, India, where thorium-bearing monazite sands produce high levels of γ-radiation [48]. With the exception of residential exposure to radon, studies of high background radiation areas have not provided convincing evidence of excess radiation-induced risk to health, largely because of rather crude study designs that have been unable to segregate the predicted radiation effects from the influence, frequently much greater, of other causal factors. However, a large cohort of residents of Kerala has been established, which together with individual dose measurements, information on other risk factors, and cancer registration data may eventually produce valuable results [49].

Current radiation risk models for leukemia imply that childhood leukemia is particularly sensitive to induction by radiation and suggest that 15–20% of cases in Great Britain, where the average annual dose to the RBM from natural background radiation is $\sim$1.3 mSv, might be caused by such exposure, although the uncertainties in this attributable fraction are considerable [50, 51]. Epidemiological studies have been unable to reliably detect this risk, although the variation of the RBM dose from natural background radiation is not great, so large studies are required to achieve sufficient statistical power. However, recently, a nationwide case–control study in Denmark found a positive association between residential radon exposure and childhood acute lymphoblastic leukemia, which the authors suggested could account for 9% (95% CI: 1%, 21%) of cases in Denmark [52]. This is an interesting finding, although conventional dosimetric modeling shows that radon produces only a small component of the RBM dose from natural background radiation [51], and the doses were assessed from a model based upon explanatory variables such as house type and geology that needs to be investigated further in terms of its predictive accuracy. In any event, the Danish findings are not inconsistent with the predictions of conventional modeling if uncertainties are taken into account [51].

Considerable effort has been devoted to the study of residential radon and lung cancer. The major difficulty is controlling the effects of tobacco smoke, and failure to adequately do so led to misleading findings in a geographical correlation study of average radon levels and lung cancer mortality rates in US counties – Cohen reported a statistically significant *negative* correlation between average radon concentration and lung cancer mortality [53], which Puskin showed was most likely to be the consequence of uncontrolled confounding by tobacco smoke [54]. Case–control studies that have gathered information on individual smoking habits and measured radon levels have been carried out around the world, and pooled analyses have been conducted in Europe [55], North America [56], and China [57]. The pooled data analyses found risk coefficients for lung cancer that are compatible with the dose–response gradients predicted by the studies of underground hard rock miners occupationally exposed to radon. The combined effect of radon and tobacco smoke upon the risk of lung cancer is more than additive, so a smoker has a greater increase in lung cancer risk per unit exposure to radon than a nonsmoker.

Atmospheric nuclear weapons tests, particularly at their height in the late 1950s and early 1960s, led to exposure to the radioactive debris of these tests and the opportunity to study potential health effects, especially those cancers that are particularly sensitive to induction by radiation, such as childhood leukemia and thyroid cancer after exposure in childhood. Thyroid cancer was found to be in excess among Marshall Islanders accidentally exposed to high levels of fallout after an unexpected change in wind direction after a nuclear test at Bikini Atoll in 1954 [58], and thyroid neoplasms (malignant plus benign) showed a highly significant trend with assessed thyroid dose in a cohort of those exposed to fallout as children living near the Nevada Test Site [59], although the number of thyroid cancers was too small to obtain reliable results. A geographical/temporal correlation study of thyroid cancer mortality and incidence across the United States found some weak evidence for an influence of fallout from the Nevada Test Site upon risk for those exposed in infancy and for those born in the 1950s [60]. Recently, thyroid tumors in communities near the Semipalatinsk Test Site in present-day Kazakhstan were the subject of a study using reconstructed thyroid doses from the test explosions [61]. A significant dose response for thyroid nodules was found for both external and internal doses, at a level compatible with that derived from previous studies of external exposure, but there was no suggestion that the comparatively small number of thyroid cancers was associated with dose.

Leukemia, mainly among children, and atmospheric nuclear weapons testing have also been the subject of investigation. A significant association between acute leukemia deaths among young people and the doses received from fallout in southwestern Utah from testing at the Nevada Test Site has been reported, at a level compatible with the predictions of conventional risk models using individually reconstructed doses [62]. An indication of an effect of fallout upon leukemia incidence in the Nordic countries has also been reported, again at a level comparable with conventional predictions [63]. Recently, childhood leukemia incidence using data from 11 cancer registries in 3 continents has been examined to determine whether any influence of the marked peak of atmospheric testing during the late 1950s and early 1960s can be detected, which could be indicative of a serious underestimation of the risk posed by man-made radionuclides released into the environment; no such influence was found providing no evidence for gross inaccuracy of conventional risk models [64].

The explosion on April 26, 1986 in the core of one of the nuclear reactors at Chernobyl in the then Soviet Union released large amounts of radionuclides into the atmosphere over a period of 10 days. Much of Europe was contaminated, certain areas of present-day Ukraine, Belarus, and Russia heavily so. Of primary interest are the effects of high doses to the thyroid gland received by a large number of children from the intake of radioisotopes of iodine, mainly ^{131}I, which has a half-life of $\sim$8 days; some children received thyroid doses of 1 Gy or more from radioiodine [65]. A cohort study of thyroid cancer among those in Ukraine who were less than 18 years of age at exposure found an ERR coefficient of 5.2 (95% CI: 1.7, 27) Gy^{-1} [66], while a case–control study of thyroid cancer in Belarus and Russia among those who were less than 15 years of age at exposure reported an ERR coefficient of 4.5 (95% CI: 1.2, 7.8) Gy^{-1} [67]; these risk estimates are compatible with that derived from a pooled analysis of thyroid cancer in those exposed to external sources of radiation while under 15 years of age of 7.7 (95% CI: 2.1, 29) Gy^{-1} [24]. More than 4000 cases of thyroid cancer have been attributed to radiation exposure from Chernobyl among those exposed as children in the heavily contaminated regions of the former Soviet Union, and thyroid cancer risk models suggest that many more radiation-related cases will be diagnosed in coming years as the exposed population ages. Adults are less sensitive to radiation induction of thyroid cancer and evidence for thyroid cancers among those exposed at an older age has yet to be found. No convincing evidence exists for an effect of Chernobyl upon thyroid cancer incidence outside the former Soviet Union, but thyroid doses were much lower here [65].

At present, little reliable evidence exists for excesses of other cancers that may be attributed to exposure to radiation from Chernobyl, including childhood leukemia, although the doses received were generally much lower than the thyroid doses in children [65]. Childhood leukemia has been paid particular attention because of its sensitivity to induction by radiation, but although a case–control study in the heavily contaminated areas of Belarus, Ukraine, and Russia found some evidence of an increased risk, this evidence depended upon cases and controls from Ukraine [68], and later results from Ukraine using a different set of controls notably decreased the risk estimate [69], raising doubts about the presence of bias in the data. A study in Europe outside the former Soviet Union found little evidence for the effect of Chernobyl upon childhood leukemia incidence, which is not surprising given the generally low doses received [65].

Contamination of the environment has also occurred as a result of routine operations and accidents at other nuclear installations. During the production of plutonium for weapons use at the Mayak complex near Chelyabinsk in the Southern Urals, discharges of radioactive materials into the environment are likely to have had a detectable effect upon exposed populations. In the late 1940s and early 1950s, highly radioactive liquid effluent was released into the Techa River, causing such high levels of contamination that riverside villages for some tens of kilometers downstream had to be evacuated. Substantial effort is being devoted to the reconstruction of the doses received by the riverside residents (mainly due to external exposure and internally deposited ^{90}Sr) and to the follow-up of health effects in the population, projects that present appreciable difficulties to those involved. Nonetheless, although doses have yet to be finalized and as such reliable risk coefficients are not yet available, indications of trends with dose for leukemia and breast cancer have been reported [70, 71]. Releases to atmosphere from Mayak will have led to enhanced levels of radiation in surrounding communities, particularly in the nearest town of Ozyorsk. An increase of thyroid diseases among a sample of those residents exposed to the greatest releases of ^{131}I has been reported [72], although a larger study and better dosimetry is desirable. Finally, an accident involving a radioactive waste tank at Mayak in 1957 led to a large release of radioactivity requiring the evacuation of $\sim$10 000 residents toward the north-east of the site, but no findings of any related epidemiological study have been published.

Other nuclear installations have discharged radioactive material into the environment, although generally not on the scale of Mayak during its early years of operations. One exception is the Hanford complex in Washington State that produced the first large-scale quantities of plutonium for use in the Trinity atomic test explosion and the atomic bombing of Nagasaki. In its early years of irradiated nuclear fuel reprocessing operations, "short-cooled" fuel was reprocessed leading to the discharge of substantial quantities of ^{131}I into the atmosphere. Considerable resources have been devoted to the examination of the effect of these releases of radioiodine on the risk of thyroid cancer and other thyroid diseases in the neighborhood of Hanford through the establishment of a historical cohort of individuals who were young children at the height of the ^{131}I discharges [73]. Individual thyroid doses were estimated in an extensive dose reconstruction program: the median thyroid dose was assessed to be 97 mGy, with the highest dose being almost 3 Gy. The ERR coefficient for thyroid cancer was nonsignificantly elevated at 0.7 Gy^{-1}, which is an order of magnitude less than the ERR coefficients obtained from the pooled study of external irradiation and from the Chernobyl studies, and for all thyroid tumors (malignant and benign), the ERR was also nonsignificantly elevated at 0.1 Gy^{-1}, which is also an order of magnitude less than those obtained from the studies of Chernobyl contamination and of exposures to fallout from the Nevada Test Site and from the Semipalatinsk Test Site. Why the Hanford study produced low thyroid tumor risk estimates is not clear, although the influence of factors such as exposure to shorter-lived radioisotopes of iodine and preexisting iodine deficiency have been suggested as being of possible relevance [74].

Steel contaminated by ^{60}Co was used as a construction material in Taiwan during 1983–1984, and the affected buildings were identified in 1992. Just over

6000 people were exposed (mean cumulative reconstructed dose, ~50 mSv) and have been the subject of study. Statistically significant trends with reconstructed ^{60}Co dose of leukemia (excluding CLL) and breast cancer incidence have been reported [75]. However, the study was based on just 117 cases of cancer, so further information will arise as follow-up of this (relatively small) cohort continues.

6.6 Conclusions

Although substantial evidence of the risk of cancer posed by exposure to ionizing radiation is available from the experience of the survivors of the atomic bombings of Hiroshima and Nagasaki, this is supplemented significantly by information from a wide range of studies of exposure under varying circumstances, such as protracted exposure to chronic low levels and internal exposure to radioactive materials. The epidemiological study of exposure to ionizing radiation – in intensity probably surpassed only by the study of exposure to tobacco smoke – has provided an impressive base of knowledge from which reliable estimates of the risk of cancer arising from such exposure have been derived. Future studies hold the prospects of further substantial information as data accumulate from the continuing follow-up of existing, and possibly new, exposed groups.

References

1 International Agency for Research on Cancer (2000) *IARC Monographs on the Evaluation of Carcinogenic Risks to Humans. Vol. 75. Ionizing Radiation, Part 1: X- and Gamma (γ)-Radiation, and Neutrons*, IARC, Lyon

2 International Agency for Research on Cancer (2001) *IARC Monographs on the Evaluation of Carcinogenic Risks to Humans. Vol. 78. Ionizing Radiation, Part 2: Some Internally Deposited Radionuclides*, IARC, Lyon

3 International Commission on Radiological Protection (2007) The 2007 recommendations of the International Commission on Radiological Protection. ICRP Publication 103. *Ann. ICRP*, **37** (2–4), 1–332.

4 United Nations Scientific Committee on the Effects of Atomic Radiation (2008) *Effects of Ionizing Radiation. UNSCEAR 2006 Report. Vol. I: Report to the General Assembly with Scientific Annexes A and B*, United Nations, New York

5 Committee to Assess Health Risks from Exposure to Low Levels of Ionizing Radiation (2006) *Health Risks from Exposure to Low Levels of Ionizing Radiation. BEIR VII Phase 2*, The National Academies Press, Washington, DC.

6 United Nations Scientific Committee on the Effects of Atomic Radiation (1994) *Sources and Effects of Ionizing Radiation. UNSCEAR 1994 Report to the General Assembly, with Scientific Annexes*, United Nations, New York

7 Wakeford, R. (2004) The cancer epidemiology of radiation. *Oncogene*, **23** (38), 6404–6428.

8 Folley, J.H., Borges, W., and Yamawaki, T. (1952) Incidence of leukemia in survivors of the atomic bomb in Hiroshima and Nagasaki, Japan. *Am. J. Med.*, **13** (3), 311–321.

9 Preston, D.L., Shimizu, Y., Pierce, D.A., Suyama, A., and Mabuchi, K. (2003) Studies of mortality of atomic bomb survivors. Report 13: solid cancer and

noncancer disease mortality: 1950–1997. *Radiat. Res.*, **160** (4), 381–407.

10 Yamada, M., Wong, F.L., Fujiwara, S., Akahoshi, M., and Suzuki, G. (2004) Noncancer disease incidence in atomic bomb survivors, 1958–1998. *Radiat. Res.*, **161** (6), 622–632.

11 Preston, D.L., Cullings, H., Suyama, A., Funamoto, S., Nishi, N., Soda, M., Mabuchi, K., Kodama, K., Kasagi, F., and Shore, R.E. (2008) Solid cancer incidence in atomic bomb survivors exposed *in utero* or as young children. *J. Natl. Cancer Inst.*, **100** (6), 428–436.

12 Schull, W.J. (2003) The children of atomic bomb survivors: a synopsis. *J. Radiol. Prot.*, **23** (4), 369–384.

13 Preston, D.L., Pierce, D.A., Shimizu, Y., Cullings, H.M., Fujita, S., Funamoto, S., and Kodama, K. (2004) Effect of recent changes in atomic bomb survivor dosimetry on cancer mortality risk estimates. *Radiat. Res.*, **162** (4), 377–389.

14 Richardson, D., Sugiyama, H., Nishi, N., Sakata, R., Shimizu, Y., Grant, E.J., Soda, M., Hsu, W.L., Suyama, A., Kodama, K., and Kasagi, F. (2009) Ionizing radiation and leukemia mortality among Japanese atomic bomb survivors, 1950–2000. *Radiat. Res.*, **172** (3), 368–382.

15 Preston, D.L., Kusumi, S., Tomonaga, M., Izumi, S., Ron, E., Kuramoto, A., Kamada, N., Dohy, H., Matsuo, T., *et al.* (1994) Cancer incidence in atomic bomb survivors. Part III. Leukemia, lymphoma and multiple myeloma, 1950–1987. *Radiat. Res.*, **137** (2 Suppl.), S68–S97.

16 Preston, D.L., Ron, E., Tokuoka, S., Funamoto, S., Nishi, N., Soda, M., Mabuchi, K., and Kodama, K. (2007) Solid cancer incidence in atomic bomb survivors: 1958–1998. *Radiat. Res.*, **168** (1), 1–64.

17 Wood, J.W., Tamagaki, H., Neriishi, S., Sato, T., Sheldon, W.F., Archer, P.G., Hamilton, H.B., and Johnson, K.G. (1969) Thyroid carcinoma in atomic bomb survivors Hiroshima and Nagasaki. *Am. J. Epidemiol.*, **89** (1), 4–14.

18 Pierce, D.A., Sharp, G.B., and Mabuchi, K. (2003) Joint effects of radiation and smoking on lung cancer risk among atomic bomb survivors. *Radiat. Res.*, **159** (4), 511–520.

19 Land, C.E. (1995) Studies of cancer and radiation dose among atomic bomb survivors. The example of breast cancer. *JAMA*, **274** (5), 402–407.

20 Weiss, H.A., Darby, S.C., and Doll, R. (1994) Cancer mortality following X-ray treatment for ankylosing spondylitis. *Int. J. Cancer*, **59** (3), 327–338.

21 Chaturvedi, A.K., Engels, E.A., Gilbert, E.S., Chen, B.E., Storm, H., Lynch, C.F., Hall, P., Langmark, F., Pukkala, E., Kaijser, M., Andersson, M., Fosså, S.D., Joensuu, H., Boice, J.D., Kleinerman, R.A., and Travis, L.B. (2007) Second cancers among 104,760 survivors of cervical cancer: evaluation of long-term risk. *J. Natl. Cancer Inst.*, **99** (21), 1634–1643.

22 Little, M.P. (2001) Cancer after exposure to radiation in the course of treatment for benign and malignant disease. *Lancet Oncol.*, **2** (4), 212–220.

23 Little, M.P., Weiss, H.A., Boice, J.D., Jr., Darby, S.C., Day, N.E., and Muirhead, C.R. (1999) Risks of leukemia in Japanese atomic bomb survivors, in women treated for cervical cancer, and in patients treated for ankylosing spondylitis. *Radiat. Res.*, **152** (3), 280–292.

24 Ron, E., Lubin, J.H., Shore, R.E., Mabuchi, K., Modan, B., Pottern, L.M., Schneider, A.B., Tucker, M.A., and Boice, J.D., Jr. (1995) Thyroid cancer after exposure to external radiation: a pooled analysis of seven studies. *Radiat. Res.*, **141** (3), 259–277.

25 Preston, D.L., Mattsson, A., Holmberg, E., Shore, R., Hildreth, N.G., and Boice, J.D., Jr. (2002) Radiation effects on breast cancer risk: a pooled analysis of eight cohorts. *Radiat. Res.*, **158** (2), 220–235.

26 Travis, L.B., Hauptmann, M., Gaul, L.K., Storm, H.H., Goldman, M.B., Nyberg, U., Berger, E., Janower, M.L., Hall, P., Monson, R.R., Holm, L.E., Land, C.E., Schottenfeld, D., Boice, J.D., Jr., and Andersson, M. (2003) Site-specific cancer incidence and mortality after cerebral angiography with radioactive thorotrast. *Radiat. Res.*, **160** (6), 691–706.

27 Wick, R.R., Atkinson, M.J., and Nekolla, E.A. (2009) Incidence of leukaemia and other malignant diseases following injections of the short-lived alpha-emitter ^{224}Ra into man. *Radiat. Environ. Biophys.*, **48** (3), 287–294.

28 Boice, J.D., Jr., Preston, D., Davis, F.G., and Monson, R.R. (1991) Frequent chest X-ray fluoroscopy and breast cancer incidence among tuberculosis patients in Massachusetts. *Radiat. Res.*, **125** (2), 214–222.

29 Howe, G.R. and McLaughlin, J. (1996) Breast cancer mortality between 1950 and 1987 after exposure to fractionated moderate-dose-rate ionizing radiation in the Canadian fluoroscopy cohort study and a comparison with breast cancer mortality in the atomic bomb survivors study. *Radiat. Res.*, **145** (6), 694–707.

30 Ronckers, C.M., Doody, M.M., Lonstein, J.E., Stovall, M., and Land, C.E. (2008) Multiple diagnostic X-rays for spine deformities and risk of breast cancer. *Cancer Epidemiol. Biomarkers Prev.*, **17** (3), 605–613.

31 Doll, R. and Wakeford, R. (1997) Risk of childhood cancer from fetal irradiation. *Br. J. Radiol.*, **70**, 130–139.

32 Wakeford, R. (2008) Childhood leukaemia following medical diagnostic exposure to ionizing radiation *in utero* or after birth. *Radiat. Prot. Dosimetry*, **132** (2), 166–174.

33 Wakeford, R. and Little, M.P. (2003) Risk coefficients for childhood cancer after intrauterine irradiation: a review. *Int. J. Radiat. Biol.*, **79** (5), 293–309.

34 Doll, R. (1995) Hazards of ionising radiation: 100 years of observations on man. *Br. J. Cancer*, **72** (6), 1339–1349.

35 Miller, R.W. (1995) Delayed effects of external radiation exposure: a brief history. *Radiat. Res.*, **144** (2), 160–169.

36 Yoshinaga, S., Mabuchi, K., Sigurdson, A.J., Doody, M.M., and Ron, E. (2004) Cancer risks among radiologists and radiologic technologists: review of epidemiologic studies. *Radiology*, **233** (2), 313–321.

37 Sigurdson, A.J. and Ron, E. (2004) Cosmic radiation exposure and cancer risk among flight crew. *Cancer Invest.*, **22** (5), 743–761.

38 Fry, S.A. (1998) Studies of U.S. radium dial workers: an epidemiological classic. *Radiat. Res.*, **150** (5 Suppl.), S21–S29.

39 Committee on Health Risks of Exposure to Radon (1999) *Health Effects of Exposure to Radon. BEIR VI*, National Academy Press, Washington, DC.

40 Romanenko, A.Y., Finch, S.C., Hatch, M., Lubin, J.H., Bebeshko, V.G., Bazyka, D.A., Gudzenko, N., Dyagil, I.S., Reiss, R.F., Bouville, A., Chumak, V.V., Trotsiuk, N.K., Babkina, N.G., Belyayev, Y., Masnyk, I., Ron, E., Howe, G.R., and Zablotska, L.B. (2008) The Ukrainian–American study of leukemia and related disorders among Chornobyl cleanup workers from Ukraine: III. Radiation risks. *Radiat. Res.*, **170** (6), 711–720.

41 Kesminiene, A., Evrard, A.S., Ivanov, V.K., Malakhova, I.V., Kurtinaitis, J., Stengrevics, A., Tekkel, M., Anspaugh, L.R., Bouville, A., Chekin, S., Chumak, V.V., Drozdovitch, V., Gapanovich, V., Golovanov, I., Hubert, P., Illichev, S.V., Khait, S.E., Kryuchkov, V.P., Maceika, E., Maksyoutov, M., Mirkhaidarov, A.K., Polyakov, S., Shchukina, N., Tenet, V., Tserakhovich, T.I., Tsykalo, A., Tukov, A.R., and Cardis, E. (2008) Risk of hematological malignancies among Chernobyl liquidators. *Radiat. Res.*, **170** (6), 721–735.

42 Muirhead, C.R., Kendall, G.M., Darby, S.C., Doll, R., Haylock, R.G., O'Hagan, J.A., Berridge, G.L., Phillipson, M.A., and Hunter, N. (2004) Epidemiological studies of UK test veterans: II. Mortality and cancer incidence. *J. Radiol. Prot.*, **24** (3), 219–241.

43 Wakeford, R. (2009) Radiation in the workplace: a review of studies of the risks of occupational exposure to ionising radiation. *J. Radiol. Prot.*, **29** (2A), A61–A79.

44 Cardis, E., Gilbert, E.S., Carpenter, L., Howe, G., Kato, I., Armstrong, B.K., Beral, V., Cowper, G., Douglas, A., Fix, J. *et al.* (1995) Effects of low doses and low dose rates of external ionizing radiation: cancer mortality among nuclear industry workers in three countries. *Radiat. Res.*, **142** (2), 117–132.

45 Cardis, E., Vrijheid, M., Blettner, M., Gilbert, E., Hakama, M., Hill, C., Howe, G., Kaldor, J., Muirhead, C.R., Schubauer-Berigan, M., Yoshimura, T., Bermann, F., Cowper, G., Fix, J., Hacker, C., Heinmiller, B., Marshall, M., Thierry-Chef, I., Utterback, D., Ahn, Y.O., Amoros, E., Ashmore, P., Auvinen, A., Bae, J.M., Bernar, J., Biau, A., Combalot, E., Deboodt, P., Diez Sacristan, A., Eklöf, M., Engels, H., Engholm, G., Gulis, G., Habib, R.R., Holan, K., Hyvonen, H., Kerekes, A., Kurtinaitis, J., Malker, H., Martuzzi, M., Mastauskas, A., Monnet, A., Moser, M., Pearce, M.S., Richardson, D.B., Rodriguez-Artalejo, F., Rogel, A., Tardy, H., Telle-Lamberton, M., Turai, I., Usel, M., and Veress, K. (2007) The 15-country collaborative study of cancer risk among radiation workers in the nuclear industry: estimates of radiation-related cancer risks. *Radiat. Res.*, **167** (4), 396–416.

46 Muirhead, C.R., O'Hagan, J.A., Haylock, R.G., Phillipson, M.A., Willcock, T., Berridge, G.L., and Zhang, W. (2009) Mortality and cancer incidence following occupational radiation exposure: third analysis of the National Registry for Radiation Workers. *Br. J. Cancer,* **100** (1), 206–212.

47 Sokolnikov, M.E., Gilbert, E.S., Preston, D.L., Ron, E., Shilnikova, N.S., Khokhryakov, V.V., Vasilenko, E.K., and Koshurnikova, N.A. (2008) Lung, liver and bone cancer mortality in Mayak workers. *Int. J. Cancer,* **123** (4), 905–911.

48 Hendry, J.H., Simon, S.L., Wojcik, A., Sohrabi, M., Burkart, W., Cardis, E., Laurier, D., Tirmarche, M., and Hayata, I. (2009) Human exposure to high natural background radiation: what can it teach us about radiation risks? *J. Radiol. Prot.*, **29** (2A), A29–A42.

49 Nair, R.R., Rajan, B., Akiba, S., Jayalekshmi, P., Nair, M.K., Gangadharan, P., Koga, T., Morishima, H., Nakamura, S., and Sugahara, T. (2009) Background radiation and cancer incidence in Kerala, India: Karanagappally cohort study. *Health Phys,* **96** (1), 55–66.

50 Wakeford, R., Kendall, G.M., and Little, M.P. (2009) The proportion of childhood leukaemia incidence in Great Britain that may be caused by natural background ionizing radiation. *Leukemia,* **23** (4), 770–776.

51 Little, M.P., Wakeford, R., and Kendall, G.M. (2009) Updated estimates of the proportion of childhood leukaemia incidence in Great Britain that may be caused by natural background ionising radiation. *J. Radiol. Prot.*, **29** (4), 467–482.

52 Raaschou-Nielsen, O., Andersen, C.E., Andersen, H.P., Gravesen, P., Lind, M., Schüz, J., and Ulbak, K. (2008) Domestic radon and childhood cancer in Denmark. *Epidemiology,* **19** (4), 536–543.

53 Cohen, B.L. (1995) Test of the linear-no threshold theory of radiation carcinogenesis for inhaled radon decay products. *Health Phys.*, **68** (2), 157–174.

54 Puskin, J.S. (2003) Smoking as a confounder in ecologic correlations of cancer mortality rates with average county radon levels. *Health Phys.*, **84** (4), 526–532.

55 Darby, S., Hill, D., Auvinen, A., Barros-Dios, J.M., Baysson, H., Bochicchio, F., Deo, H., Falk, R., Forastiere, F., Hakama, M., Heid, I., Kreienbrock, L., Kreuzer, M., Lagarde, F., Mäkeläinen, I., Muirhead, C., Oberaigner, W., Pershagen, G., Ruano-Ravina, A., Ruosteenoja, E., Rosario, A.S., Tirmarche, M., Tomásek, L., Whitley, E., Wichmann, H.E., and Doll, R. (2005) Radon in homes and risk of lung cancer: collaborative analysis of individual data from 13 European case–control studies. *BMJ,* **330** (7485), 223.

56 Krewski, D., Lubin, J.H., Zielinski, J.M., Alavanja, M., Catalan, V.S., Field, R.W., Klotz, J.B., Létourneau, E.G., Lynch, C.F., Lyon, J.I., Sandler, D.P., Schoenberg, J.B., Steck, D.J., Stolwijk, J.A., Weinberg, C., and Wilcox, H.B. (2005) Residential radon and risk of lung cancer: a combined analysis of 7 North American case–control studies. *Epidemiology,* **16** (2), 137–145.

57 Lubin, J.H., Wang, Z.Y., Boice, J.D., Jr., Xu, Z.Y., Blot, W.J., De Wang, L., and Kleinerman, R.A. (2004) Risk of lung cancer and residential radon in China: pooled results of two studies. *Int. J. Cancer,* **109** (1), 132–137.

58 Takahashi, T., Schoemaker, M.J., Trott, K.R., Simon, S.L., Fujimori, K., Nakashima, N., Fukao, A., and Saito, H. (2003) The relationship of thyroid cancer with radiation exposure from nuclear weapon testing in the Marshall Islands. *J. Epidemiol.*, **13** (2), 99–107.

59 Lyon, J.L., Alder, S.C., Stone, M.B., Scholl, A., Reading, J.C., Holubkov, R., Sheng, X., White, G.L., Jr., Hegmann, K.T., Anspaugh, L., Hoffman, F.O., Simon, S.L., Thomas, B., Carroll, R., and Meikle, A.W. (2006) Thyroid disease associated with exposure to the Nevada nuclear weapons test site radiation: a reevaluation based on corrected dosimetry and examination data. *Epidemiology*, **17** (6), 604–614.

60 Gilbert, E.S., Huang, L., Bouville, A., Berg, C.D., and Ron, E. (2010) Thyroid cancer rates and ^{131}I doses from Nevada atmospheric nuclear bomb tests: an update. *Radiat. Res.*, **173** (5), 659–664.

61 Land, C.E., Zhumadilov, Z., Gusev, B.I., Hartshorne, M.H., Wiest, P.W., Woodward, P.W., Crooks, L.A., Luckyanov, N.K., Fillmore, C.M., Carr, Z., Abisheva, G., Beck, H.L., Bouville, A., Langer, J., Weinstock, R., Gordeev, K.I., Shinkarev, S., and Simon, S.L. (2008) Ultrasound-detected thyroid nodule prevalence and radiation dose from fallout. *Radiat. Res.*, **169** (4), 373–383.

62 Stevens, W., Thomas, D.C., Lyon, J.L., Till, J.E., Kerber, R.A., Simon, S.L., Lloyd, R.D., Elghany, N.A., and Preston-Martin, S. (1990) Leukemia in Utah and radioactive fallout from the Nevada test site. A case–control study. *JAMA*, **264** (5), 585–591.

63 Darby, S.C., Olsen, J.H., Doll, R., Thakrar, B., Brown, P.D., Storm, H.H., Barlow, L., Langmark, F., Teppo, L., and Tulinius, H. (1992) Trends in childhood leukaemia in the Nordic countries in relation to fallout from atmospheric nuclear weapons testing. *BMJ*, **304** (6833), 1005–1009.

64 Wakeford, R., Darby, S.C., and Murphy, M.F. (2010) Temporal trends in childhood leukaemia incidence following exposure to radioactive fallout from atmospheric nuclear weapons testing. *Radiat. Environ. Biophys.*, **49** (2), 213–227.

65 Cardis, E., Howe, G., Ron, E., Bebeshko, V., Bogdanova, T., Bouville, A., Carr, Z., Chumak, V., Davis, S., Demidchik, Y., Drozdovitch, V., Gentner, N., Gudzenko, N., Hatch, M., Ivanov, V., Jacob, P., Kapitonova, E., Kenigsberg, Y., Kesminiene, A., Kopecky, K.J., Kryuchkov, V., Loos, A., Pinchera, A., Reiners, C., Repacholi, M., Shibata, Y., Shore, R.E., Thomas, G., Tirmarche, M., Yamashita, S., and Zvonova, I. (2006) Cancer consequences of the Chernobyl accident: 20 years on. *J. Radiol. Prot.*, **26** (2), 127–140.

66 Tronko, M.D., Howe, G.R., Bogdanova, T.I., Bouville, A.C., Epstein, O.V., Brill, A.B., Likhtarev, I.A., Fink, D.J., Markov, V.V., Greenebaum, E., Olijnyk, V.A., Masnyk, I.J., Shpak, V.M., McConnell, R.J., Tereshchenko, V.P., Robbins, J., Zvinchuk, O.V., Zablotska, L.B., Hatch, M., Luckyanov, N.K., Ron, E., Thomas, T.L., Voillequé, P.G., and Beebe, G.W. (2006) A cohort study of thyroid cancer and other thyroid diseases after the Chornobyl accident: thyroid cancer in Ukraine detected during first screening. *J. Natl. Cancer Inst.*, **98** (13), 897–903.

67 Cardis, E., Kesminiene, A., Ivanov, V., Malakhova, I., Shibata, Y., Khrouch, V., Drozdovitch, V., Maceika, E., Zvonova, I., Vlassov, O., Bouville, A., Goulko, G., Hoshi, M., Abrosimov, A., Anoshko, J., Astakhova, L., Chekin, S., Demidchik, E., Galanti, R., Ito, M., Korobova, E., Lushnikov, E., Maksioutov, M., Masyakin, V., Nerovnia, A., Parshin, V., Parshkov, E., Piliptsevich, N., Pinchera, A., Polyakov, S., Shabeka, N., Suonio, E., Tenet, V., Tsyb, A., Yamashita, S., and Williams, D. (2005) Risk of thyroid cancer after exposure to ^{131}I in childhood. *J. Natl. Cancer Inst.*, **97** (10), 724–732.

68 Davis, S., Day, R.W., Kopecky, K.J., Mahoney, M.C., McCarthy, P.L., Michalek, A.M., Moysich, K.B., Onstad, L.E., Stepanenko, V.F., Voillequé, P.G., Chegerova, T., Falkner, K., Kulikov, S., Maslova, E., Ostapenko, V., Rivkind, N., Shevchuk, V., and Tsyb, A.F. (2006) Childhood leukaemia in Belarus, Russia,

and Ukraine following the Chernobyl power station accident: results from an international collaborative population-based case–control study. *Int. J. Epidemiol.*, **35** (2), 386–396.

69 Noshchenko, A.G., Bondar, O.Y., and Drozdova, V.D. (2009) Radiation-induced leukemia among children aged 0–5 years at the time of the Chernobyl accident. *Int. J. Cancer*, **127** (2), 412–426.

70 Krestinina, L., Preston, D.L., Davis, F.G., Epifanova, S., Ostroumova, E., Ron, E., and Akleyev, A. (2010) Leukemia incidence among people exposed to chronic radiation from the contaminated Techa River, 1953–2005. *Radiat. Environ. Biophys.*, **49** (2), 195–201.

71 Ostroumova, E., Preston, D.L., Ron, E., Krestinina, L., Davis, F.G., Kossenko, M., and Akleyev, A. (2008) Breast cancer incidence following low-dose rate environmental exposure: Techa River Cohort, 1956–2004. *Br. J. Cancer*, **99** (11), 1940–1945.

72 Mushkacheva, G., Rabinovich, E., Privalov, V., Povolotskaya, S., Shorokhova, V., Sokolova, S., Turdakova, V., Ryzhova, E., Hall, P., Schneider, A.B., Preston, D.L., and Ron, E. (2006) Thyroid abnormalities associated with protracted childhood exposure to ^{131}I from atmospheric emissions from the Mayak weapons facility in Russia. *Radiat. Res.*, **166** (5), 715–722.

73 Davis, S., Kopecky, K.J., Hamilton, T.E., Onstad, L., and Hanford Thyroid Disease Study Team (2004) Thyroid neoplasia, autoimmune thyroiditis, and hypothyroidism in persons exposed to iodine 131 from the Hanford nuclear site. *JAMA*, **292** (21), 2600–2613.

74 Boice, J.D., Jr. (2006) Thyroid disease 60 years after Hiroshima and 20 years after Chernobyl. *JAMA*, **295** (9), 1060–1062.

75 Hwang, S.L., Hwang, J.S., Yang, Y.T., Hsieh, W.A., Chang, T.C., Guo, H.R., Tsai, M.H., Tang, J.L., Lin, I.F., and Chang, W.P. (2008) Estimates of relative risks for cancers in a population after prolonged low-dose-rate radiation exposure: a follow-up assessment from 1983 to 2005. *Radiat. Res.*, **170** (2), 143–148.

Part Three
Animal Studies

Cancer Risk Evaluation: Methods and Trends,
Edited by Günter Obe, Burkhard Jandrig, Gary E. Marchant, Holger Schütz, and Peter M. Wiedemann.

7
Animal Studies on RF EMF Cancer Effects

Clemens Dasenbrock and Jochen Buschmann

7.1
Introduction

Is there a causal link between radio frequency electromagnetic field (RF EMF) exposure and cancer? This basic question is yet to be answered.

As discussed in Chapter 8, the current animal research has its methodological limitations in terms of risk assessment. These are partly, but by far not exclusively, the result of poorly conducted research, and rather a consequence of a lack of valid research strategy. The availability of lifetime exposure bioassays in more than one species after exposure to relevant frequencies, including the investigation of the complete spectrum of organs (comparable to guideline studies performed under Good Laboratory Practice (GLP) conditions), could add significant information. This information, however, is not suitable to show the total absence of a carcinogenic effect of RF EMF. Should these studies not give evidence on carcinogenic effects in animals, this would, however, allow the conclusion that the uncertainty is acceptable to the same degree as it is for chemicals of ubiquitous use.

7.1.1
Carcinogenesis

Carcinogenesis (i.e., the development of cancer) is a complex, multistage process. Through the phases of initiation, promotion, and progression after a latency of years to decades, the process of carcinogenesis leads from the initial transformation of normal cells to a clinically manifest tumor (see Chapter 2). The first stage of this development (initiation) is characterized by direct damage of the cell genome (e.g., caused by ionizing radiation) or indirect alterations provoked by metabolites (e.g., radicals). If repair processes fail, immunological control mechanisms can lead to cell death (apoptosis, necrosis) or the genetic damage/alteration (in the cell) is permanent (mutation). Initially, the damaged cell is controlled by the normal tissue, but during the promotion phase, proliferation by clonal multiplication of the transformed cell(s) is stimulated. The changed genome thus disseminates to the daughter cells (preneoplastic foci) and the so-called benign tumors develop. The term progression phase

Cancer Risk Evaluation: Methods and Trends,
Edited by Günter Obe, Burkhard Jandrig, Gary E. Marchant, Holger Schütz, and Peter M. Wiedemann.

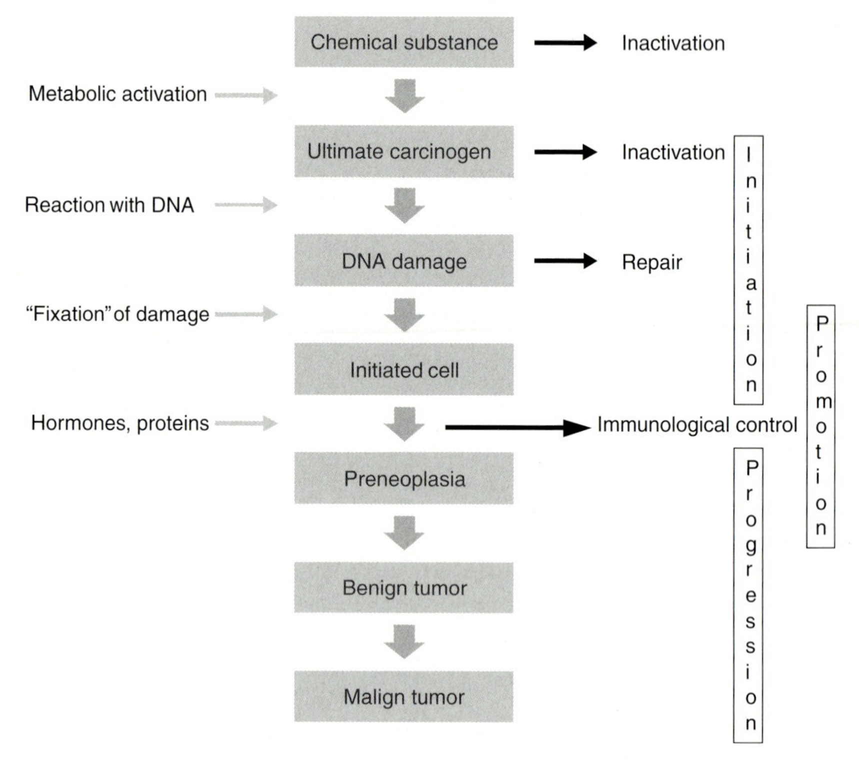

Figure 7.1 Scheme of the process of chemical carcinogenesis (adapted from Ref. [1]).

stands for the progressive transformation of benign tumors into malignant ones that are characterized, among other things, by progressive autonomy and an increased metastatic potential. A scheme of this process, adapted from Ref. [1], is provided in Figure 7.1.

7.1.2
Principles of Carcinogenicity/Toxicity Testing

As a prerequisite for a sound risk assessment of chemicals in the process of their registration, among other investigations, toxicity studies have to be performed. These studies are regulated by a complex framework of guidelines [2–6]. For the estimation of a carcinogenicity risk, the main information is delivered by a lifetime exposure test in the two rodent species rat and mouse [2, 3], supporting evidence can be collected from genotoxicity tests (see Chapter 9). The application of a comparable test strategy for the investigation of a potential carcinogenicity risk of electromagnetic fields appears to be possible and should be considered.

Several *in vivo* animal studies focusing on the determination of carcinogenic effects from RF EMF are already published. However, many of them do not fulfill the criteria that are applied to standard guideline studies. Consequently, some methodological conclusions for the performance of future investigations can be drawn from a survey of existing experimental RF EMF studies. These primarily refer to the inclusion of a sham treated and a nontreated cage or shelf control, and the simultaneous use of several groups subjected to different exposure levels (dose groups), a procedure that is well established in toxicological routine tests on chemicals, pharmaceuticals, and so on. This guarantees the detection of possible dose/effect relations. Also, the inclusion of extensive histological diagnostics (according to the IARC/WHO nomenclature) [7, 8] seems to be urgently required for covering the whole spectrum of neoplastic changes and, for example, for an improved evaluation of shifts in the dignity of neoplastic changes. Moreover, advantages and disadvantages of using freely moving versus restrained animals for exposure are to be considered in planning and performance of such studies.

Numerous "new" animal experimental studies as well as replication studies for a (re-)evaluation of available study results have been initiated worldwide. As an example of these studies, the research projects PERFORM-A within the Fifth Framework Program (FP 5) of the European Union and the CEMFEC (Combined Effects of Electromagnetic Fields with Environmental Carcinogens [9, 10]) shall be mentioned here. They are aimed at clarifying whether RF EMFs have carcinogenic or co-carcinogenic effects on the animal species of rats and mice.

Within the European study project PERFORM-A, two carcinogenicity studies were performed using B6C3F1 hybrid mice and Wistar rats exposed to RF EMF of 902 and 1747 MHz, respectively, over 2 years. In a third part project, the influence of a daily exposure to 902 MHz performed over 6 months was examined in female Sprague Dawley rats, in which mammary tumors were previously induced with DMBA (7,12-dimethylbenz[*a*]anthracene). In this animal model, the influence of RF EMF on tumor growth, incidence, latency, and dignity was explored. The fourth part project was aimed to reproduce the results of the Australian mouse study performed by Repacholi *et al.* [11] who – in 1997 – described a two- to threefold increase in lymphoma rates in Pim1 transgenic mice after 1 h exposure to an EMF of 902 MHz daily for 18 months.

The methodology and results of the (co-)carcinogenicity testing within the European project PERFORM-A are described in detail in Section 7.2.

These results, at least partly, have filled the knowledge gap(s) discussed in Section 7.3 in that the two lifetime exposure studies in rats and mice, which did not yield any evidence of a carcinogenic potential of the two investigated technologies, fulfill the criteria described for this research gap.

7.2
Exemplary Carcinogenicity Studies Testing the Possible Health Effects Related to Mobile Telephones and Base Stations (PERFORM-A)

Almost 10 years ago, the existing experimental data of *in vivo* long-term studies investigating the effects of mobile phone use and base station exposure on the organism as a whole were not sufficient for a joint risk assessment. In response to this, the European Commission (through its Fifth Framework Program), the Swiss and Austrian governments, the GSM Association, and the Mobile Manufacturers Forum supported research addressing potential long-term health implications from the use of mobile phones. Within this, the research program with the acronym PERFORM-A included two combined chronic toxicity/carcinogenicity mouse and rat studies being conducted in both sexes (according to Ref. [2]), involved three different exposure levels to achieve a radio frequency "dose" response, and used as exposure agents signals that best represented current cellular communications-based RF. Two additional studies were conducted using animals predisposed to tumor (e.g., mammary tumor or lymphoma) development. The aim of the PERFORM-A studies was to evaluate putative carcinogenic effects in mice and rats exposed to RF for up to 2 h/day, 5 days/week over a period of up to 2 years.

The applied signal simulated the exposure from GSM902 (GSM: Global System for Mobile Communication) and DCS1747 (DCS: Digital Personal Communications System) handsets, including the low-frequency amplitude modulation components, occurring during speaking (GSM Basic), listening (DTX), and moving within the environment (handovers, power control). The carrier frequency was set to the center of the system's uplink band, that is, 902 MHz for GSM and 1747 MHz for DCS. The targeted exposure levels expressed as whole-body averaged specific absorption rates ($SAR_{wb,av}$) were set at 0.4, 1.3, and 4 W/kg.

The following four separate animal studies were part of the PERFORM-A program:

PERFORM-A1: Two combined toxicity/carcinogenicity studies of 902 MHz GSM and 1747 MHz DCS wireless communication signals in B6C3F1 mice (performed by Fraunhofer ITEM, Hannover, Germany).
PERFORM-A2: Two combined toxicity/carcinogenicity studies of 902 MHz GSM and 1747 MHz DCS wireless communication signals in Wistar rats (performed by RCC Ltd, Itingen, Switzerland).
PERFORM-A3: Evaluation of 902 MHz GSM wireless communication signals on DMBA-induced mammary tumors in Sprague Dawley rats (performed by ARC Research GmbH, Seibersdorf, Austria).
PERFORM-A4: Evaluation of 902 MHz GSM wireless communication signals on lymphoma induction in Eμ-pim-1 transgenic mice (performed by Istituto di Ricerche Biomediche "A. Marxer" (RBM), Colleretto Giacosa, Italy).

The IT'IS Foundation realized the technical aspects of the RF exposure with their (sub)project *PERFORM-A5:* exposure system design, construction, and dosimetry to support studies A1–A4 (performed by IT'IS, Zurich, Switzerland).

All animal studies, including histopathology, were performed blind to all scientists involved except for the IT'IS staff controlling/monitoring the daily RF exposure. Conversely, the IT'IS staff did not know the group identifier. The key codes and identifier were not disclosed until completion of the histopathological evaluation and handover of the still blinded raw data to the representatives of the sponsors.

The studies were performed in compliance with the principles of Good Laboratory Practice and are published in separate articles (Ref. [12]: PERFORM-A1; Ref. [13]: PERFORM-A2; Ref. [14]: PERFORM-A3; Ref. [15]: PERFORM-A4).

7.2.1 Material and Methods

PERFORM-A1

The B6C3F1 mouse studies of GSM and DCS wireless communication signals comprised eight exposure groups and a cage control group, each consisting of 65 males and 65 females.

The applied signal simulated the exposure from GSM 902 MHz and DCS 1747 MHz handsets, including the low-frequency amplitude modulation components, occurring during speaking (GSM Basic), listening (DTX), and moving within the environment (handovers, power control). The carrier frequency was set to the center of the system's uplink band, that is, 902 MHz for GSM and 1747 MHz for DCS. The exposure levels were targeted at 0.4, 1.3, and 4 W/kg.

Uniform whole-body exposure was achieved by restraining the mice in tubes at fixed positions in the exposure setup. Mice were exposed to three whole-body specific absorption rate (SAR) levels or sham exposed for 2 h/day, 5 days/week during the entire 2-year exposure period.

The exposure units (wheels) consisted of two parallel stainless steel metal plates at a distance of 117 mm. A conical antenna was placed in the center between the plates. Encased between the plates, there were 65 cylindrical (plexiglass) tubes arranged radially around the antenna. The mice were placed in the exposure wheels in polycarbonate tubes with adjustable backstops for 2 h daily. Each exposure was connected with a ventilation system to provide the restraint mice with fresh air. Temperature and humidity were monitored continuously within the animal exposure units. All the exposure equipment, including necessary amplifiers, signal generator, and other electronic devices, was supplied by the IT'IS Foundation (Zurich, Switzerland) who also ensured the continuous monitoring of the exposure system, including data transmission and dosimetry.

With the exception of the daily tube restraint (RF EMF exposure: 2 h/day, 5 days/week), the animals were maintained under controlled standard animal room conditions (22 ± 2 °C, 30–70% humidity, 12 h light/dark cycle) on absorbent softwood in polycarbonate type II cages, having free access to food and water.

During the in-life phase of the studies, parameters evaluated included clinical signs, mortality, body weight gain, and food consumption. Chronic toxicity investigations with 15 male and 15 female mice per group were performed after 12 months of RF exposure, analyzing organ weights, hematology, gross pathology, and histopathology, while the remaining 50 mice per group were sacrificed after up to 24 months of RF EMF exposure.

After histopathological examination, a selection of slides, selected by an external advisor, was examined by a (small-scale) Pathology Working Group (PWG) and consensus diagnosis was reached on single questionable cases.

Finally, preserved blood samples collected at terminal sacrifices were evaluated on micronucleus formation in peripheral blood lymphocytes. The incidence of micronuclei was recorded for each mouse in 2000 polychromatic and 2000 normochromatic erythrocytes.

PERFORM-A2

In the 902 MHz GSM and 1747 MHz DCS rat studies, groups of 65 male and 65 female Wistar rats were exposed to SAR target levels of wireless communication signals of 4 W/kg (high dose), 1.33 W/kg (medium dose), 0.4 W/kg (low dose), and 0 W/kg (sham control). Exposure was performed 2 h/day, 5 days/week for 52 consecutive weeks (15 male and 15 female rats/group) or for 104 consecutive weeks (50 male and 50 female rats/group). One additional group of 65 males and 65 females was kept unexposed and served as cage control for the same intervals. At the start of the exposures, 15 male and 15 female rats were recruited from the reserve pool of animals for disease surveillance.

The exposure wheel consisted of a circular cascade of 17 sectoral waveguides, all excited by one quarter loop antenna placed in the center. The wheels were sealed by stainless steel wires on the side. During exposure, the rats were restrained in metal-free polycarbonate tubes in the horizontal position. Air from the animal room was supplied to each rat by a ventilation system. Environmental parameters of oxygen, relative humidity, and temperature were recorded. All the exposure equipment, including necessary amplifiers, signal generator, and other electronic devices, was supplied by IT'IS, Zurich (Switzerland). IT'IS also ensured the continuous monitoring of the exposure system, including data transmission and dosimetry.

Parameters evaluated included clinical (mortality, clinical signs, palpable mass observation, body weight, food consumption, ophthalmoscopic examination, and clinical pathology) and postmortem (organ weight and macroscopic and microscopic pathology) examinations.

Microscopic examination was performed on all collected tissues of all animals excluding the cage controls.

After histopathological examination, a selection of slides, selected by an external advisor, was examined by a (small-scale) Pathology Working Group and consensus diagnosis was reached on single questionable cases.

PERFORM-A3

The aim of the study "902 MHz GSM wireless communication signals: effects on DMBA-induced mammary tumors in rats" was to detect a possible modification by long-term exposure to 902 MHz GSM on incidence, nature, latency period, multiplicity, or growth of mammary tumors, which were induced by an initial administration of DMBA. The "DMBA-induced mammary tumor" model is a well-established model to investigate the influence of various factors on tumor development.

Five groups of 100 female Sprague Dawley rats were each treated once orally with 17 mg/kg DMBA at an age of 46–48 days. From the day after the DMBA dosing onward, they were treated with 902 MHz GSM signals as follows:

Sham exposure group: No exposure to RF EMF, but transferred to and kept in (like the dosed groups) the exposure setup tubes for 4 h/day. This group served as a negative control group for the exposed groups.

The low-dose group was exposed to a SAR target level of 0.44 W/kg, the medium-dose group to 1.33 W/kg, and the high-dose group to 4.00 W/kg for 4 h/day, 5 days/week for 6 months.

The exposure units (wheels) were identical to those used for PERFORM-A2, provided by IT'IS, Zurich, Switzerland.

Cage control group: No exposure to RF EMF, kept continually in their home cages. This group served to give baseline data of the animals without the restriction in the exposure units.

The investigations performed included daily animal observation, detailed weekly clinical observations, weekly examination for palpable subcutaneous tissue masses, weekly body weights, selected hematology parameters at terminal necropsy, necropsy with gross pathological examination, organ weights at terminal necropsy, and histopathology of all gross lesions, some selected organs, and 12 mammary gland sections per animal.

PERFORM-A4

The aim of the study "Lymphoma induction and carcinogenicity study in *Pim-1* transgenic mice exposed to pulsed 900 MHz electromagnetic fields" was to determine whether long-term exposure to three different levels of pulse-modulated RF fields similar to those used in digital mobile telecommunications would increase the incidence of lymphoma in *Pim-1* transgenic mice. These animals express elevated levels of the Pim-1 transgene in their lymphoid compartments, and as a result are predisposed to develop lymphomas.

This experiment was performed in order to clarify the results of a previously published study [11]. In addition, the properties of these RF fields to induce tumors other than lymphomas in this mouse model were investigated in the study, which was conducted in compliance with the main guidelines regarding carcinogenicity studies.

In the present study, groups of 50 male and 50 female Pim-1 transgenic mice were exposed to different target RF EMF levels (SAR), that is, 4 W/kg (high), 1.4 W/kg (medium), 0.5 W/kg (low level), and 0 W/kg (sham). Exposure was performed 1 h/day, 7 days/week for 18 consecutive months. An additional group of 50 males and 50 females was kept unexposed and served as cage control.

The exposure units (wheels) were identical to those used for PERFORM-A1, provided by IT'IS, Zurich, Switzerland. IT'IS also ensured the continuous monitoring of the exposure system, including data transmission and dosimetry.

Parameters evaluated included clinical (clinical signs, palpable mass observation, body weight, food consumption, and clinical pathology) and postmortem (organ weight, gross pathology, and histology) examinations.

Histology was performed on all tissues of all animals including the cage controls. After histological examination, a selection of slides was examined by a (small-scale) Pathology Working Group.

As it holds true for the other studies, this (*Pim-1*) study including histolopathology was also performed blind to all scientists involved except for IT'IS staff, and key codes were disclosed only after completion of histopathology evaluation.

7.2.2
Results

The results can be summarized as follows.

PERFORM-A1: Combined Toxicity/Carcinogenicity Studies of GSM and DCS Wireless Communication Signals in B6C3F1 Mice

Within 2 years, complete exposures (2 h/day) were performed in 98 and 99% of the expected target days in the 902 and 1747 MHz studies, respectively. The estimated exposure levels expressed as specific whole-body absorption rates (SAR_{wb}) were 4.0, 1.3, and 0.4 W/kg during the phases of GSM Basic signal modulation.

Bacteriological investigations revealed positive results (*Staphylococcus aureus, Pasteurella pneumotropica*) on the mucosal surfaces of single animals without clinical relevance. Parasitological and virological examinations during the course of the study revealed no abnormalities, indicating an undisturbed animal study.

Main clinical findings throughout this long-term study were atrophy of the hair follicles, hyperkeratosis of the hind feet, and osteoarthropathy (joint stiffening) of the knee joints (all of slight severity) that are restraint related and were observed with an increasing incidence during the course of the study in all (sham) exposure groups. Compared to the sham group, a significantly altered food intake was detected in various RF EMF exposure groups: differences (increased and decreased consumption) were limited to few and singular weeks, revealing no continuity in time or in the different RF EMF exposure levels.

Compared to the sham group, significant differences in body weight were only detected at some singular measurements and revealed no continuity in respect of time or RF EMF exposure level.

The incidence of mortality throughout the study was higher in the female groups than in the male groups. At the terminal sacrifice time point (after the 2-year exposure period), the incidence of mortality in the various tube-restrained mice was between 10 and 20% for males and between 20 and 30% for females. Comparing the mortality of the sham and three RF EMF exposure groups statistically, Kaplan–Meier test revealed no remarkable differences in the males and females of the 902 and 1747 MHz studies, respectively.

Hematological analysis after 12 months of RF EMF exposure demonstrated no abnormalities.

Relative organ weights of the brain, heart, lungs, liver, spleen, adrenals, kidneys, and gonads revealed no RF EMF exposure-related effects.

The histopathological results after 12 months of exposure illustrated a large variety of sporadic findings, all within the normal range of background alterations commonly seen in mice of this age and strain. Restraint-related lesions were detected in all exposure groups (including sham groups).

After 2 years of RF EMF exposure, among various non-neoplastic findings that did not show any exposure-related increases, a high frequency of chronic osteoarthropathy of the knee joint(s) together with pododermatitis (skin) was observed in both studies. Etiologically, these lesions were considered to be related to restraint of the animals in the exposure tubes.

Regarding the organ-related tumor incidence, the pairwise Fisher test did not show any significant increase of a specific tumor type in the radio frequency exposure groups compared to the sham exposed group, either in the 902 or in the 1747 MHz study.

Finally, there were no significant differences in the frequency of micronuclei between RF exposed, sham exposed, and cage control mice. Micronuclei were, however, significantly increased in polychromatic erythrocytes of the positive control mice [16].

PERFORM-A2: Combined Toxicity/Carcinogenicity Studies of GSM and DCS Wireless Communication Signals in Wistar Rats

Of the intended exposures, more than 98% (902 MHz) or 96.6% (1747 MHz) were successfully completed.

The estimated exposure levels in terms of SAR_{wb} were 0.44, 1.33, and 4.0 W/kg for 902 MHz groups and 0.41, 1.23, and 3.7 W/kg for the 1747 MHz groups at the low, medium, and high exposure levels, respectively.

The incidence of animals that died or were killed before the end of the exposure period did not indicate an effect related to exposure to 902 MHz GSM or 1747 MHz DCS wireless communication signals.

There were no clinical signs or palpable masses, body weight, food consumption, ophthalmoscopic findings, macroscopic findings, or organ weight changes considered to be related to exposure to 902 MHz GSM or 1747 MHz DCS wireless communication signals. There were also no changes in hematological, clinical biochemical, or urinalysis parameters considered to be related to exposure to 902 MHz GSM or 1747 MHz DCS wireless communication signals.

Finally, there were no significant differences in the incidence of primary neoplasms, the number of rats with more than one primary neoplasm, the multiplicity and latency of neoplasms, the number of rats with metastases, and the number of benign and malignant neoplasms between the rats exposed to wireless communication signals and rats that were sham exposed.

PERFORM-A3: Evaluation of GSM Wireless Communication Signals on DMBA-Induced Mammary Tumors in Sprague Dawley Rats

Of the intended exposures, 95% were successfully completed. The estimated averaged SAR_{wb} values were 0.4, 1.3, and 4.0 W/kg.

There was no RF EMF exposure-related effect on observations in life, on clinical and functional observations, and on the survival or the causes of death. All RF EMF exposed groups had, at several terms, significantly more palpable tissue masses. There was no RF EMF exposure-related effect on body weights, hematology endpoints, and organ weights. There was also no RF EMF exposure-related effect found on gross pathology at necropsy.

For histopathology endpoints, the following main findings were observed: when compared with the sham exposed groups, there were significantly

- less animals with benign neoplasms in all RF EMF dose groups,
- more animals with malignant neoplasms in the high RF EMF dose group,
- more adenocarcinomas (in total number) in the low RF EMF dose group,
- more malignant neoplasms (in total number) in the low and high RF EMF dose groups,
- more animals with adenocarcinoma (any number) in the high RF EMF dose group, and
- less animals with one (or any number of) fibroadenoma in the low and medium RF EMF dose groups.

PERFORM-A4: Evaluation of GSM Wireless Communication Signals on Lymphoma Induction in Eμ-Pim-1 Transgenic Mice

During 18 months, complete exposures (1 h/day) were performed in 98.5% of the targeted days. The estimated exposure levels were 0.5, 1.4, and 4.0 W/kg SAR_{wb}.

Before the end of the exposure period, 75 males and 92 females died or were killed. The distribution of unscheduled deaths across the groups indicated that, in comparison to controls, there was poorer survival in all the male groups exposed to RF EMF and in females exposed to 0.5 W/kg. Since the decreased survival was not exposure level related, it was considered incidental. There was no evidence from the histopathology that exposure to RF EMF was responsible for the early death of any animal.

No effects on body weight, food consumption, and palpable masses that could be related to RF EMF exposure were found.

In the hematology investigations, no changes in white blood cell differential count, total blood cell count, and related parameters were noted. No organ weight changes and no macroscopic changes that could be related to RF EMF exposure were found.

Histological investigation revealed that the exposure of *Pim-1* transgenic mice to a pulsed 900 MHz electromagnetic field for 18 months at absorption rates of 0.5 and 1.4 W/kg was not considered to have affected the incidence of any neoplastic finding. In particular, the incidence of malignant lymphoma in animals exposed to RF EMF was not significantly different from that seen in cage or sham controls.

A dose-related positive trend in the incidence of males bearing a Harderian gland adenoma ($p < 0.0028$, one-tailed test) with four of six cases occurring in males exposed to 4.0 W/kg and the remaining two in males exposed to 1.4 W/kg contrasted with the result in females where four of the six cases occurred in controls and the two in females exposed to 0.5 W/kg.

7.2.3
Discussion

PERFORM-A1

Concerning the tumor incidence in the carcinogenicity study of GSM (902 MHz) and DCS (1747 MHz) wireless communication signals in B6C3F1 mice, a comparison with historical control data from B6C3F1 mice in NTP carcinogenicity studies [17] shows that the tumor rates observed in the present studies are within the range of the NTP tumor data, although for some organs (liver, Harderian glands, and hematopoietic system) marked differences between both data sets exist. The main tumor types observed in the present studies were also well in line with the corresponding tumor incidences in untreated B6C3F1 mice from a transgeneration carcinogenicity study, conducted at Fraunhofer ITEM [18].

PERFORM-A2

The increased/decreased incidences of some non-neoplastic and neoplastic findings in the study "902 MHz GSM and 1747 MHz DCS wireless communication signals: combined chronic toxicity/carcinogenicity study in the Wistar rat" are regarded incidental rather than exposure related.

The finding of four prostate adenomas in the 1747 MHz high-dose group of this study was not statistically significant compared to the sham control (in which none was found). However, prostate adenomas are uncommon in the Wistar rat and it cannot be totally discounted that the incidence of this uncommon tumor type is associated with exposure to a high dose of wireless communication signals at the 1747 MHz frequency.

PERFORM-A3

For giving the histopathological results of this study a relative weight, a comparison of the sham exposed group versus the cage control group was added. Among the two not RF EMF exposed groups, significant differences were repeatedly noted.

Compared to the sham exposed group, the *cage control group* had

- a marginally higher mean body weight,
- higher mean adrenal gland weights,
- more animals with palpable tissue masses in the mammary gland (actually the highest number of all groups, that is, sham exposed, RF EMF exposed, and cage control groups),
- more tissue masses in the mammary gland at necropsy (actually the highest number of all groups),
- more animals with hyperplasia or neoplasia (actually the highest number of all groups),
- more animals with malignant neoplasms,
- more benign neoplasms in the mammary gland (actually the highest number of all groups),

- more malignant neoplasms in the mammary gland (and, of course, more neoplasms of any dignity, actually the highest number of all groups),
- a higher number of both adenocarcinomas and fibroadenomas (actually the highest number of all groups) in the mammary gland, and
- some other significant differences in tumor distribution, in general with the higher number than the sham exposed group.

The almost identical body weights of the two control groups indicate that the manipulation and restriction in the exposure setups did not stress the animals severely. No cause for the higher tumor formation in the cage control animals could be identified.

In the mammary glands and their neoplastic lesions, there were some significant differences between one or more of the RF EMF exposed groups and the sham exposed groups found, especially

- more palpable tissue masses,
- more animals with malignant neoplasms, and
- less animals with benign neoplasms.

This indicates that there might be an effect of the exposure to RF EMF on the number and dignity of the tumors formed, while all other aspects (e.g., time of onset, total number, multiplicity, size) remained unaffected.

In this context, it is important to include the cage control group for comparison of the effects and to establish their relative biological importance. In fact, in all the above-mentioned parameters with significant effects, the cage control group had equal or even worse results than the worst RF EMF exposed group ("worse" stands here for more tumors or a higher malignancy of the tumors).

The most pronounced differences noted in this study were those between the sham exposed group and the cage control group. After having performed an identical study in China, almost the same results were reported by Yu *et al.* [19].

PERFORM-A4

In the lymphoma induction and carcinogenicity study in *Pim-1* transgenic mice exposed to pulsed 900 MHz electromagnetic fields, the statistically significant difference in the incidence of males bearing a Harderian gland adenoma was considered to have arisen fortuitously and not as a result of exposure to RF EMF.

None of the nonneoplastic findings encountered in this study were considered to have been affected by exposure to RF EMF. The range of findings encountered was broadly similar to that commonly observed in aging mice.

In addition, the SAR_{wb} was about three times greater than the highest average SAR in the Repacholi *et al.* study [11]. Thus, the results of Repacholi *et al.* [11] were not confirmed. Also, Utteridge *et al.* [20] who exposed PIM-1 transgenic and wild-type mice for 1 h/day, 5 days/week for up to 24 months to 894 MHz GSM signals at SARs of 0.25, 1.0, 2.0, and 4.0 W/kg did not confirm the findings of Repacholi *et al.* [11].

7.2.4 Conclusions

In studies of this dimension with numerous parameters determined, isolated findings that are seemingly treatment related are to be expected, as there may be several "incidental" significant differences between the groups, even in the absence of any biologically effective agent.

Under the conditions of study **PERFORM-A1**, the exposure of male and female B6C3F1 mice to electromagnetic fields of 902 and 1747 MHz wireless communication signals at a whole-body absorption rate of up to 0.4, 1.3, or 4.0 W/kg, 2 h/day, 5 days/week over a period of up to 24 months produced no evidence that the exposure had any adverse health effect or any influence on the incidence or severity of the background non-neoplastic and neoplastic lesions observed.

Data on micronucleus formation in peripheral blood erythrocytes did not indicate RF-induced genotoxicity in mice after 2 years of exposure.

Under the conditions of study **PERFORM-A2**, Wistar rats exposed to 902 MHz GSM signals at SAR levels of up to 0.44, 1.33, or 4.0 W/kg or 1747 MHz DCS signals at SAR levels of 0.41, 1.23, or 3.7 W/kg for 2 h/day, 5 days/week over a period of up to 24 months produced no evidence that exposure had any effect on the incidence or severity of any neoplastic or nonneoplastic condition. It is not known whether the presence of four prostate adenomas in the 1747 MHz DCS wireless communication signals high-dose group can be directly attributed to treatment.

Under the conditions of study **PERFORM-A3**, it is concluded that this study produced a borderline evidence of long-term repeated exposure to 902 MHz GSM signals at SAR levels of up to 0.4, 1.3, or 4.0 W/kg affecting the DMBA-induced mammary tumor response in rats with an equivocal biological relevance.

Under the conditions of study **PERFORM-A4**, exposure of *Pim-1* transgenic mice to a pulsed 900 MHz electromagnetic field at an absorption rate of 0.5, 1.4, or 4.0 W/kg, daily for 1 h, over a period of not less than 18 months produced no evidence that exposure had any effect on the incidence or severity of any neoplastic or non-neoplastic condition.

Overall, three out of four studies produced no evidence that exposure had any effect on the incidence or severity of any neoplastic or non-neoplastic condition (with one equivocal finding in one study). The only effect observed is a borderline one in the study investigating effects on the DMBA-induced mammary tumor response.

7.3 Research Gaps

Real lifetime (*in utero* up to death) exposure bioassays after radiation to human exposure-relevant frequencies, including the investigation of the complete spectrum of organs (comparable to guideline studies performed under GLP conditions), are still missing. The current US program [21] may fill this gap.

In summary, it can be concluded that currently only a small number of chronic RF EMF studies have been published and that there still are few studies using characteristic GSM or UMTS mobile radio signals. Long-term experimental studies investigating the GSM and UMTS signal technology widely spread in Europe are still rare. "Early" studies were not always satisfactorily performed or published, so their contribution to an assessment of human health risks from chronic RF EMF is limited. Especially, the incomplete, nonstandardized histopathological evaluations found in carcinogenicity studies and the partially unclear exposure technology (RF signal) minimize the relevance of these investigations.

Based on such data, a sound scientific assessment of potential cancer risks from RF EMF will be a matter of controversial discussion; however, concrete evidence for such health concerns to date has not been found. The upcoming IARC monograph 102 meeting "Non-ionizing radiation, Part II: radiofrequency electromagnetic fields" in May 2011 will set an interim point in the ongoing discussion.

The main factors limiting the relevance of currently existing studies on carcinogenicity of RF EMF are

- the absence of any reproducible effects,
- in addition to the above-described cancer studies of PERFORM-A, the absence of further guideline type studies covering the whole organ spectrum, using standardized histopathology, a sufficiently high number of animals and more than one "dose" group,
- the academic type of studies so far published, using models that are often not sufficiently standardized and, therefore, of limited value for a sound risk assessment, and
- the investigation of different frequencies and/or modulation, where it is nearly impossible to extrapolate from one frequency range to another (in terms of toxicology, each basically and essentially different frequency and/or signal modulation would be considered a different compound requiring specific testing).

7.3.1
Exposure Assessment

In the case of toxicity testing in chemicals, the sequence of events is to apply in animal studies doses close to the maximum tolerated ones in order to determine the no observed adverse effect level (NOAEL) [4, 22]. Applying a safety factor (see Section 7.3.4), a safe dose for humans is determined based on the NOAEL from animal studies. Consequently, one of the core principles in toxicity testing in animals is overdosing. This principle is strongly limited if RF EMF is tested in animal models, since overdosing would result in heating of the organism, which should be omitted. This significantly compromises the above-described procedure for chemicals.

Unlike in the case of chemicals, for pharmaceuticals normally the intended therapeutic dose is known before testing, allowing the determination of the doses to be tested using the opposite procedure: applying a safety factor to the intended human use would determine the doses used in the experiment.

Applying this approach for RF EMF testing could solve the problem of producing (easily measurable) thermal effects in the experiments. To be able to do this, however, we need more data on worst-case exposure scenarios that could occur for humans under the conditions of the existing and intended use of RF EMF. Only if these data are known, a research strategy as described in Section 7.4.2 can be applied.

7.3.2
Mode of Action

Due to the complex nature of the process of carcinogenicity, different modes of action could contribute to the eventually observed effect of an increased tumor rate. Consequently, should such effects be observed reproducibly in valid RF EMF studies, then the investigation of the underlying mechanism is important in terms of risk assessment: if the nature of the effect allows to determine a threshold for the observed effect (mostly applicable in the case of a co-carcinogenic effect), then the above-described normal procedure of risk assessment can be applied. In the contrary situation (e.g., genotoxic mechanism of tumor induction), if no threshold of the effect can be assumed, this procedure cannot be applied.

Since guideline studies cannot normally provide such information, it is important to better interlink basic and applied research, should data suggest carcinogenic effects due to RF EMF (see Section 7.4.1).

7.3.3
Susceptibility

The susceptibility of the currently used rodent models for carcinogenicity has been shown for many chemical compounds. The striking differences in lifetime between the model animals and the target organism (human) are outweighed by differences in the repair mechanisms so that the process of carcinogenesis is comparable in these species. The question is what the situation is in the case of RF EMF. For a sound data extrapolation, a good dosimetry is of high priority (see Section 7.4.2.). While relatively good models for the absorption of RF EMF in humans are available, it is highly important to also have such models for experimental animals. This also includes the availability of such models for the developing organisms. The predictive value of animal experiments will significantly benefit if these models can be used in order to compare the exposure situation in the target organs of animals and humans (see also Section 7.3.4).

7.3.4
Extrapolation

When testing chemicals, normally a safety factor of 100 (10 for interspecies differences, 10 for differences in susceptibility in humans) in extrapolation from animal studies to the human exposure has proven to be protective. It must be further discussed whether this safety factor should be the same when testing RF EMF. On one hand, the safety factor of 100 is applied when a broad spectrum of guideline studies in at least two species can serve as a basis. This situation is not given in the case of RF EMF, suggesting a higher safety factor. On the other hand, the safety factor partly consists of metabolic differences between the animal models and humans, which does not play a role for RF EMF. The latter fact would suggest the use of a lower safety factor.

7.4 Proposed Research Strategy

In order to get more information, which can be better used in the process of risk assessment, the adoption of procedures similar to those of toxicological standard tests applied for the registration of chemicals, drugs, or pesticides appears possible and reasonable.

Since these methods have been developed and applied for a long time and fixed in permanently improved guidelines, they are scientifically well based and widely accepted for a successful risk assessment of chemical substances. Consequently, these animal models are known to be predictive for humans, and the applied methods are well standardized and validated.

When planning this type of studies, the following points should be considered beforehand:

- Selection of the best suitable and sensitive species/strain
- Selection of relevant field strengths as an analogue to dose ranges applied in (chronic) toxicity and carcinogenicity testing
- Consideration of the advantages and disadvantages of the use of free moving versus restraint animals

Although the performance of complete carcinogenicity studies in two species is complex and expensive, such studies should deliver essential information for risk assessment. An example was given in the series of PERFORM-A studies.

7.4.1 Interaction between Hypothesis-Driven Studies and "Apical" Tests

On comparing the different scientific approaches applied in the different studies accessible so far, two basic groups of studies can be distinguished:

- Studies aiming at validating a specific hypothesis (hypothesis-driven studies)
- Studies mainly serving as a means for "global" risk assessment (toxicology type of studies)

The former are strongly academic attempts at basic research level, often of high scientific value, and are performed using very sensitive model systems. Extrapolation to the human target organism is often limited, and it is very difficult (above all in case of negative results from the verification of single hypotheses) to draw conclusions on "general" health risks to humans.

The latter study group uses a markedly toxicological approach in respect of applied research: different test methods are used to examine various groups of test animals exposed to different field strengths, including as many potentially health relevant endpoints as possible. This approach, best described as "apical," due to the large number of test parameters allows a fairly good extrapolation to humans and, therefore, is highly suitable for an assessment of health risks. However, such studies, due to their "broad spectrum" characteristics, are often less profound than those of the first group, thereby not always being sensitive enough for exploring specific health risks.

A promotion of the dialogue between both groups of researchers could produce significant synergies: If, on one hand, results achieved in basic research can be reproduced in applied research and, on the other hand, findings obtained by applied studies can be confirmed by mechanistic approaches, the framework of risk assessment will be considerably improved.

7.4.2
Thermal versus Nonthermal Effects

One basic principle of toxicity testing is overdosing in animal studies in order to be able to apply safety factors in the process of extrapolation from the model (animal species) to the target organism (human). When performing animal studies on potential adverse effects of RF EMF, energy is added to biological systems. As a consequence, this leads to thermal reactions in animals, and there is no real "athermal" range. Attempts were made to introduce such a threshold at the energy flow density that leads to an increase in rectal body temperature of 1 °C, but this appears to be rather artificial.

In order to give practical guidance for future studies, it is suggested to use "dose" energy flow densities high enough to ensure the application of appropriate safety factors. The basis for such a comparison could be the exposure assessment of humans, produced by base stations and/or mobile phones using a worst-case scenario. The use of lower dose group(s) is strongly recommended in order to find potential dose–response relationships and (threshold) "doses" without effects.

When performing these animal studies, the body temperature of the animals and the cage temperature should be measured in an appropriate way, either during the study or in pilot experiments [23]. For the (rectal) body temperature, this could either be done toward the end of the daily exposure period in order to determine an expected increase in temperature or soon after switching off the exposure in order to determine a transient decrease in body temperature in response to stopping the external administration of energy [23]. By doing this, it should be shown that the selected exposure range does not cause "excessive" heating (e.g., more than 1 °C in body temperature during exposure – comparable to the known increase of body temperature in humans exposed to RF EMF at specific absorption rate of 4 W/kg for 30 min [24]).

With an exposure like this and using a study type that is based on carcinogenicity testing, either no effects will be observed or it will be possible to find (reproducible) effects. In the former case, this indicates the (relative) safety of the exposure, while in the latter it is appropriate to check whether these effects are (solely) produced by heating and/or if these effects are species specific or can be extrapolated from animals to man. But even when it can be shown that they are secondary to thermal ones, but their occurrence in humans cannot be excluded, the effects still remain relevant. Since the endpoints investigated are based on (modified) toxicity test guidelines, it can be assumed that any indication for an increase in tumor occurrence compared to the concurrent control group will be regarded as an adverse effect. The experimental database thus established can then be used to determine whether there is any indication of a carcinogenic or co-carcinogenic effect and whether a no observed adverse effect level for the given type of exposure can be determined. By applying an appropriate safety factor, a safe exposure for humans can be assessed.

7.4.3
Role of "Omics" Studies

As described above, carcinogenesis is a complex, multistage process. Assessing individual stages of this process may be of high relevance in terms of basic research. However, in terms of risk assessment, the application of this approach is (currently) of very limited value. The "omics" approach investigates the effect of a given factor on the gene and protein expression. Thus, it can also be applied to predict genotoxic effects of a given stimulus, which play an important, but not the only, role in carcinogenesis. In this situation, "omics" methods can currently be used to either determine the mechanism of an effect observed on the level of whole organism (see Section 7.3.2) or as a screening tool to determine sort of a "relative risk" (e.g., compound A is more likely to be a carcinogen than compound B). While the investigation on the mechanism of an effect is currently limited by the lack of such (reproducible) effects, the determination of a "relative risk" could deliver useful information, if in a first step the different frequency ranges (as an analogue to different compounds in the case of chemicals) could be screened using these methods in order to set priorities for further investigations (i.e., to get some information about what frequency range is most likely to cause health effects). The results of such studies could then be used for planning further studies investigating the "critical" frequency range (should there be one) and/or signal modulation with higher priority.

7.5
Summary

Until now, a causal link between RF EMF exposure (as used in mobile telephony) and cancer has not been detected in experimental carcinogenicity studies.

However, for future studies, we recommend to adopt principles of chemical testing guidelines for evaluating a potential cancer effect of RF EMF. Besides a detailed description of the (technical) exposure conditions, the following points are considered to be most important:

- Justification of species and strain
- Exposure level (dose) selection
- At least three "dose" groups, one control
- Dose level spacing (normally two- to fourfold intervals)
- Complete histopathology of all organs of all animals

The described study design applied within the PERFORM-A project is a good example of how existing data gaps relevant for risk assessment can be closed. Mechanistic toxicology with its *in vitro* and "omics" approaches, on the other hand, can deliver valuable additional information for a sound risk assessment.

References

1 Forth, W., Henschler, D., and Rummel, W. (eds) (1983) *Allgemeine und spezielle Pharmakologie und Toxikologie*, 4th edn, Bibliographisches Institut, Mannheim.

2 OECD (2009) Combined chronic toxicity/carcinogenicity studies. Test Guideline No. 453, OECD Guidelines for the Testing of Chemicals, Paris, http://www.oecd.org/env/testguidelines (retrieved August 10, 2010).

3 OECD (2009) Carcinogenicity studies. Test Guideline No. 451, OECD Guidelines for the Testing of Chemicals, Paris, http://www.oecd.org/env/testguidelines (retrieved August 10, 2010).

4 OECD (2009) Guidance document on the design and conduct of chronic toxicity and carcinogenicity studies. Series on Testing and Assessment No. 116, OECD Guidelines for the Testing of Chemicals, Paris, http://www.oecd.org/env/testguidelines (retrieved August 10, 2010).

5 OPPTS (1998) 870.4300: Combined chronic toxicity/carcinogenicity, http://www.epa.gov/ocspp/pubs/frs/home/guidelin.htm (retrieved August 10, 2010.

6 OPPTS (1998) 870.4200: Carcinogenicity, http://www.epa.gov/ocspp/pubs/frs/home/guidelin.htm (retrieved August 10, 2010).

7 WHO-IARC (1992–1997) *International Classification of Rodent Tumours. Part I: The Rat*, IARC Scientific Publication No. 122, IARC, Lyon.

8 WHO-IARC (2001) International classification of rodent tumours, in *The Mouse* (ed. U. Mohr), Springer, Heidelberg.

9 Heikkinen, P., Ernst, H., Huuskonen, H., Komulainen, H., Kumlin, T., Maki-Paakkanen, J., Puranen, L., and Juutilainen, J. (2006) No effects of radiofrequency radiation on 3-chloro-4-(dichloromethyl)-5-hydroxy-2(5*H*)-furanone-induced tumorigenesis in female Wistar rats. *Radiat. Res.*, **166**, 397–408.

10 Heikkinen, P., Ernst, H., Huuskonen, H., Komulainen, H., Kumlin, T., Maki-Paakkanen, J., Puranen, L., and Juutilainen, J. (2007) No effects of radiofrequency radiation on 3-chloro-4-(dichloromethyl)-5-hydroxy-2(5*H*)-furanone-induced tumorigenesis in female Wistar rats. *Radiat. Res.*, **167**, 124–125.

11 Repacholi, M.H., Basten, A., Gebski, V., Noonan, D., Finnie, J., and Harris, A.W. (1997) Lymphomas in Eμ-Pim1 transgenic mice exposed to pulsed 900MHz electromagnetic fields. *Radiat. Res.*, **147**, 631–640.

12 Tillmann, T., Ernst, H., Ebert, S., Kuster, N., Behnke, W., Rittinghausen, S., and

Dasenbrock, C. (2007) Carcinogenicity study of GSM and DCS wireless communication signals in B6C3F1 mice. *Bioelectromagnetics*, **28**, 173–187.

13 Smith, P., Kuster, N., Ebert, S., and Chevalier, H.J. (2007) GSM and DCS wireless communication signals: combined chronic toxicity/carcinogenicity study in the Wistar rat. *Radiat. Res.*, **168**, 480–492.

14 Hruby, R., Neubauer, G., Kuster, N., and Frauscher, M. (2008) Study on potential effects of "902-MHz GSM-type Wireless Communication Signals" on DMBA-induced mammary tumours in Sprague-Dawley rats. *Mutat. Res.*, **649**, 34–44.

15 Oberto, G., Rolfo, K., Yu, P., Carbonatto, M., Peano, S., Kuster, N., Ebert, S., and Tofani, S. (2007) Carcinogenicity study of 217Hz pulsed 900MHz electromagnetic fields in Pim1 transgenic mice. *Radiat. Res.*, **168**, 316–326.

16 Ziemann, C., Brockmeyer, H., Reddy, S.B., Vijayalaxmi, Prihoda, T.J., Kuster, N., Tillmann, T., and Dasenbrock, C. (2009) Absence of genotoxic potential of 902MHz (GSM) and 1747MHz (DCS) wireless communication signals: *in vivo* two-year bioassay in B6C3F1 mice. *Int. J. Radiat. Biol.*, **85**, 454–464.

17 Haseman, J.K., Elwell, M.R., and Hailey, R.W. (1999) Neoplasm incidences in B6C3F1 mice: NTP historical data, in *Pathology of the Mouse: Reference and Atlas* (ed. R. Maronpot), Cache River Press, Vienna, IL, pp. 679–689.

18 Dasenbrock, C., Tillmann, T., Ernst, H., Behnke, W., Kellner, R., Hagemann, G., Kaever, V., Kohler, M., Rittinghausen, S., Mohr, U., and Tomatis, L. (2005) Maternal effects and cancer risk in the progeny of mice exposed to X-rays before conception. *Exp. Toxicol. Pathol.*, **56**, 351–360.

19 Yu, D., Shen, Y., Kuster, N., Fu, Y., and Chiang, H. (2006) Effects of 900MHz GSM wireless communication signals on DMBA-induced mammary tumors in rats. *Radiat. Res.*, **165**, 174–180.

20 Utteridge, T.D., Gebski, V., Finnie, J.W., Vernon-Roberts, B., and Kuchel, T.R. (2002) Long-term exposure of Emu-Pim1 transgenic mice to 898.4MHz microwaves does not increase lymphoma incidence. *Radiat. Res.*, **158**, 357–364.

21 McCormick, D.L. (2008) Use of experimental model systems to identify possible health effects of exposure to RF fields. FGF Workshop "Open Questions in the Research on Biological & Health Effects of Low-Intensity RF-EMF", http://www.fgf.de/english/research_projects/reports/workshops/abstracts/Abstractbook-FGF-Workshop-Stuttgart-2008.pdf (retrieved August 10, 2010).

22 Rhomberg, L.R., Baetcke, K., Blancato, J., Bus, J., Cohen, S., Conolly, R., Dixit, R., Doe, J., Ekelman, K., Fenner-Crisp, P., Harvey, P., Hattis, D., Jacobs, A., Jacobson-Kram, D., Lewandowski, T., Liteplo, R., Pelkonen, O., Rice, J., Somers, D., Turturro, A., West, W., and Olin, S. (2007) Issues in the design and interpretation of chronic toxicity and carcinogenicity studies in rodents: approaches to dose selection. *Crit. Rev. Toxicol.*, **37**, 729–837.

23 Ebert, S., Eom, S.J., Schuderer, J., Apostel, U., Tillmann, T., Dasenbrock, C., and Kuster, N. (2005) Response, thermal regulatory threshold and thermal breakdown threshold of restrained RF-exposed mice at 905MHz. *Phys. Med. Biol.*, **50**, 5203–5215.

24 ICNIRP (1998) Guidelines for limiting exposure to time-varying electric, magnetic, and electromagnetic fields (up to 300GHz). *Health Phys.*, **74**, 494–522.

8
Animal Studies in Carcinogen Identification: The Example of Power Frequency (50/60 Hz) Magnetic Fields

David L. McCormick

8.1
Introduction

Identification of a chemical or physical agent as a potential human carcinogen is generally based on the results of (a) epidemiology studies, in which cancer incidence and mortality are compared in human populations (exposed versus nonexposed groups, or in groups with varying measured or imputed levels of exposure to the agent under investigation), and (b) experimental carcinogenicity bioassays, in which incidences of site-specific neoplasms are compared in groups of laboratory animals receiving different levels of exposure to the agent. Hazard identification data generated in epidemiological and experimental studies are often supplemented with, and can be considerably strengthened by, mechanistic investigations that identify potential cellular, biochemical, or molecular mechanisms of carcinogenic action.

In the past three decades, the possible carcinogenicity of power frequency (50/60 Hz) magnetic fields (EMF) has generated significant interest both in the scientific community and in the general public. This interest was initially stimulated by the 1979 study of Wertheimer and Leeper, who reported an excess of childhood cancer in homes with wiring configurations that were suggestive of high levels of EMF [1]. In 1982, Milham reported that workers in "electrical occupations," whose exposure to EMF was presumed to be higher than that of other occupational groups, demonstrated increased mortality from leukemia [2]. These early studies stimulated a large number of follow-on epidemiology studies of EMF and cancer risk (reviewed in Refs [3–5]); numerous experimental investigations have also been conducted to assess the possible carcinogenic activity of EMF exposure (reviewed in Refs [4, 5]). The results of these studies have been integrated to develop a rational, fact-based assessment of the possible risks of human cancer that may result from exposure to EMF.

Both epidemiology studies and cancer bioassays in laboratory animals have made important contributions to this hazard assessment. This chapter discusses the overall data set that has been used to develop a cancer hazard assessment for power frequency EMF, with particular emphasis on studies that have been conducted in

Cancer Risk Evaluation: Methods and Trends,
Edited by Günter Obe, Burkhard Jandrig, Gary E. Marchant, Holger Schütz, and Peter M. Wiedemann.

laboratory animals. A critical element of the rational assessment of EMF as a possible risk for human cancer is an evaluation of the strengths and weaknesses of the data on which the assessment is based.

8.2
Strengths and Limitations of Epidemiology Studies of EMF as a Cancer Hazard

By definition, epidemiology studies are designed to investigate cancer risk in the species of interest (humans) at exposure levels that are relevant to human exposure (as seen in occupational and/or residential environments). Interpretation of the results of epidemiology studies neither does require that data be extrapolated from another species to humans nor does it require that data be extrapolated from high exposure levels (as commonly used in laboratory studies) to the often much lower levels humans are actually exposed to. Because epidemiology involves evaluations of cancer risk in humans at "real-world" exposure levels, these studies are considered by some to be the "gold standard" for cancer risk assessment.

Although epidemiology studies have very obvious strengths given their direct relevance to human disease, it is important to consider several potentially critical limitations associated with their use in cancer hazard identification. Key issues include causality versus association, limitations in statistical power, timing and duration of exposure prior to hazard identification, and dosimetry and exposure assessment. Of these issues, dosimetry and exposure assessment provide especially important challenges to the use of epidemiology to evaluate the possible risks of human exposure to EMF.

An essential consideration for the use of epidemiology to identify cancer risk factors is that epidemiology data are associative: although well-designed epidemiology studies can demonstrate association between exposure and disease risk, epidemiology cannot demonstrate causality. Because spurious or noncausal associations may be identified in epidemiology studies, mechanistic data may become extremely important to support the biological rationale of any cancer risk factor identified through epidemiology studies.

Other important limitations to the use of epidemiology in the identification of cancer hazards relate to statistical power. Epidemiology is a powerful tool in the identification of major risk factors of cancer (e.g., tobacco use and lung cancer; asbestos and mesothelioma) and risk factors underlying rare malignancies, particularly in industrial environments (e.g., vinyl chloride and angiosarcoma of the liver). However, epidemiology studies often lack adequate statistical power to identify agents that may induce a modest increase in risk that is superimposed on a finite background incidence (e.g., a dietary factor that may induce a small percentage increase in the risk of colon cancer or other common neoplasm). Simply stated, weak but true causal associations may be missed due to the limited statistical power of the epidemiology study design.

The timing of hazard identification through epidemiology may also be problematic. Because the latent period for cancer development may extend for decades, identifying the relevant period of exposure for potentially carcinogenic agents presents an important challenge to the design of any epidemiology study. Is a retrospective study that extends 5 years prior to diagnosis long enough? Is 10 years sufficient for a comprehensive evaluation? Or, must the design of a comprehensive epidemiology study include exposures that may have occurred even earlier in a patient's life?

Cancer latency also presents a major public health challenge to the use of epidemiology to identify cancer hazards: reliance on epidemiology alone could mean that individuals are exposed to a carcinogenic agent for years, if not decades, prior to its identification as a risk factor. For some ubiquitous environmental agents, delays in cancer hazard identification could result in broad population exposure for many years to what was, at the time, an unrecognized hazard. EMF is clearly included in this class of ubiquitous environmental agents.

An essential factor to be considered in the interpretation of EMF epidemiology studies is the accuracy and relevance of the exposure assessment on which comparisons are based: Is the estimate of EMF exposure developed using (in order of decreasing desirability)

- measurements of EMF flux densities to which an individual is actually exposed, as obtained by a personal dosimeter or other measurements?
- measurements of EMF flux densities present at a relevant location in the home or work site?
- measurements of EMF flux densities made at a constant point outside home ("front door" measurements)? or
- an imputed or calculated level of EMF exposure that is based on job classification, residential wire code, residential distance from a power line or substation, or other metric?

It is clear that the approach used to measure or impute EMF exposures may have a substantial effect on exposure estimates [6]. Although it has been demonstrated that wire codes and job titles may provide reasonable general estimates of EMF exposure [7, 8], these classification approaches cannot accurately capture changes in exposure with time, and at best, provide only broad exposure estimates that may not support the development of accurate odds ratios and characterization of dose–response relationships [7, 8]. Data generated over two decades suggest that the reliability of exposure assessment in EMF epidemiology studies is strengthened considerably through the use of actual field measurements [6, 8].

Even if EMF exposure assessment is based on extensive personal dosimetric data, the issue of the relevant metric for biologically important interactions at the cellular level remains largely unknown. Is EMF exposure optimally assessed using a measure of *Exposure* × *Time*, as is commonly applied to chemical exposures? Or, is the relevant metric best defined on the basis of the maximum field strength to which an individual is exposed for a defined period? Alternatively, is the most relevant metric related to changes in EMF flux density to which an individual is exposed over time? Is whole-body exposure (as produced by sources such as overhead power lines or electric blankets/bed heaters) more important than local, high-dose EMF exposure that is emitted by a point source (such as the one produced by an electric razor or hair dryer)?

Clearly, such questions must be considered carefully in evaluating the strengths of association seen in epidemiology studies involving exposure to EMF.

A parallel consideration is the temporal variability of EMF measurements: Does today's measurement of an EMF flux density in a house or other location provide an accurate estimate of the magnetic field environment that was present in the same location 5 years ago, 10 years ago, or at some other earlier time that is relevant to the pathogenesis of cancer? Because EMF flux densities in any specific location can demonstrate considerable temporal variability, the ability of such measurements to support retrospective epidemiology studies remains an important issue.

Considering the potential limitations of epidemiology data for cancer risk identification, studies in experimental model systems become critically important in developing a comprehensive "sum of the evidence" approach to assess potential cancer hazards associated with human EMF exposure.

8.3 Strengths and Limitations of Experimental Studies of EMF as a Cancer Hazard

The strengths and weaknesses of laboratory animal studies designed to evaluate the possible effects of EMF exposure on cancer risk are diametrically opposed to those seen in epidemiology studies. Because animal cancer bioassays are performed in a nonhuman test system, interspecies extrapolation of study data is required. In addition to possible species differences in response, field strengths studied in animal cancer bioassays often greatly exceed those seen in human exposures, and may thereby induce artifactual responses by overwhelming physiologic systems that could maintain normal cellular, organ, or tissue homeostasis at lower, possibly more relevant, exposure levels. These two factors impose the most important limitations on the use of experimental animal data to assess the potential risk of human exposure to EMF or any other exogenous agent.

Conversely, however, several factors provide important support for the use of animal models in evaluations of the potential carcinogenicity of EMF. A number of rodent model systems that have been used to evaluate EMF carcinogenicity have been used widely in safety assessments of pharmaceuticals, environmental chemicals, and other agents to which humans receive broad population exposure. These model systems are well characterized, have extensive historical databases, and have been shown to be generally predictive of human responses [9, 10]. Bioassays in laboratory animal model systems can also be conducted under precisely controlled and monitored exposure conditions; in consideration of the challenges associated with EMF exposure assessment in epidemiology studies, the ability to conduct animal studies under rigorously controlled exposure conditions is a major strength of this approach to hazard identification. Well-designed animal cancer bioassays also incorporate statistical considerations into study designs and can identify the minimum effects required to achieve statistical significance. Finally, using genetically engineered animal models, the possible effects of EMF on susceptible subpopulations can be investigated. Studies in such susceptible subpopulations may identify hazards that cannot be identified in normal animal or human populations, but that may substantially increase the risk of disease in individuals with a genetic predisposition, preexisting condition, concomitant exposure to other risk factors, or other factor that increases disease risk.

8.4 Role of Mechanistic Studies in EMF Hazard Assessment

The results of mechanistic investigations performed in experimental animals or *in vitro* model systems may provide critical data to establish the biological framework in which any observed health effects of EMF can be interpreted. Although identification of a presumed mechanism of action is not essential to support the identification of an agent as a cancer hazard, definition of an appropriate mechanism for the presumed biophysical interaction between EMF and target cells would strengthen any finding of carcinogenic activity. Further, elucidation of one or more biochemical or molecular pathways through which EMF could modulate processes associated with carcinogenesis would provide important support to hazard assessments developed on the basis of epidemiology and animal bioassays.

Conversely, should the results of epidemiology studies and animal bioassays be negative or conflicting, the identification of a putative mechanism of action can (a) provide a rational biological context for hazard identification and (b) support the design of optimized hazard identification studies. Should the results of epidemiology studies and animal bioassays be negative, the lack of an identified biological mechanism of action further weakens the assertion that a substantive human health hazard exists. At present, no generally accepted mechanism of action has been identified through which EMF exposure could increase cancer risk. The lack of an accepted biophysical, biochemical, or molecular mechanism through which EMF may (a) interact with key cellular targets and (b) modulate critical regulatory pathways is a major weakness in the hypothesis that EMF is a significant risk factor for human carcinogenesis.

8.5 Oncogenicity Studies of EMF

8.5.1 EMF as a Possible Risk Factor for Cancer: Epidemiology Studies

Demonstration of a strong positive association between EMF exposure and cancer risk in multiple epidemiology studies would provide compelling evidence that EMF is an important risk factor for human cancer; however, no such relationship has been established so far [3]. Since 1979, nearly 200 epidemiology studies and meta-analyses designed to investigate the possible association between occupational or residential exposure to EMF and cancer risk have been published. During this period, substantial improvements in statistical power (sample size) and EMF exposure assessment have been incorporated into study designs, and large-scale studies have been conducted to evaluate the effects of EMF exposure on both total cancer incidence and cancer incidence in specific sites in both children and adults. No clear picture has emerged from this massive effort, as EMF cancer epidemiology studies have generated both positive and negative results. When considered together, the results of epidemiology studies are insufficient to either clearly support or refute the hypothesis that EMF exposure is a significant risk factor for human cancer in any organ system.

The epidemiologic literature does suggest a weak positive association (odds ratio of $\sim$2.0) for leukemia in children exposed to EMF at flux densities of $> 0.4\,\mu T$ [3–5]. Children exposed to EMF at this level represent a small percentage (estimated at 0.8%) of the overall population, and bias underlying this finding cannot be excluded [3]. However, a working group assembled by the International Agency for Research on Cancer has concluded that limited evidence exists to support the hypothesis that EMF exposure is a risk factor for childhood leukemia [5]. By contrast, no consistent association has been established between EMF exposure and cancer in any other site in children or in any site in adults (reviewed in Refs [3–5]). As such, the IARC panel concluded that suggestive evidence of EMF oncogenicity exists only for childhood leukemia and not for any other neoplasm in either children or adults. This panel did, however, suggest that additional research was required to provide a conclusive evaluation of the carcinogenicity of EMF.

8.5.2
EMF as a Possible Risk Factor for Cancer: Experimental Studies

In situations where epidemiology does not support the conclusive identification and quantitation of the potential risks associated with exposure to an environmental agent, laboratory studies conducted in appropriate experimental model systems increase in importance. Well-designed and controlled animal studies permit evaluation of biological effects *in vivo* under tightly controlled exposure and environmental conditions, and in the absence of potential confounding variables. In consideration of the conflicting results of EMF epidemiology studies, and difficulties associated with exposure assessment in such studies, animal studies may provide the best opportunity to identify effects of EMF exposure that could translate into an increased risk of human cancer.

Assessment of the Possible Oncogenic Activity of EMF as a Single Agent

Several teams of investigators have published the results of studies that were designed to investigate the possible oncogenic activity of chronic exposure to EMF alone (Table 8.1). These studies have been conducted using the chronic rodent bioassay design, a standardized study design in which rats or mice are exposed to the test agent for the majority of their natural life span (2 years), followed by a complete microscopic evaluation of approximately 50 tissues from every animal to identify induced neoplasms and nonmalignant lesions. This study design has been widely used to evaluate the potential carcinogenicity of new drugs, chemicals, and

Table 8.1 Two-year oncogenicity evaluations of power frequency EMF in laboratory animals.

Species/ strain	Group size	Exposure levels (μT)	Frequency	Exposure (h/day)	Country of performance	Reference
Mouse B6C3F1	100/sex (♂ and ♀)	0, 20, 200, 1000 (continuous) 1000 (intermittent)	60 Hz	18.5	Unites States	[11]
Rat F344	100/sex (♂ and ♀)	0, 20, 200, 1000 (continuous) 1000 (intermittent)	60 Hz	18.5	Unites States	[12]
Rat F344	50 (♀ only)	0, 2, 20, 200, 2000	60 Hz	20	Canada	[13]
Rat F344	48/sex (♂ and ♀)	0, 500, 5000	50 Hz	22.6	Japan	[14]

environmental agents, and has been demonstrated to be a reasonable predictor of human responses [9, 10]. As a result, it is generally accepted among toxicologists that the chronic bioassay in rodents provides the best available experimental approach to identify agents that may be carcinogenic in humans.

In all studies conducted using this design, chronic exposure to power frequency EMF had no effect on the incidence of leukemia, lymphoma, or other malignancies of the hematopoietic system [11–14], the site identified by epidemiology studies as the most likely target for EMF action. Furthermore, chronic EMF exposure did not induce any consistent pattern of statistically significant increases in cancer incidence in other tissues. The very close comparability of the results of chronic rodent oncogenicity bioassays conducted (a) in two different species and (b) in three independent laboratories strengthens the conclusion that exposure to magnetic fields as a single agent does not induce cancer in rodents. In consideration of the known predictive nature of the rodent 2-year bioassay [9, 10], these findings do not support the hypothesis that EMF is a significant risk factor in the etiology of human neoplasia.

Site-Specific/Multistage Rodent Oncogenicity Studies

It is generally accepted that exposure to power frequency EMF does not induce mutations or other genetic damage that can be identified using standard genetic toxicology test systems [15, 16]. Because EMF is apparently not genotoxic, it is logical to conclude that any oncogenic effects must be mediated through mechanisms that involve cocarcinogenesis or tumor promotion, rather than DNA damage. To address this possibility, a substantial number of *in vivo* studies have been conducted to identify possible cocarcinogenic or tumor promoting effects of EMF in different target tissues (reviewed in Refs [4, 5]). The design of these studies involves simultaneous or sequential exposure to EMF in combination with another agent (generally a chemical carcinogen or ionizing radiation), and is focused on neoplastic development in a specific organ. In this regard, it should be noted that most multistage tumorigenesis studies of EMF have been conducted using animal models that were designed as research tools to study the mechanisms of site-specific carcinogenesis.

As a result, the value of these assays in hazard identification, and their utility to predict human oncogenicity, is less well understood than is the chronic rodent oncogenicity bioassay.

Well-conducted studies on the effects of EMF on cancer induction in multistage models for cancer of the hematopoietic system [17–19], brain [20, 21], and liver [22, 23] have been uniformly negative: in all cases, exposure to an inducing agent + EMF induced no more neoplasms than were induced by exposure to the inducing agent alone. Conflicting data have been presented for studies of the effects of EMF on the induction of skin and breast tumors in multistage animal models: in each case, positive data have been presented by at least one laboratory [24–26], but studies conducted in one or more additional laboratories have failed to confirm the original findings [27–30].

Overall, the peer-reviewed literature provides little compelling evidence to support the hypothesis that EMF has cocarcinogenic or tumor-promoting activity in any organ in laboratory animals. Negative results have been reported from studies conducted using two different models for neoplasms originating in the hematopoietic system [19, 20]; these data confirm and extend the results of the 2-year rodent bioassays discussed above [11–14]. Similar negative data have been presented from studies in two models for brain cancer [17, 18], a putative site of EMF carcinogenicity that had been identified in several early epidemiology studies. Data from studies of EMF action in the rat mammary gland are conflicting, but do not provide compelling evidence of an increased risk of breast cancer in animals exposed to a chemical carcinogen + EMF versus the chemical carcinogen alone [30]. When considered *in toto*, the results of multistage carcinogenicity evaluations are generally consistent with the data from 2-year oncogenicity bioassays of EMF, and do not support the hypothesis that EMF exposure is a key risk factor for carcinogenesis in these target tissues.

Oncogenicity Studies in Genetically Engineered Animals

Numerous strains of genetically engineered mice (transgenic mice or knockout mice) have been developed as experimental tools for the study of molecular mechanisms of carcinogenesis; several of these genetically engineered strains may provide sensitive test systems for cancer hazard identification. Genetically engineered animals used in cancer hazard identification demonstrate an increased incidence and/or decreased latency of malignancy as the result of either the insertion of an oncogene or the deletion of a tumor suppressor gene from the germ line. As a result of their genetic predisposition to specific cancers, genetically engineered mice provide an approach to identify effects that may occur in sensitive subpopulations, but rarely or not at all in the general population. These genetically engineered animal models may demonstrate sensitivity to EMF bioeffects that may not be seen in standard inbred or outbred animal strains.

Although the total number of completed oncogenicity evaluations of EMF in genetically engineered animals is very small, the results of published studies are consistent in their failure to demonstrate any increased risk of neoplasia resulting from exposure to EMF. In studies conducted in the United States and in Australia,

EMF exposure did not increase the incidence of hematopoietic neoplasia in PIM-1 transgenic mice [31, 32]. The designs of the US and Australian studies were quite different, and as such, can be considered to represent two different models of lymphoma induction. The study performed in the United States was a 6-month bioassay in which transgenic mice were exposed to a chemical carcinogen ± EMF; EMF exposure did not increase the incidence of T-cell lymphomas in any study group and resulted in a significant *decrease* in lymphoma incidence in the high-dose (1000 μT) group. The Australian study was a long-term (18-month) bioassay of the effects of EMF alone and did not involve coexposure to a chemical carcinogen; EMF exposure had no effect on the incidence of F-cell lymphomas in this study [32].

Similar negative data were generated in a study in the p53 knockout mouse [31], an animal that is at high risk of lymphoma development as a result of the deletion of the p53 tumor suppressor gene from the germ line [31]. Because chronic exposure to EMF did not increase the incidence of hematopoietic neoplasia in any experimental group in these studies, the data from studies in genetically engineered mouse models can be considered to confirm and extend the negative results of the 2-year rodent oncogenicity bioassays of EMF.

8.6 Conclusions

Hazard assessments are optimally developed using a "sum of the evidence" approach, in which all available epidemiologic, experimental, and mechanistic data are considered. A large number of epidemiology studies have been conducted to evaluate the possible relationship between EMF exposure and cancer; no clear associations have emerged. However, data suggesting a possible link between EMF and childhood leukemia make it impossible to dismiss the hypothesis that EMF exposure may be a risk factor for a subset of human cancer.

In hazard identification scenarios where epidemiology data are unavailable, conflicting, or inconclusive, the results of well-conducted carcinogenicity bioassays in animal models increase in importance. Animal studies can be particularly useful in situations where possibly sensitive target organs have been identified through epidemiology. In consideration of the epidemiology data for childhood leukemia, the hematopoietic system takes on particular importance as a potential target for EMF action.

Four independent 2-year oncogenicity bioassays that were performed using animal models with demonstrated predictiveness for human disease generated no data to support the hypothesis that EMF exposure is an important risk factor for human cancer [11–14]. Notably, none of these studies identified increases in the incidence of leukemia, lymphoma, or any other hematopoietic neoplasm in animals receiving chronic exposure to EMF at flux densities of up to 2000 μT. Similarly, the results of studies in multistage animal models [17–19] and transgenic animal models [31, 32] for hematopoietic neoplasia failed to demonstrate any increase in the risk of malignancy in mice exposed to EMF.

Because no mechanism has been identified through which EMF may induce cancer in any organ site, cancer risk assessment for EMF must be based entirely on epidemiology and experimental data. Epidemiology suggests at most a weak association between EMF exposure and childhood leukemia and no association with cancer incidence in any other site in either children or adults. The increased risk of childhood leukemia (odds ratio) associated with EMF exposure is approximately 2.0 and is seen at an exposure threshold (0.4 μT) to which less than 1% of children are exposed in residential environments. In consideration of this modest projected increase in risk, the relatively high exposure threshold, and the relatively low incidence of childhood leukemia in the general population, epidemiology does not suggest that EMF exposure is a major contributor to the overall risk of cancer in humans. This statement is supported by the lack of genotoxicity of EMF, and by the results of animal oncogenicity bioassays, in which no elevated risk of hematopoietic neoplasia has been identified. On this basis, human exposure to magnetic fields is considered unlikely to have a substantive impact on the risk of human cancer.

References

1 Wertheimer, N. and Leeper, E. (1979) Electrical wiring configurations and childhood cancer. *Am. J. Epidemiol.*, **190**, 273–284.

2 Milham, S., Jr. (1982) Mortality from leukemia in workers exposed to electrical and magnetic fields. *N. Engl. J. Med.*, **307**, 4414–4419.

3 Ahlbom, A., Cardis, E., Green, A., Linet, M., Savitz, D., and Swerdlow, A. (2001) Review of the epidemiologic literature on EMF and health. *Environ. Health Perspect.*, **109** (Suppl. 6), 911–933.

4 Portier, C.J. and Wolfe, M.S. (eds) (1998) *Assessment of Health Effects from Exposure to Power-Line Frequency Electric and Magnetic Fields*, National Institute of Environmental Health Sciences, Research Triangle Park, NC.

5 International Agency for Research on Cancer (2003) *IARC Monographs on the Evaluation of Carcinogenic Risk to Humans, Vol. 80. Non-Ionizing Radiation, Part 1: Static and Extremely Low-Frequency (ELF) Electric and Magnetic Fields*, International Agency for Research on Cancer.

6 Kaune, W.T. (1993) Assessing human exposure to power-frequency electric and magnetic fields. *Environ. Health Perspect.*, **101** (Suppl. 4), 131–133.

7 Tworoger, S.S., Davis, S., Schwartz, S.M., and Nirick, D.K. (2002) Stability of Wertheimer–Leeper wire codes as a measure of exposure to residential magnetic fields over a 9- to 11-year period. *J. Expo. Anal. Environ. Epidemiol.*, **12**, 448–454.

8 Bracken, T.D. and Patternson, R.M. (1996) Variability and consistency of electric and magnetic field occupational exposure measurements. *J. Expo. Anal. Environ. Epidemiol.*, **6**, 355–374.

9 Rall, D.P. (2000) Laboratory animal tests and human cancer. *Drug Metab. Rev.*, **32**, 119–128.

10 Huff, J. (1999) Value, validity, and historical development of carcinogenesis studies for predicting and confirming carcinogenic risks to humans, in *Carcinogenicity Testing, Predicting, and Interpreting Chemical Effects* (ed. K.T. Kitchin), Marcel Dekker, New York, pp. 21–123.

11 McCormick, D.L., Boorman, G.A., Findlay, J.C., Hailey, J.R., Johnson, T.R., Gauger, J.R., Pletcher, J.M., Sills, R.C., and Haseman, J.K. (1999) Chronic toxicity/oncogenicity evaluation of 60Hz (power frequency) magnetic fields in B6C3F1 mice. *Toxicol. Pathol.*, **27**, 279–285.

12 Boorman, G.A., McCormick, D.L., Findlay, J.C., Hailey, J.R., Gauger, J.R., Johnson, T.R., Kovatch, R.M., Sills, R.C., and Haseman, J.K. (1999) Chronic toxicity/oncogenicity evaluation of 60Hz (power frequency) magnetic fields in F344/N rats. *Toxicol. Pathol.*, **27**, 267–278.

13 Mandeville, R., Franco, E., Sidrac-Ghali, S., Paris-Nadon, L., Rocheleau, N., Mercier, G., Desy, M., and Gaboury, L. (1997) Evaluation of the potential carcinogenicity of 60Hz linear sinusoidal continuous-wave magnetic fields in Fischer F344 rats. *FASEB J.*, **11**, 1127–1136.

14 Yasui, M., Kikuchi, T., Ogawa, M., Otaka, Y., Tsuchitani, M., and Iwata, H. (1997) Carcinogenicity test of 50Hz sinusoidal magnetic fields in rats. *Bioelectromagnetics*, **18**, 531–540.

15 McCann, J., Dietrich, F., Rafferty, C., and Martin, A.O. (1993) A critical review of the genotoxic potential of electric and magnetic fields. *Mutat. Res.*, **297**, 61–95.

16 McCann, J., Dietrich, F., and Rafferty, C. (1998) The genotoxic potential of electric and magnetic fields: an update. *Mutat. Res.*, **411**, 45–86.

17 Babbitt, J.T., Kharazi, A.I., Taylor, J.M., Bonds, C.B., Mirell, S.G., Frumkin, E., Zhuang, D., and Hahn, T.J. (2000) Hematopoietic neoplasia in C57BL/6 mice exposed to split-dose ionizing radiation and circularly polarized 60Hz magnetic fields. *Carcinogenesis*, **21**, 1379–1389.

18 Shen, Y.H., Shao, B.J., Chiang, H., Fu, Y.D., and Yu, M. (1997) The effects of 50Hz magnetic field exposure on dimethylbenz(alpha)anthracene induced thymic lymphoma/leukemia in mice. *Bioelectromagnetics*, **18**, 360–364.

19 Boorman, G.A., Rafferty, C.N., Ward, J.M., and Sills, R.C. (2000) Leukemia and lymphoma incidence in rodents exposed to low-frequency magnetic fields. *Radiat. Res.*, **153**, 627–636.

20 Kharazi, A.I., Babbitt, J.T., and Hahn, T.J. (1999) Primary brain tumor incidence in mice exposed to split-dose ionizing radiation and circularly polarized 60Hz magnetic fields. *Cancer Lett.*, **147**, 149–156.

21 Mandeville, R., Franco, E., Sidrac-Ghali, S., Paris-Nadon, L., Rocheleau, N., Mercier, G., Desy, M., Devaux, C., and Gaboury, L. (2000) Evaluation of the potential promoting effect of 60Hz magnetic fields on *N*-ethyl-*N*-nitrosourea induced neurogenic tumors in female F344 rats. *Bioelectromagnetics*, **21**, 84–93.

22 Rannug, A., Holmberg, B., Ekstrom, T., and Mild, K.H. (1993) Rat liver foci study on coexposure with 50Hz magnetic fields and known carcinogens. *Bioelectromagnetics*, **14**, 17–27.

23 Rannug, A., Holmberg, B., and Mild, K.H. (1993) A rat liver foci promotion study with 50-Hz magnetic fields. *Environ. Res.*, **62**, 223–229.

24 Stuchly, M.A., McLean, J.R., Burnett, R., Goddard, M., Lecuyer, D.W., and Mitchel, R.E. (1992) Modification of tumor promotion in the mouse skin by exposure to an alternating magnetic field. *Cancer Lett.*, **65**, 1–7.

25 Beniashvili, D.S., Bilanishvili, V.G., and Menabde, M.Z. (1991) Low-frequency electromagnetic radiation enhances the induction of rat mammary tumors by nitrosomethyl urea. *Cancer Lett.*, **61**, 75–79.

26 Loscher, W., Wahnschaffe, U., Mevissen, M., Lerchl, A., and Stamm, A. (1994) Effects of weak alternating magnetic fields on nocturnal melatonin production and mammary carcinogenesis in rats. *Oncology*, **51**, 288–295.

27 Sasser, L.B., Anderson, L.E., Morris, J.E., Miller, D.L., Walborg, E.F., Jr., Kavet, R., Johnston, D.A., and DiGiovanni, J. (1999) Lack of a co-promoting effect of a 60Hz magnetic field on skin tumorigenesis in SENCAR mice. *Carcinogenesis*, **19**, 1617–1621.

28 Anderson, L.E., Boorman, G.A., Morris, J.E., Sassar, L.B., Mann, P.C., Grumbein, S.L., Hailey, J.R., McNally, A., Sills, R.C., and Haseman, J.K. (1999) Effect of 13-week magnetic field exposures on DMBA-initiated mammary gland carcinomas in female Sprague-Dawley rats. *Carcinogenesis*, **20**, 1615–1620.

29 Ekstrom, T., Mild, K.H., and Holmberg, B. (1998) Mammary tumours in

Sprague-Dawley rats after initiation with DMBA followed by exposure to 50Hz electromagnetic fields in a promotional scheme. *Cancer Lett.*, **123**, 107–111.

30 Boorman, G.A., McCormick, D.L., Ward, J.M., Haseman, J.K., and Sills, R.C. (2000) Magnetic fields and mammary cancer in rodents: a critical review and evaluation of published literature. *Radiat. Res.*, **153**, 617–626.

31 McCormick, D.L., Ryan, B.M., Findlay, J.C., Gauger, J.R., Johnson, T.R., Morrissey, R.L., and Boorman, G.A. (1998) Exposure to 60Hz magnetic fields and risk of lymphoma in PIM transgenic and TSG-p53 (p53 knockout) mice. *Carcinogenesis*, **19**, 1649–1653.

32 Harris, A.W., Basten, A., Gebski, V., Noonan, D., Finnie, J., Bath, M.L., Bangay, M.J., and Repacholi, M.H. (1998) A test of lymphoma induction by long-term exposure of Emu-Pim1 transgenic mice to 50Hz magnetic fields. *Radiat. Res.*, **149**, 300–307.

Part Four
Genotoxicity Studies

Cancer Risk Evaluation: Methods and Trends,
Edited by Günter Obe, Burkhard Jandrig, Gary E. Marchant, Holger Schütz, and Peter M. Wiedemann.

9
Chromosomal Aberrations in Human Populations and Cancer

Günter Obe, David C. Lloyd, and Marco Durante

9.1
Introduction

Chromosomal aberrations (CA) have been studied for a long time [1], but the mechanisms of their formation are still not yet entirely clear [2, 3]. DNA double-strand breaks (DSB) are the primary lesions leading to the production of CA. DSB can be directly induced in chromosomal DNA or can occur when DNA with single-strand lesions (SSL) is replicated during the S-phase of the cell cycle [3–8].

Chromosomes in the G0/G1 phase of the cell cycle contain one DNA molecule associated with histone and nonhistone proteins. During the S-phase, DNA is replicated and in the metaphase of mitosis each chromosome contains two chromatids each with one DNA molecule.

Molecular changes in chromosomal DNA, when not properly repaired, may lead to both gene mutations and CA [9]. CA may be induced endogenously by spontaneous damage of DNA or by exposure to various exogenous environmental agents. CA can be seen in the light microscope and this makes them an important endpoint for the analysis of mutagenic activities. Cells of different types of cancers are characterized by specific types of CA, such as reciprocal translocations (TR) that play a considerable role in the development of cancer [10–13]. Therefore, elevated CA frequencies in human cells indicate an enhanced cancer risk. One highly practical application of CA in blood lymphocytes is their use for biological dosimetry.

9.2
Chromosomal Aberrations and Their Spontaneous Frequencies in Human Peripheral Lymphocytes

Chromosomal aberrations can be of chromosome (CSA) or chromatid (CTA) type. CSA result from DSB in unreplicated DNA in the G0/G1 phase of the cell cycle. CA induced in the G0/G1 phase are replicated in S-phase and appear in both chromatids at the same sites in the following metaphase. Dicentric chromosomes (DIC), ring chromosomes (RC), and TR are typical CSA. CTA occur when DSB are induced

Cancer Risk Evaluation: Methods and Trends,
Edited by Günter Obe, Burkhard Jandrig, Gary E. Marchant, Holger Schütz, and Peter M. Wiedemann.

in replicated DNA or when SSL occur in one DNA strand that during replication gives rise to DSB and to CA in the damaged chromatid. Typical CTA are chromatid breaks (B′) and chromatid exchanges (RB′). Apart from typical CA, achromatic lesions (AL), often called "gaps," occur that are unstained areas in chromosomes no longer than the diameter of the chromatid in which they are situated [14]. AL are not CA and should be tabulated separately. Other CA types are centromere-less double fragments or deletions (B″) that could be CSA or CTA.

Varying frequencies in spontaneous levels of CA have been reported in surveys using metaphases in human peripheral lymphocytes (HPL). In a series of papers, Bender *et al.* [15–17] published data on spontaneous CA in HPL of hundreds of persons and thousands of metaphases. In the 1989 paper [16], an extension of the 1988 study [15], the aberration mean values per 100 cells based on 108 950 cells from 493 blood donors were as follows (with standard errors of the means in parentheses): CTA: AL 4.147 (0.106), chromatid deletions (B′) 0.813 (0.038), isochromatid deletions (B″) 0.039 (0.006), chromatid exchanges (RB′) 0.045 (0.007); CSA: deletions (B″) 0.338 (0.027), ring chromosomes (RC) 0.02 (0.005), dicentric chromosomes (DIC) 0.160 (0.014). Isochromatid deletions (B″) are considered as CSA or as CTA on the basis of sister unions at the break sites in the CTA, which is not entirely conclusive. There were 0.044 (0.008) translocations (TR), but these cannot be properly analyzed in conventional Giemsa-stained chromosome preparations. Lloyd *et al.* [18] compiled literature data on DIC and B″ (according to the authors, terminal and interstitial deletions) from 65 publications and from their own laboratory. In lymphocytes from 2000 donors and 211 661 analyzed cells, 0.78 DIC and 3.7 B″ were found per 1000 cells with huge variability in the data from different laboratories. Bolognesi *et al.* [19] reviewed data on CA, micronuclei (MN), and sister chromatid exchanges (SCE) from 12 Italian laboratories and reported a considerable variability in the data, a general finding in analyses of these endpoints, for example, the data for CA varied from 0.51 to 10.51%. Wolf *et al.* [20] analyzed 23 healthy males and found 1.3 (0.2–4.1)% aberrant metaphases. In a control group of 41 males and 58 females aged 17–58 years, 1.1% (±0.1) aberrant cells were found [21]. Obe [22] compiled data on spontaneous frequencies of CA in HPL of hundreds of persons and thousands of cells. Per 10 000 metaphases the following numbers of CA were found: AL = 536.24, B′ = 59.81, B″ = 41.07, RB′ = 5.13, DIC = 8.00, and RC = 1.77. The frequencies of AL and CA can be summarized in descending order as follows: AL > B′ > B″ > DIC > RB′ > RC. In 1000 metaphases, each from 99 persons (41 males, 58 females; age: 17–58 years) 642 B′, 207 B″, 119 DIC (including RC), and 58 RB′ were found in 99 000 cells; there were 665 cells with CTA and 391 with CSA [21]. Wolf *et al.* [20] analyzed 1000 cells each from 23 males (age: 40–58 years) and found 250 AL, 191 B′, 63 B″ 29 DIC (including RC), and 13 RB′ in 23 000 cells, there were 184 cells with CTA and 105 with CSA. Generally, one B″ is subtracted per DIC or RC because one and sometimes even two B″ may be produced during the formation of these CA types. The remaining B″ are therefore what are sometimes referred to as "excess acentrics," taken as separate entities unrelated to the exchange type of CA (DIC and RC).

The data sets mentioned above are not very different, and certainly the relative frequencies of the different CA types are the same. AL are the most frequent events, but as discussed above, they are not considered as typical CA. If they were considered to be B′, the frequencies of this aberration type would be too high. It cannot be decided whether B″ are CSA or CTA. If B″ would represent CSA, their origin would be a break in a G0/G1 chromosome that after DNA synthesis would appear in both chromatids at the same site in the ensuing metaphase (chromosome breaks). If CTA, then B″ would result from breaks in both chromatids at the same site (isochromatid breaks). In B″ regarded as CTA but not CSA, the ends of the broken chromatids may fuse. However, in HPL this is a rare event and cannot be used to differentiate B″ as CSA and CTA. The most frequent CA types are chromatid breaks (B′) and while some of the B″ are CTA, most of them seem to be CSA. In the group of exchange type CA, DIC (CSA) are most frequent followed by RB′ (CTA), and RC (CSA) are rare and may or may not contain a centromere. A differentiation of these two types may be difficult especially in small RC and is normally not done. In DIC, the broken ends of two chromosomes fused in such a way that the two centromeric parts are joined and the two acentric fragments also join, the latter look like B″ but, because of this mechanistic process, are regarded as part of the DIC (the same holds for RC with centromeres). This is why when recording CA, one apparent B″ is deleted per DIC and centric RC, and the remaining B″ are declared as excess fragments or acentrics.

There is another way for fusions of DSB to occur in different chromosomes, namely, a centromeric part of one chromosome fuses with the acentric fragment of another and concurrently the other two parts also fuse. The result is a reciprocal translocation. These two exchange possibilities leading to either a DIC or a TR seem to be random leading to similar frequencies of DIC and TR [23, 24]. This 1 : 1 ratio can be used to estimate a more realistic TR frequency from the observed DIC frequencies in data sets derived from Giemsa-stained material. In the data set presented by Obe [22], this would lead to 16 exchanges plus 1.77 RC resulting in 17.77 per 10 000 cells. If we assume that about 50% of the B″ are CSA, which may be an underestimation, we get about 38 CSA in 10 000 cells. Still this is lower than the frequencies of CTA (RB′, B′, 50% of B″) with 85.44 in 10 000 cells.

The analysis of CA does not always produce fully reliable results. Problematic are the distinction between AL and B′, and the ambiguity of B″. RC can be so small that it is difficult or impossible to differentiate such double minutes (DM) from small B″. Even after deleting one B″ per DIC or RC, the resulting numbers of excess acentrics are not entirely reliable. One uncertainty is that DIC or RC may give rise to two B″ in cases where the fragments fail to fuse during the formation of DIC or RC. In addition to this, it has been shown by using mBAND, which is a version of fluorescence *in situ* hybridization of chromosomes (FISH), that the exchange processes for a proportion of DIC or RC may be much more complex than the simple scheme described above, making the assumption of one or two B″ associated with these CA types obsolete [25, 26].

DIC are easily recognizable in metaphase chromosomes and data on this CA type are therefore highly reliable. As discussed above, the number of DIC in 1000 cells is in the range of 0.8–1.6, and frequently a rounded generic value of 1 DIC per 1000 cells is assumed in the absence of more precise data. DIC are induced in G0/G1 HPL by ionizing radiation to the same extent *in vivo* and *in vitro*. Therefore, an appropriate reference dose–effect curve produced by *in vitro* exposure of HPL can be used to estimate dose from the DIC scored in blood taken from irradiated persons (see Section 9.7).

One may circumvent the difficulties in differentiating CA types by simply considering the frequencies of aberrant cells, irrespective of the CA types. The frequencies of such "total damaged cells" in HPL of 22 358 persons from 11 European countries varied from 1.5 to 3.6% with a mean of 2.5 and a standard deviation of 2.2 [27].

There are several reasons for the variability of spontaneous CA in HPL. One factor may be the conditions of how HPL are cultured. It has been shown that growth characteristics of HPL depend on the choice of culture medium used [28]. Since only first *in vitro* metaphases give reliable data on CA frequencies, a proper control of cell growth is mandatory. Therefore, early data, before the introduction of fluorescence plus Giemsa staining, which allows differentiation between first and following metaphases in the chromosome preparations [29], are likely to be less reliable than later data derived from guaranteed first division scoring.

Wolf and Fender [21] showed that the variability in CA frequencies in HPL from the same lymphocyte culture decreases with the number of cells scored and concluded that at least 1000 cells should be analyzed per person. Clearly, the more cells scored per data point the more statistically reliable the derived mean frequency (see Section 9.7).

A study on CA frequencies in HPL exposed to low doses of X-rays by Lloyd *et al.* [30] with the participation of six laboratories demonstrated an interlaboratory variation in scoring of CA. The chromosome preparations were made in one laboratory so that variations in laboratory techniques cannot explain this finding. Reasons for differences in scoring CA between the participating laboratories could not be pinpointed. The study also uncovered variability in CA frequencies between blood donors indicating intrinsic differences in HPL from different persons.

The data described up to now are based on chromosomes block stained with Giemsa stain. TR in which parts of chromosomes are exchanged can generally be seen only when chromosomes are Giemsa banded [31, 32] or hybridized with specific DNA probes tagged with fluorescent dyes (FISH). The latter method has been extensively used to analyze TR. There are two types of TR, complete TR in which both chromosomes reciprocally exchanged material and incomplete TR in which seemingly only one chromosome contains a translocated element. Most incomplete TR are nevertheless complete in the sense that one exchanged part is so small that it cannot be seen with the microscope [3]; moreover, no acentric fragment appears to be present. The frequencies of reciprocal TR vary between 2.5 and 8 and of incomplete TR between 3 and 11 per 1000 cells [33, 34].

An important factor influencing CA frequencies is the lifetime of HPL in the body. HPL are noncycling cells in G0 that generally tolerate CA in their genome. It can be expected that CA, mainly DIC and TR, can be seen as long as HPL survive and that the half-life of these types of CA is a measure of the half-life of HPL. This has been studied in HPL of irradiated patients [31] and of radiation accident victims [33, 35].

Time effect studies on frequencies of DIC and TR resulted in a half-life for DIC of about 2–3 years, for reciprocal TR of 7 years, and for nonreciprocal TR of 5 years [31, 33, 35, 36]. HPL are replenished by dividing stem cells in the bone marrow and this strongly depends on the hematological status of the individual. Thus, persons irradiated with doses sufficient to cause early leucopenia exhibit much faster replenishment than those with cell counts remaining within the normal range. Upon dividing, stem cells with DIC lead to nonsurviving unbalanced progeny. HPL with DIC can therefore not be replenished by stem cells and so the half-life of DIC corresponds to the half-life of HPL. Stem cells with TR may produce balanced daughter cells and so they enter the replenished HPL pool with TR. The half-life of HPL with TR is therefore longer than the half-life of HPL with DIC. This mechanism can eventually lead to clones of HPL with same types of TR such as those reported in lymphocytes of astronauts [37]. Molecular studies of TR have uncovered deletions at the breakpoint junctions, which may occur during the molecular mechanism of the TR formation [38]. These deletions should eventually lead to genetically unbalanced progeny cells and may explain the longer, but still limited, persistence of TR when compared to DIC. The shorter half-life of nonreciprocal compared to reciprocal TR points to differences between these two TR types suggesting that a proportion of the former are accompanied by an acentric fragment, even too small to be detected, rendering cell's overall genome unstable.

9.3 Micronuclei

Micronuclei result from CA or whole chromosomes that in anaphase of mitosis are not transported to the cell poles and are left behind in the cytoplasm between the separating anaphase chromosomes. These elements surround themselves with a nuclear membrane and give rise to MN separate from the main nuclei. MN are generally analyzed in the interphase following the first mitotic division after induction of CA by blocking division of cytoplasm but not the anaphase movement of chromosomes with cytochalasin B. This results in binucleated cells in which MN can be easily recognized [39]. Bolognesi *et al.* [19] reviewed data on MN frequencies from 12 Italian laboratories and reported 2.83–9.08 micronucleated cells (MNC) in 1000 binucleated cells. In the framework of the Human Micronucleus (HUMN) Project, Bonassi *et al.* [40] analyzed MN using an international database of nearly 7000 subjects. This study demonstrated a high variability of MNC in unirradiated control subjects. The overall MNC frequency was 6.5‰ (interquartile range 3–12‰), with an effect of gender (females 7.0‰, males 6.3‰) and age ($<$ 40 years: 5.9‰; $>$ 40 years: 7.6‰). Some of the variability can be explained by methodological differences in preparing and scoring MN. A major source of variability is that MN are derived from CA whose frequencies are quite variable and from whole chromosomes. Not all CA lead to MN and in most of the studies the frequencies of centromere positive and negative MN were not determined. Apart from MN in HPL, MN also occur in other cells, and exfoliated buccal epithelial cells are an example of readily obtainable material [41–43]. As with MN in HPL, there is considerable variability in the baseline frequencies of MN with a spread of 0.05–8.4 in 1000 cells [42].

Analyses using centromere probes show high frequencies of centromere containing MN in HPL [44]. Three sets of data with 58 control donors and 60 563 binucleated cells were analyzed. MN per 1000 cells (centromere positive MN in parentheses) were 11.1 ± 6.0 (8.6 ± 6.1), 4.9 ± 1.8 (2.4 ± 1.2), and 14.4 ± 8.1 (10.3 ± 7.1). A significant effect of gender and alcohol consumption and a slight age effect on MN frequencies (see Chapter 5) were found [44].

9.4 Sister Chromatid Exchanges

Sister chromatid exchange are generally induced by SSL and result from a homologous exchange repair between sister chromatids during S-phase. SCE can be seen only after differential substitution of chromosomal DNA with thymine derivates that are incorporated into replicating DNA in such a way that in metaphase chromosomes one chromatid is labeled in both DNA strands and the other in one strand (labeling for two cell cycles) or when one DNA strand remains unlabeled and the other is labeled in one DNA strand (labeling the first cell cycle and growing the cells for another cell cycle in the absence of the label). The labeling agent can be 5-bromo-2′-deoxyuridine (BrdU) or biotin-16-2′-deoxyuridine-5′-triphosphate (biotin-dUTP). After appropriate staining, SCE can be seen as color switches between sister chromatids at homologous sites [29, 45–48]. Data on spontaneous SCE frequencies obtained with BrdU labeling are quite variable. Values of 4–9 SCE per cell [15, 16, 49] up to 14 SCE per cell [50] have been reported. Females have significantly more SCE than males, an effect of age was at best weak, and smokers had significantly more SCE than nonsmokers [15, 16, 50]. In a recent review, the values of SCE per cell were reported to vary from 3.16 to 7.91 and to exhibit an age effect [19]. SCE are much more frequent than CA and MN. Assuming an average level of about 5 SCE per cell, we would expect 50 000 SCE per 10 000 cells compared to about 14.9 exchanges (RB′, DIC, RC) and 100.9 breaks (B′, B″) [22]. Even after doubling the frequencies of DIC to take TR into account (see above), the frequencies of CA are much lower compared to SCE. SCE are not considered CA, but an indicator of an error-free repair mechanism of SSL in the S-phase of the cell cycle.

9.5 Age Dependency of CA, MN, and SCE

As already mentioned, there is a positive correlation of the frequencies of CA and to a lesser extent of SCE with age [16, 19, 20, 33, 34, 51–54].

Tonomura *et al.* [54] analyzed CA in HPL of 23 newborn infants and found no DIC and rings in 15 325 metaphases, although DIC occurred in HPL of adults.

Neri *et al.* [55] provide data on percent aberrant cells (cells with CA) in Czech children. In 2 groups of 17 school children and 24 preschool children, the following values were reported: 1.37 ± 0.89, 2.24 ± 1.11, and 1.12 ± 1.05. According to Sram and Rössner [56], the percentage of aberrant cells in the Czech unexposed general population is 1.2–2.0 and the value in the 20–63-year subset is $1.59 \pm 1.34\%$. These values are in a similar range as those for children mentioned above. Neri *et al.* [55] present data on percent aberrant cells in newborns and different age groups up to 16–19 years. In a first set of data generated between 1984 and 1993, newborns had 1.11 aberrant cells, children of 4–6 years had 1.12, those of 7–15 years had 1.63, and those of 16–19 years had 2.02, showing an age effect, which was not confirmed in a second series of studies with children of the same age groups. In the second set of data generated from 1994–1999, all children apart from newborns with 1.11% aberrant cells, as in the data set from earlier times, had fewer aberrant cells and there was no age effect. This result is interpreted to indicate a change in the extent of pollution in the environment of the children. Neri *et al.* [57] performed a meta-analysis on biomarkers of genetic damage in children, including CA, MN, and SCE. The respective values varied considerably (frequencies of CA: 0.41–2.55 (in one set of data even 72); frequencies of MN: 1.2–7.8 per 1000 cells; SCE per cell: 3.83–9.6). In a study by Merlo *et al.* [58], the frequency of CA in children aged 0–19 years is reported to be 1.24% with no difference between girls (1.21%) and boys (1.24%) and no effect of age. Bender *et al.* [16] found a positive correlation of age with DIC. A clear age effect was seen in painted chromosomes for DIC, TR, insertions, and fragments [52]. In a review by Bolognesi *et al.* [19], an age effect for CA, MN, and to a lesser extent for SCE was shown. Using FISH, the frequencies of TR were found to be clearly correlated with age [33, 34, 51, 53].

A conclusive explanation for the positive correlation of CA and MN with age cannot be given. HPL have a half-life of 2–3 years, and CA and consequently MN should be efficiently removed with time and therefore a considerable accumulation of damage is not expected. An interesting source of age-related effects are losses or gains of X-chromosomes in female and loss of Y-chromosome in male stem cells leading to an age-dependent elevation of aneuploidy [36, 59]. Whether this could lead to elevated frequencies of MN is not clear, an age effect of the frequencies of centromere positive MN was not seen [44].

The frequencies of mutations at the hypoxanthine-guanine phosphoribosyltransferase gene (HPRT) are positively correlated with age in HPL [60, 61]. HPRT mutations can result from CA [62] and as discussed in Chapter 5, the frequencies of CA in HPL are age dependent. The incidence of repair of UV irradiated plasmids transfected into fibroblasts from donors of different age decreased and the frequency of mutations increased with age [9, 63]. A decrease in repair capacity could be one reason for the elevation of CA frequencies with age. SCE seem to be correlated with age too, but the effect is not very pronounced [19]. Since SCE can only be seen in dividing cells with chromosomal DNA differentially substituted before preparation, an age effect is not expected to occur.

9.6
Origin of CA in HPL

There is considerable spontaneous damage in mammalian DNA that when not properly repaired may give rise to mutations including CA [8, 9, 64–67] (Chapter 3). Superimposed on these are the CA induced by exogenous chromosome-breaking agents that comprise mankind's normal background exposure to environmental chemicals and radiation.

CA and MN in HPL or in other mammalian cells are induced by DNA-damaging agents *in vitro* and *in vivo* and are therefore important endpoints in genotoxicity test strategies [68]. Mutagenicity of compounds in exposed people can be demonstrated by analyzing CA and MN in their HPL [69–71]. Human population studies have been performed after exposure to various types of agents such as ionizing radiation [65], chemicals in occupational settings [56], and lifestyle factors, such as cigarette smoking [72–74] or alcohol consumption [75, 76]. A recent compilation of TR data from 16 laboratories showed that cigarette smoking leads to higher TR frequencies, which correlates with higher frequencies of CA other than TR in HPL of smokers compared to nonsmokers [34, 72].

9.7
Ionizing Radiation and Chromosomal Aberrations

Ionizing radiation is an important human carcinogen and long-standing questions that have exercised the radiation protection community are "What is the low dose–response relationship for radiation-induced cancers?" and "Does the risk of radiation-induced cancers at low doses extend right down to the origin of the response curve?" These are embodied in the arguments for or against the linear no threshold hypothesis (LNT). It is clear that to date the very low dose–response issue has not been resolved by cancer epidemiology (see Chapter 6) and so it may be instructive to turn to other quantifiable radiobiological endpoints that tentatively might act as surrogate indicators for cancer risk.

One of the most sensitive endpoints that may serve as a surrogate for an elevated cancer risk is the induction of CA and there is a large body of data, using many radiation qualities, that has been developed for purposes of biological dosimetry. Figure 9.1 shows some typical *in vitro* dose–response curves for the induction of DIC in HPL. Long-term follow-up studies have shown that there is a linkage between the presence of CA in HPL and the occurrence of malignancies many years later (see Section 9.8). In addition, molecular studies on the carcinogenesis processes have highlighted that genomic lesions form an important component in several of the steps from initiation to overt tumors.

However, the DIC per se, being an unstable structure, is unlikely to be directly implicated in carcinogenesis. It is nevertheless, because of its dosimetry application, the best quantified of all the types of radiation-induced alterations. Stable types of chromosomal aberrations are a prerequisite for lesions leading to malignancy

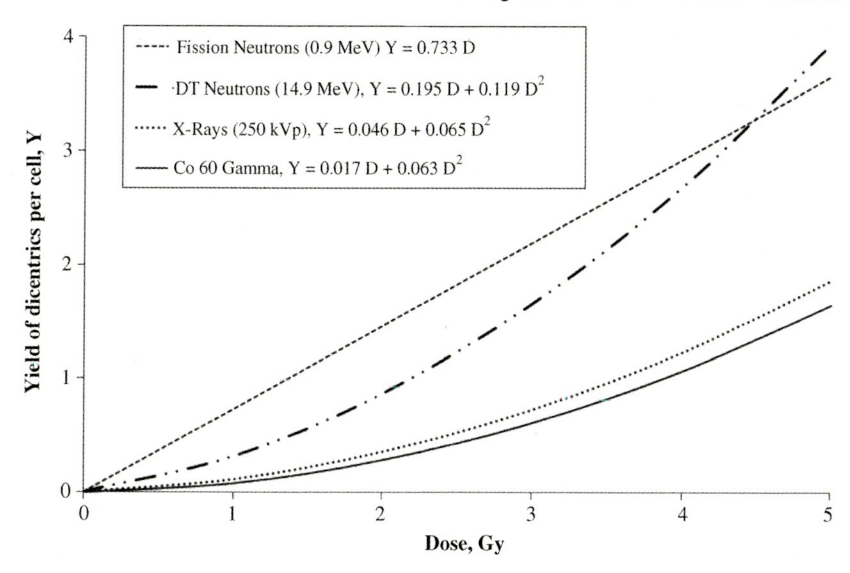

Figure 9.1 Typical linear and linear/quadratic *in vitro* DIC dose–response curves prepared with HPL for biological dosimetry.

because of their ability to pass more or less unhindered through successive cell divisions. To date, there are insufficient data on the low dose response for TR particularly in the dose range of interest in the LNT debate [76b]. As described above, the DIC frequency can nevertheless inform on stable aberration yields because there is compelling evidence from the use of FISH that the DIC and its stable homologue, the TR, are induced with equal frequencies immediately after irradiation [24]. Later, of course, the 1 : 1 ratio of DIC to TR breaks down because of their relative persistences in the cell population.

Low-dose data for the DIC do exist and Figure 9.2 shows a fit through a large, multilaboratory study (in total ~250 000 metaphases analyzed) over the range 0–50 mGy of 169 keV ISO wide series X-rays [30]. A weighted linear fit describes the data quite well down to about 10 mGy, but below this the background "noise" is such that one cannot distinguish any particular model of dose response, with or without a threshold, over the first five data points. A more recent study [77] revisited the low dose response, using Co-60 gamma radiation over a similar range, and reported an excellent linear fit with DIC yields significantly above the control frequency from 20 mGy. Probably, these two studies have demonstrated the practical limit to which low-dose DIC data can contribute to answering the dose response question. The limit is set by the background "noise."

The background frequency for DIC in healthy control subjects, based on many studies, has been reviewed above and ranges from about 1/500 to 1/2000 cells and a consensus value of 1/1000 cells is frequently taken [78]. This, together with the number of cells that is feasible to analyze, poses a practical limit when using the DIC assay as a biological dosimeter for investigating suspected or actual cases of radiation overexposure. It is customary to specify doses with 95% confidence limits [79, 80]

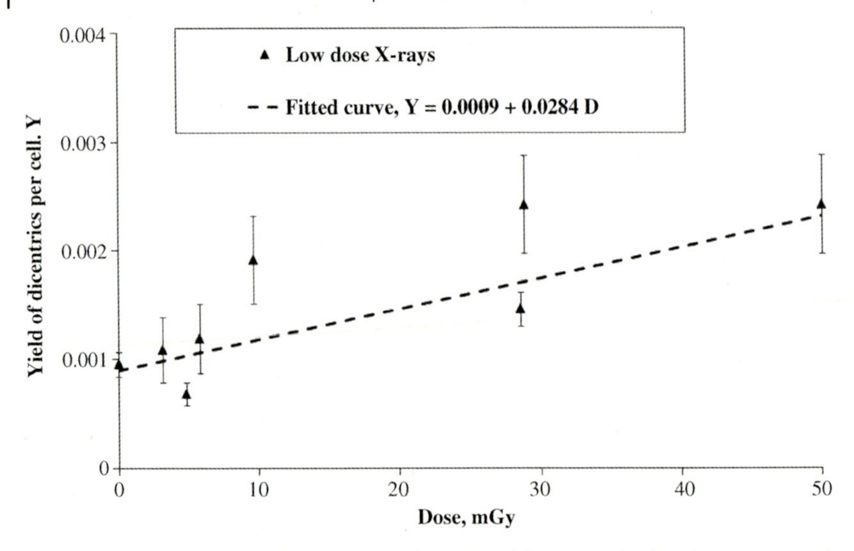

Figure 9.2 A weighted linear dose–response fit to DIC yields measured in lymphocytes exposed to low doses of X-radiation. Reproduced from Ref. [30].

and this uncertainty arises from two sources: the standard error on the fitted dose–response curve and on the measured DIC frequency in the HPL of a patient under investigation. Table 9.1 shows the upper and lower 95% limits on four dose estimates up to 1 Gy when using the gamma curve in Figure 9.1. The table shows that the uncertainties can be reduced by scoring more cells. However, with a conventional light microscope a very experienced technician can analyze about 500 metaphases per day, and thus there are logistical limits to just how many cells can be analyzed for a case investigation.

Conventional microscopy has now been replaced in some laboratories by automated systems using computer-driven microscopes that are able to locate metaphases

Table 9.1 The 95% confidence limits on dose estimates based on the numbers of metaphases scored for DIC induced by cobalt-60 gamma radiation using the dose–response curve $Y = (0.0005 \pm 0.0005) + (0.0164 \pm 0.0035)D + (0.0493 \pm 0.0028)D^2$.

Dose estimate (mGy)	Confidence limits	No. of cells examined		
		500	1000	2000
100	Upper	327	239	202
	Lower	<0	7	33
250	Upper	445	391	352
	Lower	82	130	166
500	Upper	708	651	604
	Lower	324	364	403
1000	Upper	1211	1145	1101
	Lower	820	868	905

and capture them as digitized images [81]. A recent refinement is "dicentric hunting" software that can scan the images and identify candidate DIC for operator review [82].

Technological advances are therefore now moving toward being able to remove the restraint of scoring typically about 1000 metaphases per case and so perhaps lowering the minimum detectable dose by the DIC assay. This raises issues concerning the precise definition of dose detection as opposed to dose measurement. A pragmatic suggestion is that an exposure can be regarded as having been *detected* whenever the dose estimate is positive but its lower confidence limit is zero or less. This is illustrated in Table 9.1 for the detection of 100 mGy by scoring 500 cells. If the lower limit is above zero, for example, again in Table 9.1, 100 mGy from 1000 cells, the dose estimate can be regarded as a *measurement*.

So, how might the introduction of automated DIC scoring impinge on the limitations illustrated by Table 9.1? Suppose, for example, that it would become possible to score one million metaphases. Table 9.2 shows that a dose estimate of 3.3 mGy would fulfill the criterion of a measurement with the, above zero, lower confidence limit of 0.1 mGy.

Unfortunately, the values shown in Table 9.2 are deceptive and not feasible. This is because there is statistical "noise" for any person and certainly variation in sensitivity from person to person. For radiation accident investigations, one does not have the luxury of a preaccident control sample from the patient and so the generic background has to be assumed. In reality, it could be higher and this would limit any benefit to be gained from any extra scoring above the traditional 1000 cells.

Nevertheless, given that computerized microscopes are now available, it is worth considering what realistic improvement they offer. What is the cutoff point between scoring more cells to improve the real measurable lower limit rather than simply obtaining the spurious sensitivity illustrated in Table 9.2? The answer lies in the accuracy of the linear dose–response yield coefficient that predominates at low doses. For high-energy gamma rays, such as from Co-60, a good calibration curve has an uncertainty on the linear coefficient of about 20% (1 standard error). With this being

Table 9.2 Dose estimates and their 95% confidence limits, in Gy, for various numbers of DIC scored in 1 million cells with an assumed background of 1/1000 cells.

Observed number of dicentrics	Lower confidence limit	Dose estimate	Upper confidence limit
1000	0	0	0.0031
1050	0	0.0025	0.0056
1066	0.0001	0.0033	0.0064
1070	0.0003	0.0035	0.0066
1100	0.0017	0.0049	0.0081
2000	0.0406	0.0442	0.0476

The calculations were made with a generic low LET dose–response yield equation of $Y = 0.001 + 0.02D + 0.06D^2$.

Table 9.3 Dose estimates and their 95% confidence limits, in Gy, for various numbers of DIC scored in 10 000 cells with an assumed background of 1/1000 cells.

Observed number of dicentrics	Lower confidence limit	Dose estimate	Upper confidence limit
18	0	0.038	0.098
20	0	0.046	0.103
22	0	0.053	0.109
24	0	0.060	0.113
26	0.006	0.067	0.117

The calculations were made with a generic low LET dose–response yield equation of $Y = 0.001 + (0.02 \pm 0.004)D + 0.06D^2$.

the limiting factor on converting the DIC frequency to dose, there is little to be gained by measuring the patient's DIC frequency to better than 20%, that is, scoring about 25 dicentrics. When one reworks the calculations, one obtains the values shown in Table 9.3. Thus, in practice, automation combined with a good calibration curve could reduce the minimum measurable dose from about 100 mGy based on the traditional scoring of 1000 cells to about 70 mGy from 10 000 cells. At this point, the uncertainties on the individual measurement and the linear yield coefficient contribute approximately equally to the overall uncertainty on the dose estimate. A major limitation is the spontaneous frequency of DIC that of course is unknown for any irradiated individual referred for biological dosimetry.

In comparing physical and biological dosimetry, it is important to draw a distinction between *physically absorbed dose* and *biological relevant effects*. Ionizing radiation deposits energy inside biological targets, according to its energy and charge and the chemical composition of the target. Such energy divided by the target mass represents the *physically absorbed dose* and is expressed in the unit Gy. The resultant primary damage in cellular DNA is mostly enzymatically repaired, and the CA resulting from the damage remaining after such restitution, because of misrepair, determines the biological consequences including an elevated cancer risk. The health outcome therefore depends on the efficiency of DNA repair mechanisms and within the population there is considerable variability. Correct restitution of DNA will be generally more likely in young healthy individuals than in the elderly or those carrying genetic deficiencies. A wide range of other host factors such immunosuppression and vitamin deficiency are also important. Thus, CA, as biomarkers, are a surrogate measure of the *biologically relevant effect* of exposure to radiation and to a certain extent are indicative of health consequences and, at low dose, one of the biggest concern is cancer induction.

Calibration curves prepared for biological dosimetry can be only an approximation to the circumstances of a given case. Certainly, an appropriate radiation source, for example, Co-60 or X-rays, should be selected and consideration given to dose rate. However, it is impossible to reproduce *in vitro* the exact *in vivo* absorbed energy spectrum caused, for example, by scattered radiation that will be of lower energy and hence higher LET than the incident radiation. The output should be given as dose (in Gy) because the *in vitro* calibrations will have been monitored with instruments traceable to a national physical standard. After producing a dose estimate in Gy, one would wish to relate this to risk. Risk (of stochastic diseases) employs the unit of equivalent dose in Sv that is based on arbitrary assumptions of the radiation protection community concerning the relative ranking of the radiation qualities in their effectiveness in causing cancer [83]. The ICRP [83] proposes Sv/risk coefficients derived mainly from epidemiological studies such as the one on the A-bomb survivors' life span study. One may then use these coefficients to generate advice on risk for irradiated individuals for whom dose estimates have been made by either physical or biological methods. This should be relatively easy when dealing with low LET radiation because the assumed conversion between Sv and Gy is unity. However, this is an approximation that is maintained for ease and convenience of calculation. There is ample evidence from several biological systems including CA in HPL that across the spectrum of radiations considered as "low LET" there is, particularly at low doses, a considerable variation in biological effectiveness. In general, the lower the energy the higher the effectiveness per Gy.

It may be tempting to estimate directly the equivalent dose from the biodosimetric test by using a calibration curve with a reference radiation. If, for example, a calibration curve is constructed *in vitro* by using gamma rays, then the x-axis could be expressed in Sv rather than in Gy and the equivalent dose could be directly estimated from the CA frequency. However, the oversimplification of assuming a radiation weighting factor of unity for all low LET radiations, given the biological evidence that their effectiveness varies, renders this unwise.

9.8
CA and Cancer in Human Populations

As described above, when counseling irradiated persons on their risk of developing cancer, a two-step process is adopted. First, the dose is estimated that should be based on all available information but, in the absence of reliable physics, CA can assume great importance. Second, after establishing the dose, risk is addressed using the epidemiologically derived Sv/risk coefficients proposed by international bodies such as ICRP [83]. The coefficients may be weighted where appropriate to take account of factors such as the individual's age and gender. It is a long-term objective to replace this two-step approach so that cancer risk could be estimated directly if an appropriate calibration curve of CA versus risk can be established. This would need to rely on cells such as HPL that are acceptable to sample and it has been hypothesized that chromosomal damage in HPL reflects similar damage occurring in precursor cells for carcinogenic processes in target tissues. Evidence that CA in HPL are positively correlated with cancer risk has been obtained in a large cohort study, performed by the European Study Group on Cytogenetic Biomarkers and Health (ESCH) in a group of 3541 healthy subjects screened for CA in the last three decades in five European countries [84]. Subjects were divided into three categories (low, medium, and high) based on percentiles of CA frequencies and followed up for

Table 9.4 Results from the multivariate Cox regression analyses of CA frequencies on total cancer incidence (data from Finland, Norway, and Sweden) and mortality (data from Italy)[a].

CA frequency	Incidence ratio (95% CI) (Scandinavia)	Mortality ratio (95% CI) (Italy)
Low	1.00	1.00
Medium	1.22 (0.71–2.12)	1.38 (0.70–2.71)
High	2.08 (1.26–3.40)	2.56 (1.35–4.86)

a) The multivariate models included additional factors: gender, age at test ($\leq$ 50 years, >50 years), time since test ($\leq$ 9 years, > 9 years), and country (Finland, Norway, and Sweden; not applicable for Italy). CI: confidential interval. Adapted from Ref. [84].

cancer incidence or mortality for many years. A significant increase in both endpoints was found for the "high-frequency" group (66–100 percentile), which experienced an occurrence of cancer more than double compared to those subjects classified as "low-frequency group" (Table 9.4 and Figure 9.3). The impact of CA frequency on survival is demonstrated in Figure 9.3, which shows the results of survival analyses performed on incidence (Nordic countries) and mortality data (Italy).

Many studies of this type have been performed and all came to the same conclusion, namely, that high frequencies of CA in human populations are correlated

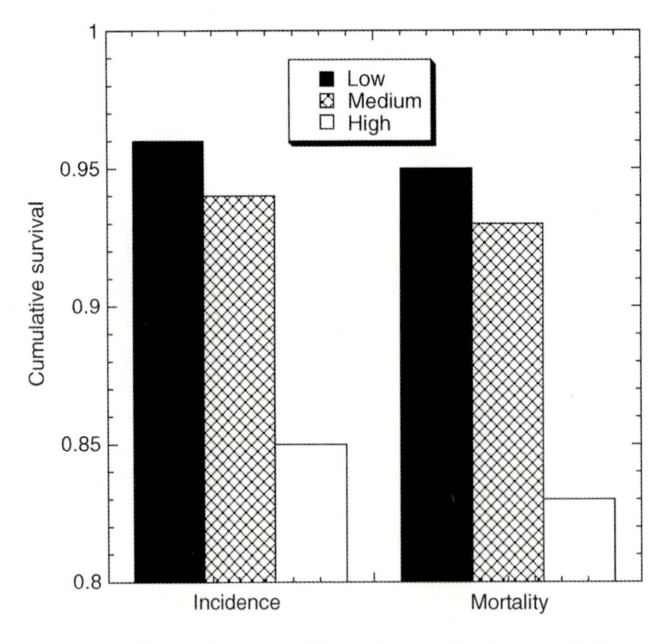

Figure 9.3 Disease-free survival for total cancer incidence (Nordic cohort) or mortality (Italian cohort) after 20 years from the blood test assessing the frequency of basal CA in lymphocytes. Subjects were divided into three groups according to the percentile of the distributions: low, medium, and high CA frequency. Adapted from Ref. [84].

with a significantly higher risk of developing cancer when compared to persons with low frequencies of CA [85–93].

These findings support the presence of a causal link between CA frequencies and cancer risk. Experimental studies and studies on A-bomb survivors demonstrated a positive relationship between CA induced by ionizing radiation and cancer risk; therefore, a correlation between CA and cancer to increase with the dose of ionizing radiation or other mutagenic agents would be expected. The results of a case–control study nested within the ESCH specifically address this aspect [87, 89]. A careful reconstruction of past exposures for all cancer cases and matched controls has been performed, showing that with few exceptions the prediction of cancer risk by CA frequencies seems not to be modified by exposure to genotoxic agents, leading to the assumption that the relationship between CA and cancer is independent of exposure to carcinogens [89]. Of course, there must be reasons for the occurrence of CA and as discussed before these are both endogenous and exogenous. In some analyses, increased risks of cancers of specific types were observed after exposure to ionizing radiation, polyvinyl chloride, and plastics, welding fumes, metals, and smoking [89, 92]. Like CA, MN, but not SCE, are correlated with cancer [88, 90]. In view of the fact that CA and consequently MN are induced by mutagenic and carcinogenic agents, positive correlations of CA and MN with cancer are not really unexpected. As discussed above, CSA are induced by DSB in G0/G1 cells and therefore it would be expected that especially CSA, such as DIC, should be correlated with cancer, which was found in some [92, 93] but not in other studies [89]. CTA result from SSL that upon transformation to DSB in S-phase may give rise to CTA in the following metaphase. CA observed spontaneously in HPL should mainly result from DSB induced *in vivo*. Apart from being induced by exogenous agents, such as ionizing radiation, substantial amounts of DSB occur spontaneously *in vivo* [7]. DSB may give rise to CSA and SSL may lead to CTA when cells with such lesions enter S-phase, which is the case when the cells are put in culture. During repair of SSL in G0/G1, DSB may eventually occur and these would add to the pool of CSA. TR in stem cells of HPL may be compatible with cell survival and by this will add to the frequencies of CSA. TR can be analyzed sufficiently only with FISH and enough data are not yet available to permit correlations between TR and cancer. These aspects would rather lead to the expectation of correlations of CSA and not of CTA with cancer. This is further supported by the finding that there is no correlation of SCE frequencies with cancer. SCE result from repair of SSL during S-phase [46, 48]. If CTA are indicative of cancer, then why is that not the case for SCE? *GSTM1* and *GSTT1* null genotypes did not change the association between CA and cancer; this is interesting since these genes produce proteins important for the detoxification of genotoxic agents [91].

Whatever the exact mechanistic connection between CA in HPL and cancer, the results clearly demonstrate that CA in lymphocytes are a broad biomarker of cancer risk in humans, reflecting both genotoxic effects of carcinogens and individual cancer susceptibility. At present, CA in HPL are used purely as biological dosimeter to verify or compensate for the absence of physical dosimetry measurements. While physical dosimetry provides the same risk for all individuals, CA take into account the

Figure 9.4 Estimates of the excess relative risk (ERR) of cancer incidence based on the measurements of CA in astronauts or from physical dosimetry coupled to a risk assessment model recommended by ICRP [83] and NCRP [100]. Data are related to 15 Russian cosmonauts who were in space for 115–380 days in single missions on board of the International Space Station or MIR. Estimated doses ranged from 55–365 mSv. ERR estimates from CA are performed from the ratio of pre- and postflight CA yields in HPL, using the risk coefficient derived from the ESCH epidemiological study. Physical estimates are based on the absorbed dose in Gy, an average quality factor in low Earth orbit of 2.4, and the cancer risk coefficient recommended by ICRP [83] and NCRP [100]. For details see Ref. [96].

biological responses of the individual, which depend on radiation plus many other factors, and this could provide a personalized estimate of the associated risk.

This idea has been applied to estimate risk in populations exposed to mixed radiation fields in cases where the uncertainty is particularly high, such as astronauts. Within the limits of experimental uncertainty in the measurements of CA, and the different techniques used, the results of these comparisons (Figure 9.4) generally support the current models used to estimate space radiation risk in long-term manned missions, such as to Mars [94–98].

Figure 9.4 shows data on cancer risks based on CA in HPL of 15 cosmonauts flying for more than 3 months on MIR or the International Space Station, that is, in low Earth orbit (LEO). Excess relative risk (ERR) for cancer was calculated either from induced CA in HPL of cosmonauts (ERRch) or from physical measurements of radiation on board the space station (ERRphys). ERRch values estimated using the ESCH database (see above) vary between –0.08 and 0.42 (average of 0.09). ERRphys is calculated on the basis of radiation dose measured on board, an average quality factor, a factor derived from cancer incidence in A-bomb survivors, and a dose and dose rate effectiveness factor. The ERRphys values vary between 0.02 and 0.13 (average 0.05) [96, 98]. Data based on CA are more variable than data calculated from radiation dose. Reasons for this may be statistical errors in determining CA frequencies and interindividual variability of the extent of CA induction in different astronauts. Irrespective of the great variability of CA data, ERRch and

ERRphys are relatively similar indicating that cytogenetic biomarkers are quite reliable endpoints to measure radiation-induced cancer risk. Compared to astronauts exposed to radiation in LEO, estimation of cancer risk for astronauts traveling to the moon or Mars will be more difficult because of the highly complex radiation exposure scenarios especially due to high-energy protons and high-energy heavy (HZE) ions [94, 96, 97, 99].

The new knowledge on the human genome structure paves the way for studies addressing correlations between specific genetic polymorphisms and cancer. Given the proven association between CA and cancer risk, CA and MN can also be used as surrogate endpoints to test the risk associated with specific polymorphisms [91, 101]. These studies are still preliminary but clearly very promising to reach the goal of a genetic screening of individuals at elevated cancer risk.

References

1 Natarajan, A.T. (2002) Chromosome aberrations: past, present and future. *Mutat. Res.*, **504**, 3–16.

2 Bryant, P.E. (2007) Origin of chromosome aberrations: mechanisms, in *Chromosomal Alterations: Methods, Results and Importance in Human Health* (eds G. Obe and Vijayalaxmi), Springer, Berlin, pp. 177–199.

3 Obe, G., Pfeiffer, P., Savage, J.R.K., Johannes, C., Goedecke, W., Jeppesen, P., Natarajan, A.T., Martinez-Lopez, W., Folle, G.A., and Drets, M.E. (2002) Chromosomal aberrations: formation, identification and distribution. *Mutat. Res.*, **504**, 17–36.

4 Christmann, M., Roos, W.P., and Kaina, B. (2007) DNA methylation damage: formation, repair and biological consequences, in *Chromosomal Alterations: Methods, Results and Importance in Human Health* (eds G. Obe and Vijayalaxmi), Springer, Berlin, pp. 99–121.

5 Goedecke, W. (2007) Double strand break repair mechanisms in mammalian cells, in *Chromosomal Alterations: Methods, Results and Importance in Human Health* (eds G. Obe and Vijayalaxmi), Springer, Berlin, pp. 55–65.

6 Iliakis, G., Wu, W., Wang, M., Terzoudi, G.I., and Pantelias, G.E. (2007) Backup pathways of nonhomologous end joining may have a dominant role in the formation of chromosomal aberrations, in *Chromosomal Alterations: Methods, Results and Importance in Human Health* (eds G. Obe and Vijayalaxmi), Springer, Berlin, pp. 67–85.

7 Pfeiffer, P., Goedecke, W., and Obe, G. (2000). Mechanisms of DNA double-strand break repair and their potential to induce chromosomal aberrations. *Mutagenesis*, **15**, 289–302.

8 Vilenchik, M.M. and Knudson, A.G. (2003) Endogenous DNA double-strand breaks: production, fidelity of repair, and induction of cancer. *Proc. Natl. Acad. Sci. USA*, **100**, 12871–12876.

9 Friedberg, E.C., Walker, G.C., Siede, W., Wood, R.D., Schultz, R.A., and Ellenberger, T. (2006) *DNA Repair and Mutagenesis*, 2nd edn, ASM Press, Washington, DC.

10 Adhvaryu, S.G. (2007) Chromosome analysis in cancer patients: applications and limitations, in *Chromosomal Alterations: Methods, Results and Importance in Human Health* (eds G. Obe and Vijayalaxmi), Springer, Berlin, pp. 479–493.

11 Mitelman, F., Mertens, F., and Johansson, B. (1997) A breakpoint map of recurrent chromosomal rearrangements in human neoplasia. *Nat. Genet*, **15**, 417–474.

12 Mitelman, F., Johansson, B., and Mertens, F. (eds) (2006) Mitelman Database of Chromosome Aberrations in Cancer. http://cgap.nci.nih.gov/Chromosomes/Mitelman.

13 Mitelman, F., Johansson, B., and Mertens, F. (2007) The impact of translocations and gene fusions on cancer causation. *Nat. Rev. Cancer*, **7**, 233–245.

14 Savage, J.R.K. (2004). On the nature of visible chromosomal gaps and breaks. *Cytogenet. Genome Res.*, **104**, 46–55.

15 Bender, M.A., Preston, R.J., Leonard, R.C., Pyatt, B.E., Gooch, P.C., and Shelby, M.D. (1988) Chromosomal aberration and sister-chromatid exchange frequencies in peripheral blood lymphocytes of a large human population sample. *Mutat. Res.*, **204**, 421–433.

16 Bender, M.A., Preston, R.J., Leonard, R.C., Pyatt, B.E., and Gooch, P.C. (1989) Chromosomal aberration and sister-chromatid exchange frequencies in peripheral blood lymphocytes of a large human population sample. *Mutat. Res.*, **212**, 149–154.

17 Bender, M.A., Preston, R.J., Leonard, R.C., Pyatt, B.E., and Gooch, P.C. (1990) On the distributions of chromosomal aberrations in human peripheral blood lymphocytes in culture. *Mutat. Res.*, **244**, 215–220.

18 Lloyd, D.C., Purrott, R.J., and Reeder, E.J. (1980) The incidence of unstable chromosome aberrations in peripheral blood lymphocytes from unirradiated and occupationally exposed people. *Mutat. Res.*, **72**, 523–532.

19 Bolognesi, C., Abbondandolo, A., Barale, R., Casalone, R., Dalpra, L., DeFerrari, M., Degrassi, F., Forni, A., Lamberti, L., Lando, C., Migliore, L., Padovani, P., Pasquini, R., Puntoni, R., Sbrana, I., Stella, M. and Bonassi, S. (1997) Age-related increase of baseline frequencies of sister chromatid exchanges, chromosome aberrations, and micronuclei in human lymphocytes. *Cancer Epidemiol. Biomarkers Prev.*, **6**, 249–256.

20 Wolf, G., Arndt, D., Kotschy-Lang, N., and Obe, G. (2004) Chromosomal aberrations in uranium and coal miners. *Int. J. Radiat. Biol.*, **80**, 147–153.

21 Wolf, G. and Fender, H. (2002). Chromosomal analysis in human population monitoring, in *Some Aspects of Chromosome Structure and Functions* (eds R.C. Sobti, G. Obe, and R.S. Athwal), Narosa Publishing House, New Delhi, pp. 133–148.

22 Obe, G. (1986) Spontaneous level of somatic chromosome aberrations in man, in *Monitoring of Occupational Genotoxicants* (eds M. Sorsa and H. Norppa), Alan R. Liss, New York, pp. 25–37.

23 Hansmann, I., Meyding, U., and Virsik, R.P. (1983) X-ray induced reciprocal translocations and dicentrics in human G0 lymphocytes. *Int. J. Radiat. Biol.*, **43**, 91–93.

24 Edwards, A.A., Lindholm, C., Darroudi, F., Stephan, G., Romm, H., Barquinero, J., Barrios, L., Caballin, M-R., Roy, L., Whitehouse, C.A., Tawn, E.J., Moquet, J.E., Lloyd, D.C., and Voisin, P. (2005) Review of translocations detected by FISH for retrospective biological dosimetry applications. *Radiat. Prot. Dosimetry*, **113**, 396–402.

25 Johannes, C., Chudoba, I., and Obe, G. (1999) Analysis of X-ray-induced aberrations in human chromosome 5 using high-resolution multicolour banding FISH (mBAND). *Chromosome Res.*, **7**, 625–633.

26 Johannes, C., Horstmann, M., Durante, M., Chudoba, I., and Obe, G. (2004) Chromosome intrachanges and interchanges detected by multicolour banding in lymphocytes: searching for clastogen signatures in the human genome. *Radiat. Res.*, **161**, 540–548.

27 Bonassi, S. (2007) Chromosomal aberration in peripheral blood lymphocytes of healthy subjects and risk of cancer, in *Chromosomal Alterations: Methods, Results and Importance in Human Health* (eds G. Obe and Vijayalaxmi), Springer, Berlin, pp. 495–504.

28 Obe, G. and Beek, B. (1982) The human leukocyte test system, in *Chemical Mutagens. Principles and Methods for Their Detection*, vol. **7** (eds F.J. De Serres and A. Hollaender), Plenum Press, New York, pp. 307–400.

29 Perry, P. and Wolff, S. (1974). New Giemsa method for the differential staining of sister chromatids. *Nature*, **251**, 156–158.

30 Lloyd, D.C., Edwards, A.A., Leonard, A., Deknudt, G.L., Verschaeve, L., Natarajan, A.T., Darroudi, F., Obe, G., Palitti, F., Tanzarella, C., and Tawn, E.J. (1992) Chromosomal aberrations in human lymphocytes induced *in vitro* by very low doses of X-rays. *Int. J. Radiat. Biol.*, **61**, 335–343.

31 Buckton, K.E., Hamilton, G.E., Paton, L., and Langlands, A.O. (1978) Chromosome aberrations in irradiated ankylosing spondylitis patients, in *Mutagen-Induced Chromosome Damage in Man* (eds H.J. Evans and D.C. Lloyd), Edinburgh University Press, Edinburgh, pp. 142–150.

32 Buckton, KE., Court Brown, WM., and Smith, PG. (1967) Lymphocyte survival in men treated with X-rays for ankylosing spondylitis. *Nature*, **214**, 470–473.

33 Pressl, S., Romm, H., Ganguly, B.B., and Stephan, G. (2000) Experience with FISH-detected translocations as an indicator in retrospective dose reconstructions. *Radiat. Prot. Dosimetry*, **88**, 45–49.

34 Sigurdson, A.J., Ha, M., Hauptmann, M., Bhatti, P., Sram, R.J., Beskid, O., Tawn, E.J., Whitehouse, C.A., Lindholm, C., Nakano, M., Kodama, Y., Nakamura, N., Vorobtsova, I., Oestreicher, U., Stephan, G., Yong, L.C., Bauchinger, M., Schmid, E., Chung, H.W., Darroudi, F., Roy, L., Voisin, P., Barquinero, J.F., Livingston, G., Blakey, D., Hayata, I., Zhang, W., Wang, C.Y., Bennett, L.M., Littlefield, L.G., Edwards, A.A., Kleinerman, R.A., and Tucker, J.D. (2008) International study of factors affecting human chromosome translocations. *Mutat. Res.*, **652**, 112–121.

35 Lindholm, C., Romm, H., Stephan, G., Schmid, E., Moquet, J., and Edwards, A. (2002) Intercomparison of translocation and dicentric frequencies between laboratories in a follow-up of the radiological accident in Estonia. *Int. J. Radiat. Biol.*, **78**, 883–890.

36 Evans, H.J. (1986) What has been achieved with cytogenetic monitoring? in *Monitoring of Occupational Genotoxicants* (eds M. Sorsa and H. Norppa), Alan R Liss, New York, pp. 3–23.

37 George, K., Durante, M., Willigham, V., and Cucinotta, F.A. (2004) Chromosome aberrations of clonal origin are present in astronaut's blood lymphocytes. *Cytogenet. Genome Res.*, **104**, 245–251.

38 Povirk, L.F. (2006). Biochemical mechanisms of chromosomal translocations resulting from DNA double-strand breaks. *DNA Repair*, **5**, 1199–1212.

39 Fenech, M. (2007) Cytokinesis-block micronucleus assay: a comprehensive "cytome" approach for measuring chromosomal instability, mitotic dysfunction and cell death simultaneously in one assay, in *Chromosomal Alterations: Methods, Results and Importance in Human Health* (eds G. Obe and Vijayalaxmi), Springer, Berlin, pp. 241–255.

40 Bonassi, S., Fenech, M., Lando, C., Lin, Y.P., Ceppi, M., Chang, W.P., Holland, N., Kirsch-Volders, M., Zeiger, E., Ban, S.Y., Barale, R., Bigatti, M.P., Bolognesi, C., Jia, C., Di Giorgio, M., Ferguson, L.R., Fucic, A., Lima, O.G., Hrelia, P., Krishnaja, A.P., Lee, T.K., Migliore, L., Mikhalevich, L., Mirkova, E., Mosesso, P., Muller, W.U., Odagiri, Y., Scarfi, M.R., Szabova, E., Vorobtsova, I., Vral, A., and Zijno, A. (2001) HUman MicroNucleus Project: international database comparison for results with the cytokinesis-block micronucleus assay in human lymphocytes: I. Effect of laboratory protocol, scoring criteria, and host factors on the frequency of micronuclei. *Environ. Mol. Mutagen.*, **37**, 31–45.

41 Bonassi, S., Biscotti, B., Kirsch-Volders, M., Knasmueller, S., Zeiger, E., Burgaz, S., Bolognesi, C., Holland, N., Thomas, P., and Fenech, M. (2009) State of the art survey of the buccal micronucleus assay: a first stage in the HUMNxl project initiative. *Mutagenesis*, **24**, 295–302.

42 Holland, N., Bolognesi, C., Kirsch-Volders, M., Bonassi, S.,

Zeiger, E., Knasmueller, S., and Fenech, M. (2008) The micronucleus assay in human buccal cells as a tool for biomonitoring DNA damage: HUMN project perspective on current status and knowledge gaps. *Mutat. Res.* doi: 10.1016/j.mrrev.2008.03.007

43 Thomas, P., Holland, N., Bolognesi, C., Kirsch-Volders, M., Bonassi, S., Zeiger, E., Knasmüller, S., and Fenech, M. (2009). Buccal micronucleus cytome assay. *Nat. Protoc.*, **4**, 825–837.

44 Iarmarcovai, G., Bonassi, S., Sari-Minodier, I., Bachiucka-Palmaro, M., Botta, A., and Osiere, T. (2007) Exposure to genotoxic agents, host factors, and lifestyle influence the number of centromeric signals in micronuclei: a pooled re-analysis. *Mutat. Res.*, **615**, 18–27.

45 Bruckmann, E., Wojcik, A., and Obe, G. (1999) X-irradiation of G1 CHO cells induces SCE which are both true and false in BrdU-substituted cells but only false in biotin-dUTP-substituted cells. *Chromosome Res.*, **7**, 277–288.

46 Wilson, D.M., III and Thompson, L.H. (2007). Molecular mechanisms of sister-chromatid exchange. *Mutat. Res.*, **616**, 11–23.

47 Wojcik, A. and Obe, G. (2007). Sister-chromatid exchanges, in *Chromosomal Alterations: Methods, Results and Importance in Human Health* (eds G. Obe and Vijayalaxmi), Springer, Berlin, pp. 271–283.

48 Wojcik, A., Bruckmann, E., and Obe, G. (2004). Insights into the mechanisms of sister chromatid exchange formation. *Cytogenet. Genome Res.*, **104**, 304–309.

49 Carrano, A.V. and Moore, D.H., II (1982) The rationale and methodology for quantifying sister chromatid exchanges in humans, in *Mutagenicity: New Horizons in Genetic Toxicology* (ed. J.A. Heddle), Academic Press, New York, pp. 267–304.

50 Anderson, D., Dewdney, R.S., Jenkinson, P.C., Lovell, D.P., Butterworth, K.R., and Cunning, D.M. (1986) in *Monitoring of Occupational Genotoxicants* (eds M. Sorsa and H. Norppa), Alan R Liss, New York, pp. 39–58.

51 Pressl, S., Edwards, A., and Stephan, G. (1999). The influence of age, sex and smoking on the background level of FISH-detected translocations. *Mutat. Res.*, **442**, 89–95.

52 Ramsey, M.J., Moore, D.H., II, Briner, J.F., Lee, D.A., Olsen, L.A., Senft, J.R., and Tucker, JD. (1995) The effect of age and lifestyle factors on the accumulation of cytogenetic damage as measured by chromosome painting. *Mutat. Res.*, **338**, 95–106.

53 Sorokine-Durm, I., Whitehouse, C., and Edwards, A. (2000). The variability of translocation yields amongst control populations. *Radiat. Protect. Dosimetry.*, **88**, 93–99.

54 Tonomura, A., Kishi, K., and Saito, F. (1983). Types and frequencies of chromosome aberrations in peripheral lymphocytes of general populations, in *Radiation-Induced Chromosome Damage in Man* (eds T.M. Ishihara and M.S. Sasaki), Alan R. Liss, Inc., New York, pp. 605–616.

55 Neri, M., Bonassi, S., Knudsen, L.E., Sram, R.J., Holland, N., Ugolini, D., and Merlo, D.F. (2006) Children's exposure to environmental pollutants and biomarkers of genetic damage I. Overview and critical issues. *Mutat. Res.*, **612**, 1–13.

56 Sram, R.J. and Rössner, P. (2007). Cytogenetic analysis and occupational health, in *Chromosomal Alterations: Methods, Results and Importance in Human Health* (eds G. Obe and Vijayalaxmi), Springer, Berlin, pp. 325–340.

57 Neri, M., Ugolini, D., Bonassi, S., Fucic, A., Holland, N., Knudsen, L.E., Sram, R.J., Ceppi, M., Boccini, V., and Merlo, D.F. (2006) Children's exposure to environmental pollutants and biomarkers of genetic damage II. Results of a comprehensive literature search and meta-analysis. *Mutat. Res.*, **612**, 14–39.

58 Merlo, D.F., Ceppi, M., Stagi, E., Boccini, V., Sram, R.J., and Rössner, P. (2007) Baseline chromosome aberrations in children. *Toxicol. Lett.*, **172**, 60–67.

59 Evans, H.J. (1975) Environmental agents and chromosome damage in somatic cells in man: *in vivo* studies, in *Chromosome Variations in Human Evolution* (ed. A.J. Boyce), Taylor & Francis, London, pp. 63–82.

60 Cole, J., Green, M.H.L., James, S.E., Henderson, L., and Cole, H. (1988) A further assessment of factors influencing measurements of thioguanine-resistant mutant frequency in circulating T-lymphocytes. *Mutat. Res.*, **204**, 493–507.

61 King, C.M., Gillespie, E.S., McKenna, P.G., and Barnett, Y.A. (1994) An investigation of mutation as a function of age in humans. *Mutat. Res.*, **316**, 79–90.

62 Obe, G., Von der Hude, W., Scheutwinkel-Reich, M., and Basler, A. (1986) The restriction endonuclease Alu I induces chromosomal aberrations and mutations in the hypoxanthine phosphoribosyltransferase locus, but not in the Na^+/K^+ ATPase locus in V79 hamster cells. *Mutat. Res.*, **174**, 71–74.

63 Moriwaki, S.-I., Ray, S., Tarone, R.E., Kraemer, K.H., and Grossman, L. (1996) The effect of donor age on the processing of UV-damaged DNA by cultured human cells: reduced DNA repair capacity and increased DNA mutability. *Mutat. Res.*, **364**, 117–123.

64 Barnes, D.E. and Lindahl, T. (2004) Repair and genetic consequences of endogenous DNA base damage in mammalian cells. *Annu. Rev. Genet.*, **38**, 445–476.

65 BEIR, VII, Phase 2 (2006) *Health Risks from Exposure to Low Levels of Ionizing Radiation*, The National Academies Press, Washington, DC.

66 Klungland, A. and Bjelland, S. (2007). Oxidative damage to purines in DNA: role of mammalian Ogg1. *DNA Repair*, **6**, 481–488.

67 Lindahl, T. (1993) Instability and decay of the primary structure of DNA. *Nature*, **362**, 709–715.

68 Kirkland, D., Aardema, M., Henderson, L., and Müller, L. (2005) Evaluation of the ability of a battery of three *in vitro* genotoxicity tests to discriminate rodent carcinogens and non-carcinogens I. Sensitivity, specificity and relative predictivity. *Mutat. Res.*, **584**, 1–256.

69 Albertini, R.J., Anderson, D., Douglas, G.R., Hagmar, L., Hemminki, K., Merlo, F., Natarajan, A.T., Norppa, H., Shuker, D.E.G., Tice, R., Waters, M.D., and Aitio, A. (2000) IPCS guidelines for the monitoring of genotoxic effects of carcinogens in humans. *Mutat. Res.*, **463**, 111–172.

70 Bonassi, S., Ugolini, D., Kirsch-Volders, M., Strömberg, U., Vermeulen, R., and Tucker, J.D. (2005) Human population studies with cytogenetic biomarkers: review of the literature and future prospects. *Environ. Mol. Mutagen.*, **45**, 258–270.

71 Natarajan, A.T. and Obe, G. (1980) Screening of human populations for mutations induced by environmental pollutants: use of human lymphocyte system. *Ecotoxicol. Environ. Saf.*, **4**, 468–481.

72 IARC (1986) *IARC Monographs on the Evaluation of Carcinogenic Risks to Humans: Tobacco Smoking*, IARC, Lyon.

73 Obe, G., Vogt H-J., Madle, S., Fahning, A., and Heller, W-D. (1982). Double-blind study on the effect of cigarette smoking on the chromosomes of human peripheral blood lymphocytes *in vivo*. *Mutat. Res.*, **92**, 309–319.

74 Obe, G., Heller, W-D., and Vogt, H-J. (1984). Mutagenic activity of cigarette smoke, in *Mutations in Man* (ed. G. Obe), Springer, Berlin, pp. 223–246.

75 IARC (1988) *IARC Monographs on the Evaluation of Carcinogenic Risks to Humans: Alcohol Drinking*, IARC, Lyon.

76 (a) Obe, G. and Anderson, D. (1987) Genetic effects of ethanol. *Mutat. Res.*, **186**, 177–200. (b) Tucker, J.D. (2008) Low-dose ionizing radiation and chromosome translocations: a review of the major considerations for human biological dosimetry. *Mutat. Res.*, **659**, 211–220.

77 Iwasaki, T., Takashima, Y., Suzuki, T., Yoshida, M.A., and Hayata, I. (2011) The dose response of chromosome aberrations in human lymphocytes induced *in vitro* by very low-dose γ rays, *Radiat. Res.*, **175**, 208–213.

78 IAEA (2001) Cytogenetic Analysis For Radiation Dose Assessment: A Manual. Technical Report Series 405, IAEA, Vienna.

79 Szluinska, M., Edwards, A., and Lloyd, D. (2007). Presenting statistical uncertainty on cytogenetic dose estimates. *Radiat. Prot. Dosimetry*, **123**, 443–449.

80 ISO 19238 (2004) *Radiation Protection: Performance Criteria for Service Laboratories Performing Biological Dosimetry by Cytogenetics*, ISO, Geneva.

81 Weber, J., Scheid, W., and Traut, H. (1992). Time-saving in biological dosimetry by using the automatic metaphase finder Metafer2. *Mutat. Res.*, **272**, 31–34.

82 Schunck, C., Johannes, T., Varga, D., Lörch, T., and Plesch, A. (2004). New developments in automated cytogenetic imaging: unattended scoring of dicentric chromosomes, micronuclei, single cell gel electrophoresis, and fluorescence signals. *Cytogenet. Genome Res.*, **104**, 383–389.

83 ICRP (International Commission on Radiological Protection) (2003) ICRP Publication 92: relative biological effectiveness (RBE), quality factor (QF), and radiation weighting factor (WR). A report of the International Commission on Radiological Protection. *Ann. ICRP*, **33**, 1–117.

84 Hagmar, L., Bonassi, S., Strömberg, U., Brøgger, A., Knudsen, L., Norppa, H., and Reuterwall, C. (1998) Chromosomal aberrations in lymphocytes predict human cancer: a report from the European Study Group on Cytogenetic Biomarkers and Health (ESCH). *Cancer Res.*, **58**, 4117–4121.

85 Boffetta, P., van der Hel, O., Norppa, H., Fabianova, E., Fucic, A., Gundy, S., Lazutka, J., Cebulska-Wasilewska, A., Puskailerova, D., Znaor, A., Kelecsenyi, Z., Kurtinaitis, J., Rachtan, J., Forni, A., Vermeulen, R., and Bonassi, S. (2007) Chromosomal aberrations and cancer risk: results of a cohort study from central Europe. *Am. J. Epidemiol.*, **165**, 36–43.

86 Bonassi, S. (1999) Combining environmental exposure and genetic effect measurements in health outcome assessment. *Mutat. Res.*, **428**, 177–185.

87 Bonassi, S., Hagmar, L., Strömberg, U., Montagud, A.H., Tinnerberg, H., Forni, A., Heikkila, P., Wanders, S., Wilhardt, P., Hansteen, I.L., Knudsen, L.E., and Norppa, H. (2000) Chromosomal aberrations in lymphocytes predict human cancer independently of exposure to carcinogens. *Cancer Res.*, **60**, 1619–1625.

88 Bonassi, S., Znaor, A., Ceppi, M., Lando, C., Chang, W.P., Holland, N., Kirsch-Volders, M., Zeiger, E., Ban, S., Barale, R., Bigatti, M.P., Bolognesi, C., Cebulska-Wasilewska, A., Fabianova, E., Fucic, A., Hagmar, L., Joksic, G., Martelli, A., Migliore, L., Mirkova, E., Scarfi, M.R., Zijno, A., Norppa, H., and Fenech, M. (2007) An increased micronucleus frequency in peripheral blood lymphocytes predicts the risk of cancer in humans. *Carcinogenesis*, **28**, 625–631.

89 Bonassi, S., Norppa, H., Ceppi, M., Strömberg, U., Vermeulen, R., Znaor, A., Cebulska-Wasilewska, A., Fabianova, E., Fucic, A., Gundy, S., Hansteen, I.L., Knudsen, L.E., Lazutka, J., Rossner, P., Sram, R.J., and Boffetta, P. (2008) Chromosomal aberration frequency in lymphocytes predicts the risk of cancer: results from a pooled cohort study of 22 358 subjects in 11 countries. *Carcinogenesis*, **29**, 1178–1183.

90 Norppa, H., Bonassi, S., Hansteen, I.L., Hagmar, L., Stromberg, U., Rossner, P., Boffetta, P., Lindholm, C., Gundy, S., Lazutka, J., Cebulska-Wasilewska, A., Fabianova, E., Sram, R.J., Knudsen, L.E., Barale, R., and Fucic, A. (2006) Chromosomal aberrations and SCEs as biomarkers of cancer risk. *Mutat. Res*, **600**, 37–45.

91 Rossi, A.M., Hansteen, I.L., Skjelbred, C.F., Ballardin, M., Maggini, V., Murgia, E., Tomei, A., Viarengo, P., Knudsen, L.E., Barale, R., Norppa, H., and Bonassi, S. (2009) Association between frequency of chromosomal aberrations and cancer risk is not influenced by genetic polymorphisms in GSTM1 and GSTT1. *Environ. Health Perspect.*, **117**, 203–208.

92 Rössner, P., Boffetta, P., Ceppi, M., Bonassi, S., Smerhovsky, Z., Landa, K., Juzova, D., and Sram, R.J. (2005) Chromosomal aberrations in lymphocytes of healthy subjects and risk of cancer. *Environ. Health Perspect.*, **113**, 517–520.

93 Liou, S.H., Lung, J.C., Chen, Y.H., Yang, T., Hsieh, L.L., Chen, C.J. and Wu, T.N. (1999) Increased chromosome-type chromosome aberration frequencies as biomarkers of cancer risk in a blackfoot endemic area. *Cancer Res.*, **59**, 1481–1484.

94 Cucinotta, F.A. and Durante, M. (2006) Cancer risk from exposure to galactic cosmic rays: implications for space exploration by human beings. *Lancet Oncol.*, **7**, 431–435.

95 Durante, M. (2002) Biological effects of cosmic radiation in low-earth orbit. *Int. J. Mod. Phys.*, **17**, 1713–1721.

96 Durante, M. (2005) Biomarkers of space radiation risk. *Radiat. Res.*, **164**, 467–473.

97 Durante, M. and Cucinotta, F.A. (2008) Heavy ion carcinogenesis and human space exploration. *Nat. Rev. Cancer*, **8**, 465–472.

98 Durante, M., Bonassi, S., George, K., and Cucinotta, FA. (2001) Risk estimation based on chromosomal aberrations induced by radiation. *Radiat. Res.*, **156**, 662–667.

99 Obe, G., Facius, R., Reitz, G., Johannes, I., and Johannes, C. (1999). Manned mission to Mars and chromosome damage. *Int. J. Radiat. Biol.*, **75**, 429–433.

100 NCRP (National Council on Radiation Protection and Measurements) (2000) Radiation Protection Guidance for Activities in Low-Earth Orbit. Report No. 132. NCRP, Bethesda, MD.

101 Iarmarcovai, G., Bonassi, S., Botta, A., Baan, R.A., and Orsière, T. (2008) Genetic polymorphisms and micronucleus formation: a review of the literature. *Mutat. Res.*, **658**, 215–233.

10
Cytogenetic Studies in Mammalian Somatic Cells Exposed to Radio Frequency Radiation: A Meta-Analysis

Vijayalaxmi and Thomas J. Prihoda

10.1
Introduction

Genotoxicity investigations are important since excess DNA damage in somatic cells can lead to the development of cancer and/or cell death. Hence, during the past several decades researchers have investigated the extent of genetic damage in mammalian somatic cells exposed *in vitro* and/or *in vivo* to radio frequency radiation (RFR, 300 MHz to 300 GHz range) using different techniques/experimental protocols, but the data reported were controversial [1]. Meta-analysis is widely used in biomedical research, especially when the outcomes in different investigations are controversial. This chapter describes a meta-analysis of the data reported for two genetic damage endpoints, namely, chromosomal aberrations (CA) and micronuclei (MN), in peer-reviewed scientific journals during the years 1990–2005. Similar meta-analysis data for other genotoxicity endpoints have been published earlier [2]. The objectives were to obtain (a) a good overall "quantitative" estimate of the damage reported in RFR-exposed compared to that in control cells that were sham exposed and/or unexposed, (b) the correlation between certain specific RFR exposure characteristics and increased genotoxicity that is larger than the random variability, and (c) whether the damage indices in RFR-exposed cells were within the spontaneous levels reported in historical database.

10.2
Materials and Methods

The methods employed for the meta-analysis were based on the recommendations made in several standard textbooks [3–5]. All publications in peer-reviewed scientific journals from 1990 to 2005 were assembled using a combination of key/search words in Medline, PubMed, and Ovid databases. The compiled list contained 63 publications [2]. Each publication was examined to record the detailed information in an Excel spreadsheet (Microsoft, WA); only CA and MN are discussed in this chapter. When the investigators have examined different numbers of cells in the same or

Cancer Risk Evaluation: Methods and Trends,
Edited by Günter Obe, Burkhard Jandrig, Gary E. Marchant, Holger Schütz, and Peter M. Wiedemann.

different experiments (e.g., MN in 500, 1000, 2000, or 4000 cells), a standardized "unit" was obtained and used as a more homogeneous measure. The standardized unit for CA was in 100 cells and MN was in 1000 cells. The "units" in RFR-exposed cells were integrated to obtain overall mean and standard deviation (SD) to designate to the "RFR-exposed group" while similar data on controls were assigned to the "control group." These are the "descriptive" data for standardized units from which the meta-analysis was conducted.

10.2.1 Meta-Analysis

Vijayalaxmi and Obe [1] identified several variables in the experimental protocols used by different investigators in different countries. Only three specific variables related to RFR exposure characteristics, namely, (i) frequency, (ii) specific absorption rate (SAR), and (iii) exposure as continuous wave (CW), pulsed wave (PW), and occupationally exposed/mobile phone users (CP), were considered in the meta-analysis to determine their "potential" influence on CA and MN. Each characteristic was further classified into subgroups. RFR frequency as (a) all frequencies, (b) $\leq$2000 MHz, and (c) >2000 MHz. SAR was subclassified as (a) information not reported (NR), (b) all together, (c) $\leq$2 W/kg, (d) $\leq$5 W/kg, and (e) >5 W/kg. The exposure was subclassified into (a) all together, (b) CW, (c) PW, and (d) CP. The Statistical Analysis System, Version 9.1 for Windows (SAS 2006), was used for all analyses described below.

10.2.2 Magnitude of Difference between RFR-Exposed and Controls (*E–C*)

Fixed effects model that assume a single "fixed" effect was used to calculate pooled estimates of *E–C* [3]. This is a conservative approach that provides very "narrow" confidence intervals (CIs), more likely to find significant differences between RFR-exposed and control groups. The variability in sample size and the results obtained in different experiments were taken into consideration to obtain a "weight." Separate statistical analyses were performed for CA and MN because of the differences in their standardized units. The pooled weighted standard error (SE) was obtained from all publications and it was then used to compute the 95% CI to obtain a quantitative estimate of *E–C*. The method is described in detail earlier [2].

10.2.3 Effect Size or Standardized Mean Difference

Another method regularly used in the meta-analysis is to determine "unitless" measure called "effect size" (ES) or "standardized mean difference" (ES) between RF radiation-exposed and control groups in each publication. Random effects models suggested by Hunter and Schmidt [4] were used to calculate the ES. These models are more accurate than the traditional random effects models and have several advantages and are recommended by the National Research Council [6]. Thus, the data reported in each publication were considered an independent random sample with some degree of variability. The method also corrected "bias" in the estimated ES and the provided "weights" for the data in each publication [2].

10.2.4
Multiple Regression Analysis

Since the meta-analysis considered the influence of several subgroups in RFR exposure characteristics on CA and MN, the percent contributions of each subgroup for the outcomes in *E–C* were examined using the standard output of weighted multiple regression analysis with adjustments as described in Ref. [7]. Seven predictor variables in RFR exposure characteristics (RFR frequency subgroups ($\leq$2000 and >2000 MHz) as one predictor variable; SAR subgroups (all unreported SARs, $\leq$2 W/kg, 2–5 W/kg, and >5 W/kg) as three predictor variables; CW/PW/CP subgroups as two predictor variables) adjusted for each other provided "weighted" regression coefficients and sums of squares for *E–C* [3]. The weighted regression coefficients and sums of squares for each predictor variable, for residual variability, and for total variability in the regression were obtained from SAS software [8]. The SE of these regression coefficients from SAS was adjusted [7] for the weighted meta-analysis of subgroup effects. The percent variance due to the predictor variables was calculated from each of the weighted sums of squares as percentage of their total. The percent contribution of each subgroup to the effect/outcome observed on *E–C* and ES on each genotoxicity endpoint was estimated.

10.2.5
Heterogeneity

The degree of homogeneity or heterogeneity in the results reported in the publications was examined in the weighted multiple regression analysis [7] using the random error for testing heterogeneity of effects to verify the validity of the models used for both *E–C* and ES. The residual weighted sums of squares were used in the chi-square "goodness-of-fit" test (heterogeneity in *E–C* obtained for CA and MN) with appropriate degrees of freedom [7]. When the test gave significant results, the data indicated heterogeneity, that is, factors that were not considered in this meta-analysis had an influence on the differences between RFR-exposed and control conditions. Such data were further examined to explain which subgroup RFR exposure characteristic contributed to the heterogeneity.

10.2.6
Historical Database

To provide a proper perspective in the evaluation of potential "adverse" effects of RFR exposure, the genotoxicity indices reported in RFR research investigations were compared with the "spontaneous" indices in normal cells published in a large historical database. A simple descriptive meta-analysis was performed by pooling the spontaneous incidence of CA, MN, or SCE reported in normal cells in several studies [9–23] and weighted by the sample size and variance. The spontaneous indices obtained for CA and MN were compared with those in RFR-exposed and control groups in the meta-analysis database.

10.3 Results

A chronological list of publications included in the meta-analysis is presented in Table 10.1 (see the details of each publication in Ref. [2]). Although several genotoxicity endpoints were investigated, only the meta-analysis results for CA and MN are presented here.

The *E–C* and ES data obtained for CA and MN (total sample size, mean, SE, and 95% CI) are presented in Table 10.2. The multiple regression analysis data for *E–C* and percent contribution of each RFR exposure characteristic to CA and MN are presented in Table 10.3. The heterogeneity test results are presented in Table 10.4.

The weighted mean *E–C* and 95% CI obtained were very small for CA (ranging from 0.10 to 0.92, that is, <1 aberration in 100 cells) and MN (ranging from 0.06 to 6.13, that is, <6.5 MN/1000 cells) and not significantly different between RFR-

Table 10.1 List of publications included in the meta-analysis in chronological order.

No.	First author, year	No.	First author, year	No.	First author, year
1	Garaj-Vrhovac, 1990a	22	Vijayalaxmi, 1997a	43	Trosic, 2002
2	Garaj-Vrhovac, 1990b	23	Vijayalaxmi, 1997b	44	Gadhia, 2003
3	Kerbacher, 1990	24	Malyapa, 1998	45	Koyama, 2003
4	Ciaravino, 1991	25	Phillips, 1998	46	McNamee, 2003
5	Garaj-Vrhovac, 1991	26	Garaj-Vrhovac, 1999	47	Mashevich, 2003
6	Garson, 1991	27	Maes, 2000	48	Vijayalaxmi, 2003a
7	Fucic, 1992	28	Vijayalaxmi, 2000	49	Vijayalaxmi, 2003b
8	Garaj-Vrhovac, 1992	29	Zotti-Martelli, 2000	50	Zeni, 2003
9	Garaj-Vrhovac, 1993	30	Lalic, 2001	51	Hook, 2004
10	Maes, 1993	31	Li, 2001	52	Koyama, 2004
11	Sarkar, 1994	32	Maes, 2001	53	Lagroye, 2004a
12	d'Ambrosio, 1995	33	Sykes, 2001	54	Lagroye, 2004b
13	Lai, 1995	34	Vijayalaxmi, 2001a	55	Trosic, 2004
14	Maes, 1995	35	Vijayalaxmi, 2001b	56	Baohong, 2005
15	Lai, 1996	36	Vijayalaxmi, 2001c	57	Diem, 2005
16	Maes, 1996	37	d'Ambrosio, 2002	58	Gandhi, 2005a
17	Antonopoulos, 1997	38	Bisht, 2002	59	Gandhi, 2005b
18	Lai, 1997	39	McNamee, 2002a	60	Gorlitz, 2005
19	Maes, 1997	40	McNamee, 2002b	61	Komatsubara, 2005
20	Malyapa, 1997a	41	Mei-Bian, 2002	62	Zeni, 2005
21	Malyapa, 1997b	42	Tice, 2002	63	Zotti-Martelli, 2005

For details of each publication, see Ref. [2].

Table 10.2 Meta-analysis of magnitude of difference (*E–C*) and effect size between RFR-exposed and control groups based on sample size and variance for chromosomal aberrations in 100 cells and micronuclei in 1000 cells.

RFR characteristics	Endpoint	*E–C*					ES		
		Total *N*	Mean	SE	CI (95%)	*p*-value	ES	SE	CI (95%)
All frequencies	CA	348	0.32	0.4	−0.4 to 1.1	0.198	0.3	0.1	0.1–0.6
≤2000 MHz	CA	136	0.15	0.4	−0.7 to 1.0	0.363	0.1	0.2	−0.2 to 0.5
>2000 MHz	CA	212	0.48	0.5	−0.6 to 1.5	0.186	0.4	0.2	0.1–0.8
All SARs	CA	302	0.28	0.3	−0.4 to 0.9	0.201	0.2	0.1	−0.1 to 0.5
NR-SARs[a)]	CA	164	0.92	0.6	−0.2 to 2.1	0.057	0.2	0.2	−0.1 to 0.6
≤2 W/kg	CA	40	0.16	0.2	−0.3 to 0.6	0.233	0.3	0.3	−0.4 to 0.5
≤5 W/kg	CA	58	0.11	0.2	−0.2 to 0.4	0.255	0.3	0.3	−0.4 to 0.9
>5 W/kg	CA	80	0.27	0.2	−0.1 to 0.7	0.099	0.1	0.3	−0.4 to 0.6
All CW-PW	CA	348	0.32	0.4	−0.4 to 1.1	0.198	0.3	0.1	0.1–0.6
CW	CA	62	0.54	1.4	−2.2 to 3.3	0.350	0.9	0.3	0.3–1.5
PW	CA	122	0.10	0.2	−0.3 to 0.3	0.486	0.1	0.2	−0.3 to 0.5
CP	CA	164	0.92	0.6	−0.2 to 2.1	0.057	0.2	0.2	−0.1 to 0.6
All frequencies	MN	3993	0.15	0.2	−0.1 to 0.4	0.155	0.2	0.0	0.1–0.2
≤2000 MHz	MN	3298	0.26	0.1	0.1–0.5	0.003	0.1	0.0	0.0–0.2
>2000 MHz	MN	695	0.11	0.7	−1.3 to 1.5	0.438	0.6	0.1	0.4–0.8
All SAR	MN	3981	0.15	0.2	−0.1 to 0.4	0.154	0.2	0.0	0.1–0.2
NR-SARs[a)]	MN	389	0.06	1.4	−2.6 to 2.7	0.482	0.9	0.1	0.6–1.2
≤2 W/kg	MN	1670	0.36	0.1	0.2–0.5	0.000	0.2	0.1	0.1–0.3
≤5 W/kg	MN	2384	0.24	0.1	0.0–0.5	0.014	0.1	0.0	0.0–0.2
>5 W/kg	MN	1208	0.46	0.2	0.2–0.8	0.001	0.0	0.1	−0.1 to 0.2
All CW-PW	MN	3993	0.15	0.2	−0.1 to 0.4	0.155	0.2	0.0	0.1–0.2

(*Continued*)

Table 10.2 *(Continued)*

RFR characteristics	Endpoint	*E–C*					ES		
		Total *N*	Mean	SE	CI (95%)	*p*-value	ES	SE	CI (95%)
CW	MN	1062	0.09	0.2	−0.4 to 0.5	0.356	0.5	0.1	0.4–0.6
PW	MN	2770	0.22	0.1	0.1–0.4	0.004	−0.0	0.0	−0.1 to 0.1
CP	MN	161	6.13	2.8	0.6–11.7	0.015	2.1	0.3	1.6–2.6

N: sample size; SE: standard error; CI: 95% confidence interval.

a) NR-SARs: specific absorption rates were not reported in some publications.Incidence of CA in the historical database: mean 1.5/100 cells (SD = 3.7; n = 15 594). Incidence of MN in the historical database: mean 9.0/1000 cells (SD = 8.0; n = 8667).

Table 10.3 Multiple regression analysis of the effects of RFR exposure characteristics on the magnitude of difference between RFR-exposed and control groups (*E–C*, based on sample size and variance) and effect size (ES) observed for chromosomal aberrations (CA) and micronuclei (MN).

		Number of effects examined	Percentage contribution due to			
			Frequency (MHz)	SAR (W/kg)	CW/PW/CP	Regression "goodness-of-fit" test[a]
CA	*E–C*	17	0.3	41.3[e]	14[h]	44.4[b]
CA	ES		5.4	14.2	15.2	65.3
MN	*E–C*	174	7.3[c]	1.7[f],[g]	27.2[i],[j]	63.8[b]
MN	ES		4.2[d]	0.6	23.9[k]	71.2

a) Contribution due to factors other than the RFR characteristics.
b) $p \le 0.05$ (heterogeneity in the predictors' effects observed in RFR exposure characteristics). Multiple regression coefficients for significant effects ($p < 0.05$).

Frequency (MHz) effects:

c) Change in effect due to <2000 MHz RFR frequency was smaller than >2000 MHz (-2.67 ± 0.16).
d) Change in effect due to <2000 MHz RFR frequency was smaller than >2000 MHz (-0.56 ± 0.18).

Specific absorption rates – W/kg effects:

e) Change in effect due to nonreported SAR was greater than <2 W/kg SAR (3.17 ± 0.77).
f) Change in effect due to >5 W/kg was greater than <2 W/kg SAR (0.30 ± 0.06).
g) Change in effect due to nonreported SAR was greater than 5 W/kg SAR (31.03 ± 8.63).

CW/PW/CP effects:

h) Change in effect due to CW was greater than PW (0.49 ± 0.17).
i) Change in effect due to CP was greater than PW (6.36 ± 0.23).
j) Change in effect due to CW was smaller than PW (-2.35 ± 0.17).
k) Change in effect due to CP was greater than PW (3.48 ± 0.47).

exposed and control groups for CA ($p > 0.05$). However, for MN, there were significant differences between the two groups at ≤2000 MHz frequency, several SARs, and PW/CP exposures ($p < 0.05$). The weighted mean ES and 95% CI obtained for CA and MN were very small and ranged from 0.0 to 0.9. The only exception was for MN with maximum mean of 2.1. The pattern of larger/smaller ES values was similar to the corresponding large/small *E–C* values (Table 10.2).

10.3.1 Multiple Regression Analysis and Heterogeneity

The overall percent contribution to the variability observed in *E–C* and ES for all endpoints due to frequency, SAR, and CW/PW/CP was of smaller magnitude compared to that obtained for goodness-of-fit tests. Nonetheless, some of them were significant ($p < 0.05$) and are explained in detail by the regression coefficients given in the footnote (Table 10.3). For CA, the variability in *E–C* due to frequency (0.3%) was not significant ($p > 0.05$), while that due to SAR (41.3%) and CW/PW/CP

Table 10.4 Heterogeneity in RFR exposure characteristics on the effects observed in chromosomal aberrations and micronuclei.

		Number of *E–C* effects examined	Sample size (*N*)	Controls (*C*)	RFR-exposed (*E*)
CA/100 cells	Mean		174	1.5	2.8
	Standard deviation			2.6	4.2
	Upper limit[a]			5.3	
	E–C range[b]	17		−0.5 to 7.7 (3 of 17 = 17.6%)	
	Complete dosimetry[c]			−0.5 to 0.8 (0 of 12 = 0%)	
	Incomplete dosimetry[c]			2.05 to 7.68 (3 of 5 = 60%)	
MN/1000 cells	Mean		1940	5.0	5.7
	Standard deviation			2.4	6.2
	Upper limit[a]			4.7	
	E–C range[b]	174		−9.2 to 31.4 (10 of 174 = 5.8%)	
	Complete dosimetry[c]			−9.2 to 9.3 (2 of 161 = 1.2%)	
	Incomplete dosimetry[c]			−5.0 to 31.4 (8 of 13 = 61.5%)	

N: number of RF radiation exposed – control (*E–C*) values examined in the multiple regression analysis.
The percentages in parenthesis are the number of studies with differences (*E–C*) greater than the upper limit values.
a) Upper limit is 2 × standard deviation above control mean, that is, 97.5 percentile.
b) RFR-exposed – control (*E–C*) range is the minimum and maximum for all *E–C* values used in the multiple regression.
c) The description of dosimetry, that is, complete or incomplete, was considered an example of high or poor quality of publication, respectively.

(14.0%) was significant ($p < 0.05$). However, the variability in ES due to RFR frequency (5.4%), SAR (14.2%), and CW/PW/CP (15.2%) was not significant ($p > 0.05$). For MN, the variability in *E–C* and ES due to RFR frequency (7.3 and 4.2%), SAR (1.7 and 0.6%), and CW/PW/CP (27.2 and 23.9%) was all significant ($p < 0.05$) (except the 0.6% ES due to SAR).

Random error accounted for a maximum variability observed in *E–C* and ES values for each endpoint (goodness-of-fit data, last column in Table 10.4). When the data were significant ($p < 0.05$), the indication was that factors other than the three RFR characteristics might explain the variability in all endpoints. Further details for the goodness-of-fit tests for *E–C* values are presented in Table 10.4. Although almost all the effects of *E–C* for CA and MN were within normal range of the controls, there was some heterogeneity of effects for these regression models; 3 out of 17 effects in CA (17.6%) and 10 out of 174 effects in MN (5.8%). The details entered in the Excel spreadsheet revealed the fact that all CA and most of the MN publications did not provide complete information on dosimetry and this might explain the larger *E–C* effects.

10.3.2 Comparison of Meta-Analysis Data with Those in Historical Database for CA, MN, and SCE

Historically, the spontaneous indices reported in the peripheral blood lymphocytes from normal individuals [9–23] were as follows: CA – 1.5/100 cells (standard deviation 3.7; $n = 15\,594$); MN – 9.0/1000 cells (standard deviation 8.0; $n = 8667$). The maximum indices obtained in RFR-exposed and control groups in the meta-analysis were similar to the above indices in the historical database.

10.4 Cytogenetic Endpoints as Biomarkers for Cancer Risk

Several researchers have conducted a systematic analysis of the spontaneous incidence of CA as a biomarker to predict carcinogenic risk in humans and reported that increased incidence of CA is the most reliable biomarker to predict increased cancer risk in humans [24–28] and that the aberration frequencies were increased even prior to the clinical manifestation of disease. Similarly, preliminary evidence has been presented that an increased incidence of MN predicts enhanced risk of cancer in humans [29].

10.5 Perspective from Meta-Analysis and Conclusions

The importance of conducting cytogenetic investigations comes from the fact that most genotoxic agents are also carcinogens. Organizations such as the ICNIRP [30] and IEEE [31] have recommended safety guidelines (based on 4 W/kg SAR that is the threshold for heat generation) to protect the general public and personnel who are occupationally exposed to RFR.

Overall, investigations that were conducted under the recommended safety guidelines indicated no significant differences in CA and MN between RFR-exposed and control groups, and the mean differences (*E*–*C*) as well as the effect size due to RFR exposure were small. Also, the mean indices for CA and MN in RFR-exposed and sham-/unexposed controls were within the spontaneous levels reported in the historical database. Since no single genotoxicity endpoint, by itself, is capable of determining the genotoxic potential and the consequent cancer risk from occupational and environmental agents [27], it is relevant to include more than one genotoxicity endpoint for DNA damage assessment in future RFR research investigations.

References

1 Vijayalaxmi and Obe, G. (2004) Controversial cytogenetic observations in mammalian somatic cells exposed to radiofrequency radiation. *Radiat. Res.*, **162**, 481–496.

2 Vijayalaxmi and Prihoda, T.J. (2008) Genetic damage in mammalian somatic cells exposed to radiofrequency radiation: a meta-analysis of the data from 63 publications (1990–2005). *Radiat. Res.*, **169**, 561–574.

3 Lipsey, M.W. and Wilson, D.B. (2001) *Practical Meta-Analysis*, Sage Publications, London.

4 Hunter, J.E. and Schmidt, F.L. (2004) *Methods of Meta-Analysis: Correcting Error and Bias in Research Findings*, 2nd edn, Sage Publications.

5 Lang, T.A. and Secic, M. (2006) Synthesizing the results of related studies: reporting systematic reviews and meta-analyses, in *How to Report Statistics in Medicine: Annotated Guidelines for Authors, Editors, and Reviewers*, 2nd edn., American College of Physicians, Philadelphia, PA, Chapter 17, pp. 255–279.

6 National Research Council (1992) *Combining Information: Statistical Issues and Opportunities for Research*, National Academy of Science Press, Washington, DC.

7 Hedges, L.V. and Olkin, I. (1985) *Statistical Methods for Meta-Analysis*, Academic Press, New York.

8 SAS (2006) Statistical Analysis System Version 9.1 for Windows, SAS Institute, Cary, NC.

9 Lloyd, D.C., Purrott, R.J., and Reeder, E.J. (1980) The incidence of unstable chromosome aberrations in peripheral blood lymphocytes from unirradiated and occupationally exposed people. *Mutat. Res.*, **72**, 523–532.

10 Vijayalaxmi and Evans, H.J. (1982) In *vivo* and *in vitro* effects of cigarette smoke on chromosomal damage and sister-chromatid exchange in human peripheral blood lymphocytes. *Mutat. Res.*, **92**, 321–332.

11 Fenech, M. and Morley, A.A. (1985) The effect of donor age on spontaneous and induced micronuclei. *Mutat. Res.*, **148**, 99–105.

12 Obe, G. (1986) Spontaneous levels of somatic chromosome aberrations in man, in *Monitoring of Occupational Genotoxicants* (eds M. Sorsa and H. Norppa), Alan R. Liss, New York.

13 Bender M.A., Preston, R.J., Leonard, R.C., Pyatt, B.E., Gooch, P.C., and Shelby, M.D. (1988) Chromosomal aberrations and sister chromatid exchange frequencies in peripheral blood lymphocytes of a large human population sample. *Mutat. Res.*, **204**, 421–433.

14 Bender, M.A., Preston, R.J., Leonard, R.C., Pyatt, B.E., and Gooch, P.C. (1989) Chromosomal aberration and sister

chromatid exchange frequencies in peripheral blood lymphocytes of a large human population sample. II. Extension of age range. *Mutat. Res.*, **212**, 149–154.

15 Bender, M.A., Preston, R.J., Leonard, R.C., Pyatt, B.E., and Gooch, P.C. (1992) On the distribution of spontaneous SCE in human peripheral blood lymphocytes. *Mutat. Res.*, **281**, 227–232.

16 Bonassi, S., Bolognesi, C., Abbondandolo, A., Barale, R., Bigatti, P., Camurri, L., Dalpra, L., De Ferrari, M., Forni, A., and Puntoni, R. (1995) Influence of sex on cytogenetic end points: evidence from a large human sample and review of the literature. *Cancer Epidemiol. Biomarkers Prev.*, **4**, 671–679.

17 Bolognesi, C., Abbondandolo, A., Barale, R., Casalone, R., Dalpra, L., De Ferrari, M., Degrassi, F., Forni, A., Lamberti, L., and Bonassi, S. (1997) Age-related increase of baseline frequencies of sister chromatid exchanges, chromosome aberrations, and micronuclei in human lymphocytes. *Cancer Epidemiol. Biomarkers Prev.*, **6**, 249–256.

18 Stephan, G. and Pressl, S. (1999) Chromosomal aberrations in peripheral lymphocytes from healthy subjects as detected in first cell division. *Mutat. Res.*, **446**, 231–237.

19 Bonassi, S., Fenech, M., Lando, C., Lin, Y., Ceppi, M., Chang, W.P., Holland, N., Kirsch-Volders, M., Zeiger, E., and Zijno, A. (2001) Human micronucleus project: international database comparison for results with cytokinesis-block micronucleus assay in human lymphocytes. I. Effect of laboratory protocol, scoring criteria, and host factors on the frequency of micronuclei. *Environ. Mol. Mutagen.*, **37**, 31–45.

20 Fenech, M., Bonassi, S., Turner, J., Lando, C., Ceppi, M., Chang, W.P., Holland, N., Kirsch-Volders, M., Zeiger, E., and Zijno, A. (2003) Intra- and inter-laboratory variation in the scoring of micronuclei and nucleoplasmic bridges in binucleate human lymphocytes. Results of an international slide-scoring exercise by the HUMN project. *Mutat. Res.*, **534**, 45–64.

21 Hagmar, L., Stromberg, U., Bonassi, S., Hansteen, L., Knudsen, L.E., Lindholm, C., and Norppa, H. (2004) Impact of types of lymphocyte chromosomal aberrations on human cancer risk: results from Nordic and Italian cohorts. *Cancer Res.*, **64**, 2258–2263.

22 Neri, M., Ceppi, M., Knudsen, L.E., Merlo, D.F., Barale, R., Puntoni, R., and Bonassi, S. (2005) Baseline micronuclei frequency in children: estimates from meta- and pooled analyses. *Environ. Health Perspect.*, **113**, 1226–1229.

23 Rossner, P., Boffetta, P., Ceppi, M., Bonassi, S., Smerhovsky, Z., Landa, K., Juzova, D., and Sram, R.J. (2005) Chromosomal aberrations in lymphocytes of healthy subjects and risk of cancer. *Environ. Health Perspect.*, **113**, 517–520.

24 Sasaki, Y.F., Sekihashi, K., Izumiyama, F., Nishidate, E., Saga, A., Ishida, K., and Tsuda, S. (2000) The comet assay with multiple mouse organs: comparison of comet assay results and carcinogenicity with 208 chemicals selected from the IRRC monographs and U.S. NTP carcinogenicity data base. *Crit. Rev. Toxicol.*, **30**, 629–799.

25 Hagmar, L., Bonassi, S., Stromberg, U., Brogger, A., Knudsen, L., Norppa, H., and Reuterwall, C. (1998) Chromosomal aberrations in lymphocytes predict human cancer: a report from the European Study Group on Cytogenetic Biomarkers and Health (ESCH). *Cancer Res.*, **58**, 4117–4121.

26 Tucker, J.D., Eastmond, D., and Littlefield, G.L. (1997) *Cytogenetic End-Points as Biological Dosimeters and Predictors of Risk in Epidemiological Studies*, IARC, Lyon, pp. 185–200.

27 Bonassi, S., Ugolini, D., Kirsch-Volders, M., Stromberg, U., Vermeulen, R., and Tucker, J.D. (2005) Human population studies with cytogenetic biomarkers: review of the literature and future prospective. *Environ. Mol. Mutagen.*, **45**, 258–271.

28 Norppa, H., Bonassi, S., Hansteen, I., Hagmar, L., Stromberg, U., Rossner, P., Boffetta, P., Lindholm, C., Gundy, S., and Fucic, A. (2006) Chromosomal aberrations and SCEs as biomarkers of cancer risk. *Mutat. Res.*, **600**, 37–45.

29 Bonassi, S., Znaor, A., Ceppi, M., Lando, C., Chang, W.P., Holland, N., Kirsch-Volders, M., Zeiger, E. *et al.* (2007) An increased micronucleus frequency in peripheral blood lymphocytes predicts the risk of cancer in humans. *Carcinogenesis*, **28**, 625–631.

30 ICNIRP (International Commission on Non-Ionizing Radiation Protection) (1998) Guidelines for limiting exposure to time-varying electric, magnetic, and electromagnetic fields (up to 300 GHz). *Health Phys.*, **74**, 494–522.

31 IEEE (Institute of Electric and Electronics Engineers) (1999) Standard for Safety Levels with Respect to Human Exposure to Radio Frequency Electromagnetic Fields, 3 kHz to 300 GHz. Institute of Electrical and Electronics Engineers, Piscataway, NJ, http://www.standards.ieee.org/, C95.1.

Part Five
Omics: A New Tool for Cancer Risk Assessment?

Cancer Risk Evaluation: Methods and Trends,
Edited by Günter Obe, Burkhard Jandrig, Gary E. Marchant, Holger Schütz, and Peter M. Wiedemann.

11
Genomics and Cancer Risk Assessment

Michal R. Schweiger and Bernd Timmermann

11.1
Introduction

So far, monogenetic diseases are the gold standard for the identification of human disease-causing mutations, and most of the time experiments are done on single-gene levels with clear gene–phenotype correlations [1, 2]. In comparison to monogenetic diseases, complex genetic diseases such as cancer follow much more difficult paths, and with conventional technologies only a limited access to underlying (epi)-genetic changes can be generated. For these diseases, systems biological approaches are the *non plus ultra* and hope arises that with systematic analyses it will be possible to understand tumor pathology in greater detail. In addition to the understanding of mechanisms, whole-genome approaches allow the identification of significant tumor markers that will facilitate the diagnosis and treatment of cancer patients.

In this regard, systematic and global data sets covering DNA, RNA, proteins, and metabolites make it possible to characterize the biology of a disease process in individual patients to an extent by far surpassing the extent of our knowledge about human biology a few years ago. While it took 10 years and US$ 1 billion to complete the sequence of the human genome, we are now able to generate a similar amount of information with next-generation sequencing (NGS) technologies within days and predictions calculate that within a few years the genome of one patient will be generated in a few minutes at a cost of US$ 1000 or even less [3].

Similar, though maybe less dramatic, improvements have been made in many other technology areas, providing a basis for combining the information on pathways gained in decades of biomedical research worldwide with a genome (and proteome) scale analysis of the disease process in individual patients, to establish models able to predict, in detail, individual response of patients to different therapies.

In oncology, targeted chemotherapies have been under development and in clinical use only in recent years. These drugs aim at on inhibition or activation of specific cellular target proteins, most frequently protein kinases. Parallel to this development, the recognition emerged that only a subset of patients significantly benefit from those chemotherapies. The others are needlessly treated, suffering from side effects without benefiting from the treatment. The ability to identify in advance

Cancer Risk Evaluation: Methods and Trends,
Edited by Günter Obe, Burkhard Jandrig, Gary E. Marchant, Holger Schütz, and Peter M. Wiedemann.

the subpopulation of patients most likely to respond to a particular therapy is summarized in the concept of personalized cancer therapy. The fundamental principle underlying personalized therapy is the enormous heterogeneity present among tumors, even among tumors of the same class. Using newest high-throughput technologies, we have learned that a wide spectrum of different genetic alterations ranging from mutations to copy number and structural variations are found in each tumor. This complexity is aggravated even more by epigenetic variations found in many tumor entities.

To establish individual cancer models and to optimize the treatment for individual patients, many critical points have to be considered. The following chapter gives an overview of technologies and strategies that can contribute to this goal, starting from the recording of the medical history over the generation of high-throughput genetic data. The primary goal is twofold: generated data can be integrated into systems biological approaches, which then can be used for the analysis of underlying pathogenetic carcinogenic mechanisms and for the identification of biomarkers to facilitate diagnosis and treatment stratifications of patients.

11.2 Tissue Material

New system-wide technologies raise a broad range of clinically or molecular biologically relevant questions including the problem of tumor resistance to therapy, the different growth rates and progression of tumors, properties required for metastatic spread, or the tumor evolution from single-tumor stem cells. However, one of the key factors for systems biological approaches is to have access to tissue material of high quality that is combined with extensive clinical information. This poses an additional burden on clinical collaborators who are required to not only follow their routine diagnosis and treatment regimens but also take care of the extraction of tissue material and the right asservation technique, for example, storage in liquid nitrogen or formalin fixation. One way to avoid the shortage of clinical material is to take advantage of already established cancer cell lines. These cell lines bear the advantage that they are more or less a homogeneous resource mainly due to selection processes during cultivation. In addition, DNA can be generated in nearly unlimited amounts, which is especially advantageous for extensive validation experiments that are still required for NGS data [4, 5]. On the other side, these cell lines are simply models of tumors and might therefore harbor alterations that are acquired during propagation and maintenance of cell lines. Closer to the *in vivo* situation are primary tumor materials [6–8]. However, tumors are heterogeneous and consist of normal cells intermingled within tumor cells. Even micro- or macrodissections of tumors do not result in 100% tumor cell content and thus only percentage of tumor tissues can be approximated. Since all alignment and single nucleotide variation (SNV)-calling algorithms are based on clear homo- and heterozygous distributions of mutated and wild-type sequences, SNV detection in cancer tissues is more complicated than that in monogenetic diseases and a second method for validation is urgently required.

Next to the bioinformatics difficulties for data analyses access to high-quality material poses a bottleneck. Even though large biobanks have been established, they mainly store material as formalin fixed and paraffin embedded tissues. With this asservation technology, specialized protocols are required that permit the extraction and conversion of DNA from formalin-fixed and paraffin-embedded (FFPE) tissue samples [9]. FFPE preparations are incompatible with many downstream molecular biology techniques, because they suffer from the problem of cross-linked and fragmented DNA [10]. In contrast, NGS technologies are applicable to samples of even more than 20-year-old FFPE material without major modifications to the DNA extraction protocols [11]. The use of FFPE DNA material opens up access to a variety of population stratifiers for clinical trials and enables routine diagnostic workups of patients. Another bottleneck in cancer research is the limited amount of tissue material available. New technologies are in sight for single-strand sequencing approaches that then circumvent the need for amplification and preparation of large tissue fractions. However, until these technologies are available for routine use, DNA analysis can be performed both on a small number of cells obtained by laser capture microdissection and on DNA extracted from diverse body fluids such as blood, urine, or sputum using advanced sample NGS preparation technologies [12] (see also Chapter 12).

11.3 Analysis Technologies

11.3.1 DNA Microarrays

Global cellular changes can be measured using microarray technologies. In standard microarrays, nucleotide probes are immobilized on solid surfaces. Single-stranded DNA or cDNA samples of interest are distributed over the array and complementary fragments can hybridize (targets) to the probes. Unbound fractions are washed off. Probe–target hybridizations are usually detected and quantified using fluorophore-, silver-, or chemiluminescence-labeled targets. A wide variety of DNA microarrays offer the possibilities to measure changes in gene expression, to genotype single nucleotide polymorphisms (SNPs), to resequence DNA regions, or to determine the copy number of DNA segments [13]. Gene expression analysis in cells or tissues is used for the determination of specific cell stages, or overexpression and knockdown of genes is used as functional readout of the proteins. Another important area for gene expression arrays is the stratification of tumors in regard to distinct biological properties or individual prognoses (see Chapter 12). The expression levels of thousands of distinct genes are measured in tissue preparations. Subsequent bioinformatics correlation analyses make it possible to identify a subset of genes, called signature genes, which are characteristic for a specific tumor phenotype, for example, drug resistance, progression, or recurrence. Pioneering work has been carried out on breast cancers to distinguish those tumors that become metastatic from those that will remain locally restricted. Using gene expression arrays has made it possible to predict the clinical course with more than 90% accuracy. Since so far all women are treated with an aggressive chemotherapy irrespective of the tumor characteristics, the distinction between aggressive and less aggressive would spare many women unnecessary treatment (up to 85% of the tumors will not metastasise) [14]. Another field of application of arrays is comparative genome hybridization, called array CGH. This technique is commonly used to detect microdeletions and duplications in patients with congenital abnormalities. It is also used to detect copy number alterations in cancer samples, and comparison with next-generation sequencing data show, as expected, good overlaps between the two methods.

11.3.2
DNA Sequencing

DNA sequencing was first described by Maxam and Gilbert and Sanger *et al.* in 1977 [15, 16]. Since the establishment of the technology, the efficiency and cost effectiveness of sequencing have improved continuously. At each step, more ingenious DNA analyzers, enzymes, and analysis algorithm have provided better accuracy and higher throughput (see Figure 11.1).

Advancements in automated capillary electrophoresis led to the start of the Human Genome Project (HGP) that was launched in 1990 to decipher the entire human genome. In an international effort, over 1000 scientists sequenced the genome and announced its completion in 2001 [17, 18]. Diagnostic sequencing is still performed by semiautomated, capillary-based Sanger sequencing. The target region is usually amplified by PCR and then sequenced by cycle sequencing that involves several rounds of denaturation, primer annealing, and extension. This approach can achieve

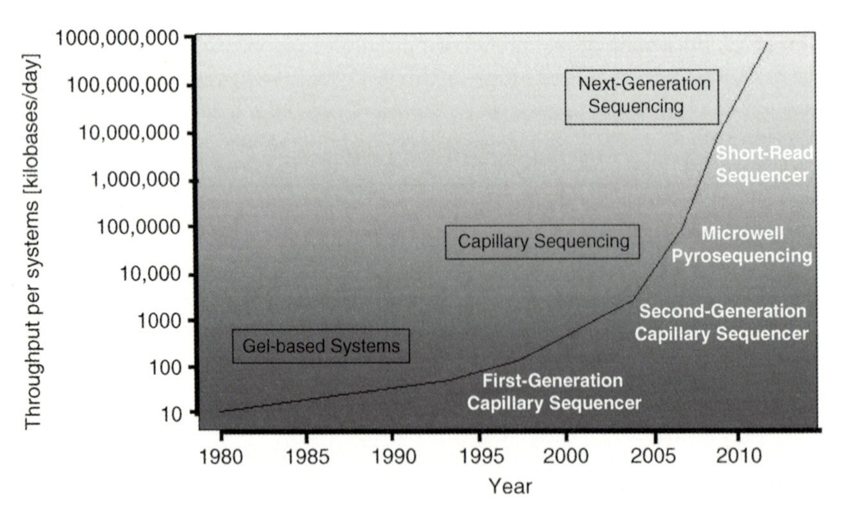

Figure 11.1 Development of sequencing throughput. Adapted from M.R. Stratton *et al. Nature*, 458, 719–724 (2009).

a read length of ~1 kb with a high accuracy. More recently, the first clinical genetics laboratories have acquired massively parallel sequencing systems to allow simultaneous screening for variations in a large number of targets and samples.

11.3.3
Next-Generation Sequencing

The continually growing demand of high-throughput technologies led to the development of next-generation sequencing technologies a few years ago [19, 20]. With these technologies sequencing of the whole human genome is feasible within a few days and offers exciting possibilities for a systematic genome analysis. The worldwide Personal Genome Project (PGP) and the 1000 Genomes Project aim at sequencing over 1000 individual genomes to gain insight into genomic variability. The development of NGS technologies has initiated a real revolution in genomics analyses. With these technologies, an enormous parallel analysis of genomic DNA has become possible within a few days. Key features of these technologies include the spatial immobilization of millions of short DNA fragments followed by a massively parallel sequencing process [21]. Progress with these technologies has mainly been achieved by analysis of germ line variations where the material (blood) is more or less homogeneous by nature and major challenges have evolved when first complete cancer genomes were analyzed. So far, a limited number of cancer genomes has been analyzed and the first one was a cytogenetically normal acute myeloid leukemia genome that has revealed eight validated somatic mutations [6]. Within a similar range with 32 nonsynonymous somatic mutations lies the profile of a sequenced breast tumor [8]. Finally, recently the complete genomes of lung cancer and melanoma cell lines have been published and indicate correlations between DNA repair mechanisms and mutational spectra [4, 5].

11.3.4
Current Technologies

The 454 Genome Sequencer (454 Life Sciences, Roche Applied Sciences) was the first commercially available next-generation sequencing system on the market [22]. This instrument carries out pyrosequencing reactions in parallel by loading hundreds of thousands of beads coated with homogeneous DNA fragments into individual wells of a reaction plate (Pico Titer Plate) [23, 24]. Beads are generated by the immobilization of fragmented DNA at a concentration that, on average, results in the addition of one fragment per bead. The fragments are then independently amplified in an oil emulsion mixture that creates separate microreactors for each bead (emPCR). The sequencing process uses an enzymatic cascade to generate light from inorganic phosphate molecules released by incorporation of nucleotides as the polymerase replicates the template DNA strand. As present, the 454 FLX system produces up to 660 Mb sequence information in 10 h with a median read length between 400 and 500 bp (Table 11.1). The raw accuracy per nucleotide sequenced on the FLX system (99.5%) is low (like all massively parallel sequencers) in relation to Sanger sequenc-

Table 11.1 Comparison of next-generation sequencing platforms.

Platform	Sequencing principle	Read length (bases)	Compartments	Run time (days)	Gb per run	Gb per day
Roche/454 GS FLX (titanium)	Pyrosequencing	400 (average)	2, 4, 8, 16	0.5	0.66	1.3
Roche/454 GS Junior (titanium)	Pyrosequencing	400 (average)	1	0.5	0.04	0.08
Illumina GA IIe	Sequencing by synthesis	25–150	8	1–14	38	2.7
Illumina GA IIx	Sequencing by synthesis	25–150	8	1–14	95	6.8
Illumina HiScanSQ	Sequencing by synthesis	25–100	8	8	50	6.25
Illumina HiSeq 2000	Sequencing by synthesis	25–100	16	1–9	200	22
SOLiD V4	Sequencing by ligation	35, 50, 75	16	4–11	100	9

Gb: gigabases.

ing. However, extremely high confidence base calls can be achieved via a high sequencing depth.

Illumina offers now four different sequencing systems. The Illumina HiSeq 2000 system (Illumina Inc., San Diego, CA) is the platform with the highest output and can produce 20 Gb sequence output per day with 100 bp in length (GAIIx system: 150 bp). The Illumina sequencer achieves parallelization by the so-called *bridge amplification* of DNA fragments immobilized onto the flow cell of the instrument at a concentration that promotes a dense array of nonoverlapping fragment colonies. Each fragment colony is then sequenced one base at a time by the cyclical addition of fluorescently labeled nucleotides that are conjugated with a reversible terminator [25] (see Figure 11.2).

The Applied Biosystems SOLiD System (now Life Technologies, Foster City, CA) sequences by multiple cycles of hybridization and ligation [26]. The sequencing reaction is initiated via the ligation of a universal anchoring primer to an 8-mer sequence derived from a population of fluorescently labeled 8-mers. After ligation of the first 8-mer, the emitted fluorescence is read followed by cleavage of the three downstream universal bases and another cycle of ligation. The ability of DNA ligase to discriminate between populations of fluorescently labeled oligonucleotides facilitates the sequencing reactions that generate approximately 9 Gb of sequence per day and read lengths up to 75 bp (Table 11.1). The main advantage of the SOLiD technology is that each base is interrogated twice, resulting in very accurate raw reads that require a lower amount of oversampling to reach a threshold value of confidence for base calling.

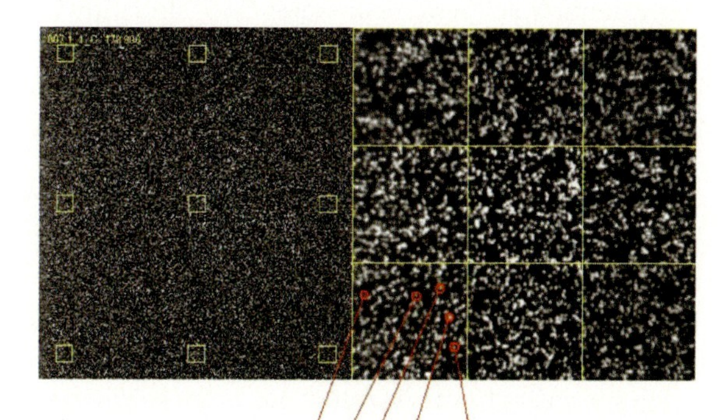

Figure 11.2 Sequencing by synthesis (Illumina). Each light signal represents the incorporation of fluorescently labeled nucleotides. Sequencing reads are generated through the cyclical addition of complementary nucleotides. In the case of a resequencing experiment, these reads are mapped to the known reference sequence. In this example, a single nucleotide change (G/C) is detected.

All these technologies provide digital information on DNA sequences of the fragments that are assembled and aligned to reference genomes using bioinformatics tools. Digital information is the basis for resequencing approaches and quantification modules for gene expression analyses or chromatin immunoprecipitation experiments. The parallel sequencing of millions of DNA molecules is especially useful for sequencing heterogeneous material, as is the case with cancer tissues. However, this is not only advantageous but also poses challenges to SNP calling algorithms because clear homo- and heterozygous SNP ratios are washed out by wild-type sequences from the normal cells [27, 28].

While next-generation sequencing technologies have increased the sequencing throughput enormously, several enrichment, amplification, and labeling steps still cause the performance to be relatively time consuming and expensive. Costs are, however, dropping rapidly for most second-generation sequencing systems (www.illumina.com, www.454.com, www.solid.appliedbiosystems.com), down to a few thousand dollars per genome announced by Complete Genomics (www.completegenomics.com), a recent start-up company developing an optimized automated second-generation sequencing approach to be run in service mode.

In comparison, future "third-generation sequencing" approaches, relying on detection of binding of nucleotide triphosphate to polymerase in real time (Pacific Biosciences, Visigen), nanopore (e.g., Oxford Nanopore) and scanning probe sequencing approaches are directed toward sequencing of single DNA molecules without any prior amplification or labeling [29–32]. PacBio and Visigen use optical techniques to detect the triphosphate about to be incorporated. In the procedure developed by Pacific Biosciences, multiple "zero mode waveguide" structures on a chip define minute volumes containing single polymerase molecules. DNA sequences are read out by a series of desoxynucleotide triphosphates that are labeled with different fluorescent dyes and illuminated during the incorporation step. During incorporation of the triphosphate analogue, the fluorescent label is cleaved off together with the pyrophosphate group, allowing the next incorporation step. In addition to determining the sequence, this procedure has been shown to be also able to detect base modifications in the DNA due to its influence on the kinetics of incorporation. The protocol under development at Visigen is conceptually quite similar. In this case, however, binding of the fluorescence-labeled desoxynucleotide triphosphate is detected by energy transfer from a quantum dot on the polymerase to the fluorescence label on the triphosphate group. The basic principle of nanopore sequencing is that a DNA strand or a cleaved nucleotide is passed through a nanopore and induces changes in the current applied [30]. The use of electrical currents for nucleotide identification promises to distinguish all four nucleotides and in addition identify 5-methyl cytosines (5mC). This would imply that during one sequencing process, all "five" nucleotides (A, T, C, G, and 5mC) could be distinguished and no additional manipulation of DNA would be required for the construction of methylation patterns. In scanning probe sequencing, another approach under development, the DNA molecule is immobilized and the scanning instrument records the nucleotides. A quite new player on the third-generation market is Ion Torrent. The Ion Torrent technology relies on the fact that when nucleotides are added to a growing

DNA strand, hydrogen ions are released. The platform immobilizes DNA strands in tiny wells within a semiconductor chip and then washes the four nucleotides one by one over the wells. As each base is incorporated, it releases hydrogen ions that can be detected as they pass through a pore at the base of each well. The downside is that the technology is susceptible to the same kind of homopolymer stretch problem that plagued the 454 technology. If there are several nucleotides of the same type in the DNA strand, all will be incorporated at once – and distinguishing the signal corresponding to six nucleotides (nts) versus the one produced by 7 nts will be very challenging. Ion Torrent showed data suggesting that it was possible to accurately read across homopolymers up to six bases long, but it is unclear how well it will do for longer stretches. The main advantage of this technology would be very low instrument price and running costs.

11.3.5
Targeted DNA Sequencing

Sequencing of entire genomes is an important application of NGS. Although all types of genetic polymorphisms can be identified using whole-genome resequencing approaches, this method is still too expensive to be conducted routinely. Instead, many research and diagnostic goals might be achieved by sequencing only a fraction of the genome.

For disease analysis, specific sets of genes involved in the pathomechanism or implicated by whole genome association studies are of major interest. PCR amplification using a high-fidelity thermostable polymerase has proven to be a reliable method for isolating genomic areas of interest in preparation for Sanger sequencing. However, given its limited multiplex capability, traditional PCR is an impractical method of genomic enrichment for next-generation sequencers.

Several targeted sequence enrichment techniques to reduce DNA sequence complexities have been established. In particular, microarray-based genomic selection (MGS), multiplex exon capture, or bead-based enrichment methods are already commercially available and used for targeted sequencing approaches [33–37]. Main differences are the amount of input DNA, the ease of performance, and if they are hybridization or synthesis based. In addition, as sequencing capacities per run continue to increase and are predicted to result in throughputs of up to 300 Gb per run by the end of the year (according to Illumina and SOLiD announcements), the ability to easily multiplex multiple samples will get increasingly important.

11.3.6
Copy Number Variations

Genome rearrangements resulting in aberrant transcriptional events are common features in human cancer. Thus, besides point mutations, extended genome rearrangements are implicated in tumorigenesis, including translocations, inversions, small insertions/deletions (indels), and copy number variations (CNVs) (see Chapter 2). Indels are most often defined as deletions or insertions below 1 kb (thousand base pairs) of DNA, whereas CNVs comprise alterations larger than 1 kb of DNA. Recent analyses using genome-wide approaches have uncovered the importance of structural genomic variations in health and disease. Furthermore, genetic association studies have implicated CNVs in cancer. The connection of changes in CNVs to several diseases has boosted the development of new technologies to investigate these rearrangements. First methods involve microscopic examination of chromosome banding, PCR, fluorescence *in situ* hybridization (FISH), and microarrays. More recently, NGS technologies have been utilized for these analyses. These approaches offer important advantages over conventional methods such as microarrays or array comparative genomic hybridization. In particular, in addition to quantitative information, they provide data about qualitative mechanisms, for example, balanced rearrangements such as reciprocal translocations and inversions that would have been otherwise overseen. Moreover, since the sequencing is based on digital modes, they are able to detect variants that are present in a subpopulation of cells. The same rearrangements can be found in minute amounts in blood samples and can be used to monitor the course of the disease [38]. Given the short read lengths of most NGS technologies, paired end sequencing approaches have been developed where short DNA segments, which are many hundreds of bp separate from each other, are brought together and sequenced. Besides the technology advances, new algorithms have been developed that identify "chimeric" reads. Chimeric reads are those fragments that incorporate DNA segments from different parts of the genome and need to be split before correct alignments can be determined.

11.3.7
DNA Methylation

The nucleotide sequence is the primary level of genetic information and the basic principle of genetic inheritance. Another level of complexity in genomic DNA arises from epigenetic variations in DNA segments that are also underlying the inheritance of phenotypes both from generation to generation and from cell to cell during cell division. In humans, cytosine methylation was the first mark discovered. It is required both for the regulation of gene expression and for the silencing of transposons and other repetitive sequences [39]. The chemical modification occurs predominantly via a covalent attachment of a methyl group to the C5 position of the cytosine ring (5mC) in CpG dinucleotides. The structure of cytosine is thereby altered without changing its base-pairing properties. Altered methylation patterns have been reported in a diverse array of complex human diseases such as cancer, in systemic autoimmune and psychiatric diseases, and in monogenic epigenetic diseases [40]. Most research has been performed on cancer epigenetics and has indicated that cytosine methylations are among the earliest events in tumorigenesis. Thus, the first biomarkers have been developed on the basis of these modifications [41].

For the detection of methylated nucleotides, the DNA is marked through a "bisulfite (BS) conversion" reaction or by methylation-specific restriction analyses. For "bisulfite conversion," genomic DNA is treated with sodium bisulfite under denaturing conditions. Cytosine residues are deaminated and converted to uracil leaving methylated cytosine moieties unaffected. During following amplification reactions, uracil is converted to thymine, and subsequent technologies rely on SNP detection methods distinguishing between cytosines and thymines resulting from bisulfite conversions. Common to all technologies are pitfalls leading to false positive base callings due to incomplete conversion reactions, degraded DNA caused by harsh conversion conditions, and methylation in pseudogenes. For human genomes, up to 1% or 28 million CpG sites are estimated. Using conventional PCR amplification and sequencing strategies in order to investigate all these means that these undertakings are extremely time consuming and expensive. Only with the aid of NGS technologies do whole human genome m5C patterns become feasible. The first human genome-wide epigenetic maps after bisulfite treatment have been constructed and show that more than 93% of all CpGs can be targeted [42].

However, the sequencing capacities and costs required for whole genome analyses are still relatively high. Thus, limited analyses of parts of the genome are more practical to gain insight into methylation patterns of mammals, especially if large numbers of samples need to be analyzed. Similar to genomic analyses, targeted enrichment and analysis technologies have been developed for epigenetic investigations that significantly reduce the sequencing costs. One of the major advantages of NGS technologies in epigenetic research is that analyses can be performed genome-wide without restriction to specific preselected regions as required for array-based technologies. The completeness of the data is especially advantageous in generating methylation profiles outside CpG islands and promoter regions, for example, in gene bodies where DNA methylation changes have recently been shown to occur [43, 44]. Taken together, the number of different NGS epigenetic technologies is large, and each has its own advantages and disadvantages. The selection of the right technology for the research question investigated is a key feature in making the most of the enormous power NGS has for basic and clinical directions of research.

11.3.8 Transcriptome Analyses

First candidate gene approaches used Northern blot techniques to analyze expression levels of single genes. Furthermore, the development of microarray techniques allowed the simultaneous examination of thousands of genes in one experiment. Due to their high power and relatively low costs, microarrays are now commonly used in laboratories. However, due to a preselection of examined genes, these methods do not allow detection of novel transcripts. As an alternative, NGS technologies are used for the analysis of transcriptomes [45, 46]. With this technology, not only gene expression profiles can be established but also alternative splicing and translocations can be detected. Using long reads from 454 or PacBio technologies, splice patterns can be directly detected, providing information that is otherwise difficult to reconstruct from short read data sets.

For the sequencing of RNA species, mRNA is reverse transcribed into cDNA and fragmented, and adapters are ligated before the sequencing is performed. Protocols for sequencing on all three commonly used NGS platforms are available. In addition to expression profiling, the data can be used for the detection of new transcripts, for the discovery of new splice variants, and, if the sequencing effort is large enough, for mutational profiling. Transcriptome sequencing has been performed on a wide variety of organisms: from *Arabidopsis thaliana* and *Drosophila melanogaster* to human cell lines and tissues. Most of the experiments for gene expression profiling are performed at the total cellular RNA level. Since these experiments are performed at an endpoint, temporal information is lost. However, for modeling the kinetics of cellular processes, especially this kind of information is required. These problems can be addressed by using nascent RNA that is identified by metabolic labeling of newly generated RNA. Elegant approaches have been developed that use incorporation of 4-thiouridine (4sU) into nascent RNA followed by a thiol-specific biotinylation and magnetic separation of nascent RNA [47]. Using this approach, kinetic pictures of RNA metabolism are emerging. In combination with NGS technologies and by combining several samples in one sequencing run, the throughput can be increased and detailed time curves can be obtained.

In addition to RNA analysis, small RNA molecules can also be easily enriched, detected, and analyzed using NGS technologies. MicroRNAs (miRNA) have recently emerged as crucial regulators of development and cell fate determination, which are essential elements in cancer progression (see also Chapter 2). In this field, NGS technologies are extremely powerful for the discovery of novel miRNA genes on a genome-wide scale. The combination of both sequencing approaches – transcriptome sequencing and miRNA sequencing – enables the prediction of miRNA–target RNA pairs adding a functional viewpoint to the discovery of new and already known miRNAs.

11.4
Outlook for Individualized Cancer Treatment

Taken together, recent improvements in high-throughput technologies make it possible to generate global data sets covering the genome, the transcriptome, the proteome, and the metabolome. The amount of data makes it possible for the first time to draw phenotype–genotype conclusions and to optimally stratify patients for the most appropriate cancer treatments. However, appropriate handling of these large data sets requires specific bioinformatics tools that primarily aim at directly identifying the correlation between experimental and clinical data. Next, the data are integrated into large computer models of cellular pathways. The adjustments of these cellular computer models to complex diseases (such as cancer) enable the establishment of cancer computer models that in turn can be used as predictive models for the calculation of optimal chemotherapeutic combinations. More and more the picture evolves that not only alterations in single genes but also modifications in pathways are essential for tumor pathogenesis. These computer models can facilitate the global view of tumor processes and in the future will be used to calculate both the influence of single genes on complete pathways and the influence of targeted chemotherapies on the overall cellular behavior. With this advancement from personal genomics to individualized cancer treatment, we have approached the next important step in cancer risk assessment.

References

1 Krawitz, P.M., Schweiger, M.R., Rodelsperger, C., Marcelis, C., Kolsch, U., Meisel, C., Stephani, F., Kinoshita, T., Murakami, Y., Bauer, S., Isau, M., Fischer, A., Dahl, A., Kerick, M., Hecht, J., Kohler, S., Jager, M., Grunhagen, J., de Condor, B.J., Doelken, S., Brunner, H.G., Meinecke, P., Passarge, E., Thompson, M.D., Cole, D.E., Horn, D., Roscioli, T., Mundlos, S., and Robinson, P.N. (2010) Identity-by-descent filtering of exome sequence data identifies PIGV mutations in hyperphosphatasia mental retardation syndrome. *Nat. Genet.* doi: 10.1038/ng.653.

2 Ng, S.B., Turner, E.H., Robertson, P.D., Flygare, S.D., Bigham, A.W., Lee, C., Shaffer, T., Wong, M., Bhattacharjee, A., Eichler, E.E., Bamshad, M., Nickerson, D.A., and Shendure, J. (2009) Targeted capture and massively parallel sequencing of 12 human exomes. *Nature*, **461**, 272–276.

3 Branton, D., Deamer, D.W., Marziali, A., Bayley, H., Benner, S.A., Butler, T., Di Ventra, M., Garaj, S., Hibbs, A., Huang, X., Jovanovich, S.B., Krstic, P.S., Lindsay, S., Ling, X.S., Mastrangelo, C.H., Meller, A., Oliver, J.S., Pershin, Y.V., Ramsey, J.M., Riehn, R., Soni, G.V., Tabard-Cossa, V., Wanunu, M., Wiggin, M., and Schloss, J.A. (2008) The potential and challenges of nanopore sequencing. *Nat. Biotechnol.*, **26**, 1146–1153.

4 Pleasance, E.D., Cheetham, R.K., Stephens, P.J., McBride, D.J., Humphray, S.J., Greenman, C.D., Varela, I., Lin, M.L., Ordonez, G.R., Bignell, G.R., Ye, K., Alipaz, J., Bauer, M.J., Beare, D., Butler, A., Carter, R.J., Chen, L., Cox, A.J., Edkins, S., Kokko-Gonzales, P.I., Gormley, N.A., Grocock, R.J., Haudenschild, C.D., Hims, M.M., James, T., Jia, M., Kingsbury, Z., Leroy, C., Marshall, J., Menzies, A., Mudie, L.J., Ning, Z., Royce, T., Schulz-Trieglaff, O.B., Spiridou, A., Stebbings, L.A., Szajkowski, L., Teague, J., Williamson, D., Chin, L., Ross, M.T., Campbell, P.J., Bentley, D.R., Futreal, P.A., and Stratton, M.R. (2010) A comprehensive catalogue of somatic mutations from a human cancer genome. *Nature*, **463**, 191–196.

5 Pleasance, E.D., Stephens, P.J., O'Meara, S., McBride, D.J., Meynert, A., Jones, D., Lin, M.L., Beare, D., Lau, K.W., Greenman, C., Varela, I., Nik-Zainal, S., Davies, H.R., Ordonez, G.R., Mudie, L.J., Latimer, C., Edkins, S., Stebbings, L., Chen, L., Jia, M., Leroy, C., Marshall, J., Menzies, A., Butler, A., Teague, J.W., Mangion, J., Sun, Y.A., McLaughlin, S.F., Peckham, H.E., Tsung, E.F., Costa, G.L., Lee, C.C., Minna, J.D., Gazdar, A., Birney, E., Rhodes, M.D., McKernan, K.J., Stratton, M.R., Futreal, P.A., and Campbell, P.J. (2010) A small-cell lung cancer genome with complex signatures of tobacco exposure. *Nature*, **463**, 184–190.

6 Ley, T.J., Mardis, E.R., Ding, L., Fulton, B., McLellan, M.D., Chen, K., Dooling, D., Dunford-Shore, B.H., McGrath, S., Hickenbotham, M., Cook, L., Abbott, R., Larson, D.E., Koboldt, D.C., Pohl, C., Smith, S., Hawkins, A., Abbott, S., Locke, D., Hillier, L.W., Miner, T., Fulton, L., Magrini, V., Wylie, T., Glasscock, J., Conyers, J., Sander, N., Shi, X., Osborne, J.R., Minx, P., Gordon, D., Chinwalla, A., Zhao, Y., Ries, R.E., Payton, J.E., Westervelt, P., Tomasson, M.H., Watson, M., Baty, J., Ivanovich, J., Heath, S., Shannon, W.D., Nagarajan, R., Walter, M.J., Link, D.C., Graubert, T.A., DiPersio, J.F., and Wilson, R.K. (2008) DNA sequencing of a cytogenetically normal acute myeloid leukaemia genome. *Nature*, **456**, 66–72.

7 Shah, S.P., Kobel, M., Senz, J., Morin, R.D., Clarke, B.A., Wiegand, K.C., Leung,

G., Zayed, A., Mehl, E., Kalloger, S.E., Sun, M., Giuliany, R., Yorida, E., Jones, S., Varhol, R., Swenerton, K.D., Miller, D., Clement, P.B., Crane, C., Madore, J., Provencher, D., Leung, P., DeFazio, A., Khattra, J., Turashvili, G., Zhao, Y., Zeng, T., Glover, J.N., Vanderhyden, B., Zhao, C., Parkinson, C.A., Jimenez-Linan, M., Bowtell, D.D., Mes-Masson, A.M., Brenton, J.D., Aparicio, S.A., Boyd, N., Hirst, M., Gilks, C.B., Marra, M., and Huntsman, D.G. (2009) Mutation of FOXL2 in granulosa-cell tumors of the ovary. *N. Engl. J. Med.*, **360**, 2719–2729.

8 Shah, S.P., Morin, R.D., Khattra, J., Prentice, L., Pugh, T., Burleigh, A., Delaney, A., Gelmon, K., Guliany, R., Senz, J., Steidl, C., Holt, R.A., Jones, S., Sun, M., Leung, G., Moore, R., Severson, T., Taylor, G.A., Teschendorff, A.E., Tse, K., Turashvili, G., Varhol, R., Warren, R.L., Watson, P., Zhao, Y., Caldas, C., Huntsman, D., Hirst, M., Marra, M.A., and Aparicio, S. (2009) Mutational evolution in a lobular breast tumour profiled at single nucleotide resolution. *Nature*, **461**, 809–813.

9 Bian, Y.S., Yan, P., Osterheld, M.C., Fontolliet, C., and Benhattar, J. (2001) Promoter methylation analysis on microdissected paraffin-embedded tissues using bisulfite treatment and PCR-SSCP. *Biotechniques*, **30**, 66–72.

10 Gilbert, M.T., Haselkorn, T., Bunce, M., Sanchez, J.J., Lucas, S.B., Jewell, L.D., Van Marck, E., and Worobey, M. (2007) The isolation of nucleic acids from fixed, paraffin-embedded tissues: which methods are useful when? *PLoS One*, **2**, e537.

11 Schweiger, M.R., Kerick, M., Timmermann, B., Albrecht, M.W., Borodina, T., Parkhomchuk, D., Zatloukal, K., and Lehrach, H. (2009) Genome-wide massively parallel sequencing of formaldehyde fixed-paraffin embedded (FFPE) tumor tissues for copy-number- and mutation-analysis. *PLoS One*, **4**, e5548.

12 Kerjean, A., Vieillefond, A., Thiounn, N., Sibony, M., Jeanpierre, M., and Jouannet, P. (2001) Bisulfite genomic sequencing of microdissected cells. *Nucleic Acids Res.*, **29**, E106–E116.

13 Ragoussis, J. (2009) Genotyping technologies for genetic research. *Annu. Rev. Genomics Hum. Genet.*, **10**, 117–133.

14 van de Vijver, M.J., He, Y.D., van't Veer, L.J., Dai, H., Hart, A.A., Voskuil, D.W., Schreiber, G.J., Peterse, J.L., Roberts, C., Marton, M.J., Parrish, M., Atsma, D., Witteveen, A., Glas, A., Delahaye, L., van der Velde, T., Bartelink, H., Rodenhuis, S., Rutgers, E.T., Friend, S.H., and Bernards, R. (2002) A gene-expression signature as a predictor of survival in breast cancer. *N. Engl. J. Med.*, **347**, 1999–2009.

15 Maxam, A.M. and Gilbert, W. (1977) A new method for sequencing DNA. *Proc. Natl. Acad. Sci. USA*, **74**, 560–564.

16 Sanger, F., Nicklen, S., and Coulson, A.R. (1977) DNA sequencing with chain-terminating inhibitors. *Proc. Natl. Acad. Sci. USA*, **74**, 5463–5467.

17 International Human Genome Sequencing Consortium (2004) Finishing the euchromatic sequence of the human genome. *Nature*, **431**, 931–945.

18 Lander, E.S., Linton, L.M., Birren, B., Nusbaum, C., Zody, M.C., Baldwin, J., Devon, K., Dewar, K., Doyle, M., FitzHugh, W., Funke, R., Gage, D., Harris, K., Heaford, A., Howland, J., Kann, L., Lehoczky, J., LeVine, R., McEwan, P., McKernan, K., Meldrim, J., Mesirov, J.P., Miranda, C., Morris, W., Naylor, J., Raymond, C., Rosetti, M., Santos, R., Sheridan, A., Sougnez, C., Stange-Thomann, N., Stojanovic, N., Subramanian, A., Wyman, D., Rogers, J., Sulston, J., Ainscough, R., Beck, S., Bentley, D., Burton, J., Clee, C., Carter, N., Coulson, A., Deadman, R., Deloukas, P., Dunham, A., Dunham, I., Durbin, R., French, L., Grafham, D., Gregory, S., Hubbard, T., Humphray, S., Hunt, A., Jones, M., Lloyd, C., McMurray, A., Matthews, L., Mercer, S., Milne, S., Mullikin, J.C., Mungall, A., Plumb, R., Ross, M., Shownkeen, R., Sims, S., Waterston, R.H., Wilson, R.K., Hillier, L.W., McPherson, J.D., Marra, M.A., Mardis, E.R., Fulton, L.A., Chinwalla, A.T., Pepin, K.H., Gish, W.R., Chissoe, S.L., Wendl, M.C., Delehaunty, K.D., Miner,

T.L., Delehaunty, A., Kramer, J.B., Cook, L.L., Fulton, R.S., Johnson, D.L., Minx, P.J., Clifton, S.W., Hawkins, T., Branscomb, E., Predki, P., Richardson, P., Wenning, S., Slezak, T., Doggett, N., Cheng, J.F., Olsen, A., Lucas, S., Elkin, C., Uberbacher, E., Frazier, M., Gibbs, R.A., Muzny, D.M., Scherer, S.E., Bouck, J.B., Sodergren, E.J., Worley, K.C., Rives, C.M., Gorrell, J.H., Metzker, M.L., Naylor, S.L., Kucherlapati, R.S., Nelson, D.L., Weinstock, G.M., Sakaki, Y., Fujiyama, A., Hattori, M., Yada, T., Toyoda, A., Itoh, T., Kawagoe, C., Watanabe, H., Totoki, Y., Taylor, T., Weissenbach, J., Heilig, R., Saurin, W., Artiguenave, F., Brottier, P., Bruls, T., Pelletier, E., Robert, C., Wincker, P., Smith, D.R., Doucette-Stamm, L., Rubenfield, M., Weinstock, K., Lee, H.M., Dubois, J., Rosenthal, A., Platzer, M., Nyakatura, G., Taudien, S., Rump, A., Yang, H., Yu, J., Wang, J., Huang, G., Gu, J., Hood, L., Rowen, L., Madan, A., Qin, S., Davis, R.W., Federspiel, N.A., Abola, A.P., Proctor, M.J., Myers, R.M., Schmutz, J., Dickson, M., Grimwood, J., Cox, D.R., Olson, M.V., Kaul, R., Shimizu, N., Kawasaki, K., Minoshima, S., Evans, G.A., Athanasiou, M., Schultz, R., Roe, B.A., Chen, F., Pan, H., Ramser, J., Lehrach, H., Reinhardt, R., McCombie, W.R., de la Bastide, M., Dedhia, N., Blocker, H., Hornischer, K., Nordsiek, G., Agarwala, R., Aravind, L., Bailey, J.A., Bateman, A., Batzoglou, S., Birney, E., Bork, P., Brown, D.G., Burge, C.B., Cerutti, L., Chen, H.C., Church, D., Clamp, M., Copley, R.R., Doerks, T., Eddy, S.R., Eichler, E.E., Furey, T.S., Galagan, J., Gilbert, J.G., Harmon, C., Hayashizaki, Y., Haussler, D., Hermjakob, H., Hokamp, K., Jang, W., Johnson, L.S., Jones, T.A., Kasif, S., Kaspryzk, A., Kennedy, S., Kent, W.J., Kitts, P., Koonin, E.V., Korf, I., Kulp, D., Lancet, D., Lowe, T.M., McLysaght, A., Mikkelsen, T., Moran, J.V., Mulder, N., Pollara, V.J., Ponting, C.P., Schuler, G., Schultz, J., Slater, G., Smit, A.F., Stupka, E., Szustakowski, J., Thierry-Mieg, D., Thierry-Mieg, J., Wagner, L., Wallis, J., Wheeler, R., Williams, A., Wolf, Y.I., Wolfe, K.H., Yang, S.P., Yeh, R.F., Collins, F., Guyer, M.S., Peterson, J., Felsenfeld, A., Wetterstrand, K.A., Patrinos, A., Morgan, M.J., de Jong, P., Catanese, J.J., Osoegawa, K., Shizuya, H., Choi, S., and Chen, Y.J. (2001) Initial sequencing and analysis of the human genome. *Nature*, **409**, 860–921.

19 Mardis, E.R. and Wilson, R.K. (2009) Cancer genome sequencing: a review. *Hum. Mol. Genet.*, **18**, R163–R168.

20 Shendure, J. and Ji, H. (2008) Next-generation DNA sequencing. *Nat. Biotechnol.*, **26**, 1135–1145.

21 Metzker, M.L. (2010) Sequencing technologies: the next generation. *Nat. Rev. Genet.*, **11**, 31–46.

22 Rothberg, J.M. and Leamon, J.H. (2008) The development and impact of 454 sequencing. *Nat. Biotechnol.*, **26**, 1117–1124.

23 Margulies, M., Egholm, M., Altman, W.E., Attiya, S., Bader, J.S., Bemben, L.A., Berka, J., Braverman, M.S., Chen, Y.J., Chen, Z., Dewell, S.B., Du, L., Fierro, J.M., Gomes, X.V., Godwin, B.C., He, W., Helgesen, S., Ho, C.H., Irzyk, G.P., Jando, S.C., Alenquer, M.L., Jarvie, T.P., Jirage, K.B., Kim, J.B., Knight, J.R., Lanza, J.R., Leamon, J.H., Lefkowitz, S.M., Lei, M., Li, J., Lohman, K.L., Lu, H., Makhijani, V.B., McDade, K.E., McKenna, M.P., Myers, E.W., Nickerson, E., Nobile, J.R., Plant, R., Puc, B.P., Ronan, M.T., Roth, G.T., Sarkis, G.J., Simons, J.F., Simpson, J.W., Srinivasan, M., Tartaro, K.R., Tomasz, A., Vogt, K.A., Volkmer, G.A., Wang, S.H., Wang, Y., Weiner, M.P., Yu, P., Begley, R.F., and Rothberg, J.M. (2005) Genome sequencing in microfabricated high-density picolitre reactors. *Nature*, **437**, 376–380.

24 Ronaghi, M. (2001) Pyrosequencing sheds light on DNA sequencing. *Genome Res.*, **11**, 3–11.

25 Adessi, C., Matton, G., Ayala, G., Turcatti, G., Mermod, J.J., Mayer, P., and Kawashima, E. (2000) Solid phase DNA amplification: characterisation of primer attachment and amplification mechanisms. *Nucleic Acids Res.*, **28**, E87.

26 Shendure, J., Porreca, G.J., Reppas, N.B., Lin, X., McCutcheon, J.P., Rosenbaum, A.M., Wang, M.D., Zhang, K., Mitra, R.D., and Church, G.M. (2005) Accurate multiplex polony sequencing of an evolved bacterial genome. *Science*, **309**, 1728–1732.

27 Ding, L., Wendl, M.C., Koboldt, D.C., and Mardis, E.R. (2010) Analysis of next generation genomic data in cancer: accomplishments and challenges. *Hum. Mol. Genet.* doi: 10.1093/hmg/ddq391.

28 Meyerson, M., Gabriel, S., and Getz, G. (2010) Advances in understanding cancer genomes through second-generation sequencing. *Nat. Rev. Genet.*, **11**, 685–696.

29 Blow, N. (2008) DNA sequencing: generation next-next. *Nat. Methods*, **5**, 267–274.

30 Clarke, J., Wu, H.C., Jayasinghe, L., Patel, A., Reid, S., and Bayley, H. (2009) Continuous base identification for single-molecule nanopore DNA sequencing. *Nat. Nanotechnol.*, **4**, 265–270.

31 Greenleaf, W.J. and Block, S.M. (2006) Single-molecule, motion-based DNA sequencing using RNA polymerase. *Science*, **313**, 801.

32 Sugiyama, S. (2006) Application of scanning probe microscopy to genetic analysis. *Jpn. J. Appl. Phys.*, **45**, 2305–2309.

33 Albert, T.J., Molla, M.N., Muzny, D.M., Nazareth, L., Wheeler, D., Song, X., Richmond, T.A., Middle, C.M., Rodesch, M.J., Packard, C.J., Weinstock, G.M., and Gibbs, R.A. (2007) Direct selection of human genomic loci by microarray hybridization. *Nat. Methods*, **4**, 903–905.

34 Choi, M., Scholl, U.I., Ji, W., Liu, T., Tikhonova, I.R., Zumbo, P., Nayir, A., Bakkaloglu, A., Ozen, S., Sanjad, S., Nelson-Williams, C., Farhi, A., Mane, S., and Lifton, R.P. (2009) Genetic diagnosis by whole exome capture and massively parallel DNA sequencing. *Proc. Natl. Acad. Sci. USA*, **106**, 19096–19101.

35 Hodges, E., Xuan, Z., Balija, V., Kramer, M., Molla, M.N., Smith, S.W., Middle, C.M., Rodesch, M.J., Albert, T.J., Hannon, G.J., and McCombie, W.R. (2007) Genome-wide *in situ* exon capture for selective resequencing. *Nat. Genet.*, **39**, 1522–1527.

36 Gnirke, A., Melnikov, A., Maguire, J., Rogov, P., LeProust, E.M., Brockman, W., Fennell, T., Giannoukos, G., Fisher, S., Russ, C., Gabriel, S., Jaffe, D.B., Lander, E.S., and Nusbaum, C. (2009) Solution hybrid selection with ultra-long oligonucleotides for massively parallel targeted sequencing. *Nat. Biotechnol.*, **27**, 182–189.

37 Porreca, G.J., Zhang, K., Li, J.B., Xie, B., Austin, D., Vassallo, S.L., LeProust, E.M., Peck, B.J., Emig, C.J., Dahl, F., Gao, Y., Church, G.M., and Shendure, J. (2007) Multiplex amplification of large sets of human exons. *Nat. Methods*, **4**, 931–936.

38 McBride, D.J., Orpana, A.K., Sotiriou, C., Joensuu, H., Stephens, P.J., Mudie, L.J., Hamalainen, E., Stebbings, L.A., Andersson, L.C., Flanagan, A.M., Durbecq, V., Ignatiadis, M., Kallioniemi, O., Heckman, C.A., Alitalo, K., Edgren, H., Futreal, P.A., Stratton, M.R., and Campbell, P.J. (2010) Use of cancer-specific genomic rearrangements to quantify disease burden in plasma from patients with solid tumors. *Genes Chromosomes Cancer*, **49**, 1062–1069.

39 Beck, S. and Rakyan, V.K. (2008) The methylome: approaches for global DNA methylation profiling. *Trends Genet.*, **24**, 231–237.

40 Feinberg, A.P. (2007) Phenotypic plasticity and the epigenetics of human disease. *Nature*, **447**, 433–440.

41 Banerjee, H.N. and Verma, M. (2009) Epigenetic mechanisms in cancer. *Biomarkers Med.*, **3**, 14.

42 Lister, R., Pelizzola, M., Dowen, R.H., Hawkins, R.D., Hon, G., Tonti-Filippini, J., Nery, J.R., Lee, L., Ye, Z., Ngo, Q.M., Edsall, L., Antosiewicz-Bourget, J., Stewart, R., Ruotti, V., Millar, A.H., Thomson, J.A., Ren, B., and Ecker, J.R. (2009) Human DNA methylomes at base resolution show widespread epigenomic differences. *Nature*, **462**, 315–322.

43 Ball, M.P., Li, J.B., Gao, Y., Lee, J.H., LeProust, E.M., Park, I.H., Xie, B., Daley, G.Q., and Church, G.M. (2009) Targeted and genome-scale strategies reveal gene-body methylation signatures in human cells. *Nat. Biotechnol.*, **27**, 361–368.

44 Rakyan, V.K., Down, T.A., Thorne, N.P., Flicek, P., Kulesha, E., Graf, S., Tomazou, E.M., Backdahl, L., Johnson, N., Herberth, M., Howe, K.L., Jackson, D.K., Miretti, M.M., Fiegler, H., Marioni, J.C., Birney, E.,

Hubbard, T.J., Carter, N.P., Tavare, S., and Beck, S. (2008) An integrated resource for genome-wide identification and analysis of human tissue-specific differentially methylated regions (tDMRs). *Genome Res.*, **18**, 1518–1529.

45 Morozova, O., Hirst, M., and Marra, M.A. (2009) Applications of new sequencing technologies for transcriptome analysis. *Annu. Rev. Genomics Hum. Genet.*, **10**, 135–151.

46 Sultan, M., Schulz, M.H., Richard, H., Magen, A., Klingenhoff, A., Scherf, M., Seifert, M., Borodina, T., Soldatov, A., Parkhomchuk, D., Schmidt, D., O'Keeffe, S., Haas, S., Vingron, M., Lehrach, H., and Yaspo, M.L. (2008) A global view of gene activity and alternative splicing by deep sequencing of the human transcriptome. *Science*, **321**, 956–960.

47 Friedel, C.C. and Dolken, L. (2009) Metabolic tagging and purification of nascent RNA: implications for transcriptomics. *Mol. Biosyst.*, **5**, 1271–1278.

12
Transcriptomics and Cancer Risk Assessment

Wolfgang Kemmner

12.1
Introduction

After completion of the Human Genome Project, the main tasks are the characterization of the organization of the human genome and the uncovering of the functions of individual genes. Moreover, the information gathered by the Human Genome Project now makes it possible to perform gene expression profiling studies covering the whole human genome. Gene expression describes the activity of a particular gene and therefore gives a measurement of how many times a gene is expressed or transcribed. Using gene expression, researchers are trying to find out which genes are expressed ubiquitously in all human tissues and which are expressed in a tissue-specific manner, for example, only in tumor tissues.

The latest release of the H-Invitational Database (H-InvDB; http://www.h-invitational.jp/), which is a comprehensive annotation resource for human genes and transcripts, reported a number of 34 699 human gene clusters defining 34 057 protein-coding and 642 nonprotein-coding loci [1]. Although a reliable calculation remains a challenge due to the low percentage of coding sequences, the existence of many short exons and long introns, and the high diversity of alternate transcript forms, this result fits very well with numbers suggested already in 2001 by the Human Genome Project consortium [2].

There are several ways of conducting transcriptomics or gene expression profiling. On the one hand, there are sequencing-based approaches that can be further subdivided into methods that involve sequencing of full-length RNAs or sequencing of short portions of RNAs, typically derived from the 3′ or 5′ end of the corresponding RNA. One example is the serial analysis of gene expression (SAGE) method that is based on the massive sequential analysis of short cDNA sequence tags [3]. Each tag with a size of about 14 base pairs is derived from a defined position within a transcript. Although small, this size is sufficient to identify the corresponding gene, and the number of times each tag binds to the gene of interest provides an accurate measurement of the gene expression level.

Cancer Risk Evaluation: Methods and Trends,
Edited by Günter Obe, Burkhard Jandrig, Gary E. Marchant, Holger Schütz, and Peter M. Wiedemann.

On the other hand, there are approaches based on hybridization of a specific nucleic acid probe to target molecules derived from the sample material. Here, a probe would detect all molecules that contain regions of complementarity to that probe. Good examples for such an approach are DNA microarrays. Here, DNA molecules (probes) are immobilized at defined places on a solid support (chip). These probes hybridize with complementary and fluorescently labeled RNA (or cDNA) molecules prepared from the sample material, such as cells, tissues, or blood. DNA microarrays very often are used, for example, to measure changes in messenger RNA expression levels or to detect single nucleotide polymorphisms (SNPs). However, microarrays have also been developed for the detection of micro-RNA expression, for DNA methylation analysis, for comparative genomic hybridization (CGH), and for location analysis of DNA binding proteins by pairing chromatin immunoprecipitation (ChIP).

In the following, we will focus on gene expression profiling studies carried out using such DNA microarrays. Today, this is still the method of choice for cancer research. However, this may change in the next few years when the so-called "deep sequencing" approach will become cheaper and more easy to perform [4].

12.2
Sample Preparation, Technical Issues, and Data Analysis

Gene expression is regulated by complex genetic networks with a variety of interactions at different levels (DNA, RNA, and protein), on many different timescales (seconds to years), and at various locations (nucleus, cytoplasma, and tissue). Thus, microarray analysis gives a comprehensive view of the gene expression profile at a given time point under specific conditions; this means it gives a snapshot of the messenger RNA population in a sample of interest.

Transcriptional profiling using microarray analysis or other techniques is sensitive to the manner in which the samples are collected and handled. Although we discuss gene expression as if it were occurring *in vivo*, in reality we are measuring the representation of RNA in tissues *ex vivo*. This measurement is influenced by alterations in gene expression that are due to a variety of factors such as RNA degradation and the presence of contaminating tissue.

Even the surgical manipulation can influence what we observe as gene expression patterns. This becomes clearer when we imagine what happens to the tumor during surgery. One of the first steps during surgical removal of malignant tumor is the ligation of the arterial blood supply of the tumor. From this moment on, hypoxemia of the tumor at body temperature occurs. The duration between ligation and final removal of the tumor can vary considerably and, under clinical conditions, may not be reduced to a constant interval. After removal of the tumor, logistical constraints may lead to further considerable delay before the tumor is finally transferred to $-80\,^{\circ}$C. Therefore, this lengthy process could lead to a considerable extent of RNA degradation.

After successful sampling of the tumor specimen, biopsy, or blood sample, a complex process has to be performed to prepare high-quality RNA or cDNA for

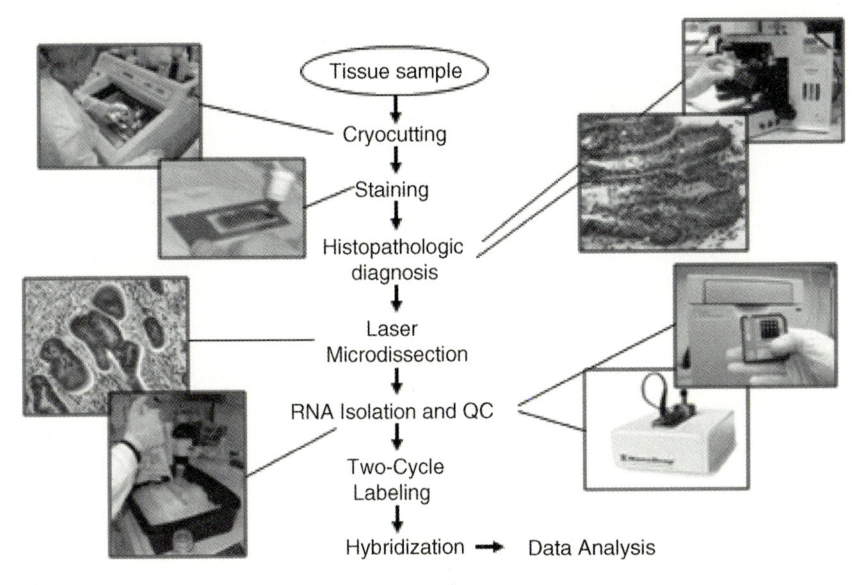

Figure 12.1 From tissue samples to microarray data.

hybridization on the microarray (Figure 12.1). The first step is the cryocutting of the frozen sample material to prepare several sections, which can be used for further processing and in parallel for histological analysis of the sample material by an experienced pathologist. Histological analysis is mandatory to be sure of what you are working with. Furthermore, samples with a high amount of necrotic areas, fatty tissue, and contaminating cells should be excluded to reduce signal noise. A further improvement could be laser microdissection of the tissue of interest. Because of the complex process of sample preparation and hybridization during microarray analysis, artifacts may be generated that lead to a high signal noise. Therefore, one should try to work with a sample material that is as pure as possible. Of course, the process of exclusion of everything else than, for example, the pure tumor tissue sharpens the analysis. A disadvantage is that molecular processes or alterations within the surrounding stroma can no longer be detected. The last steps include RNA extraction, RNA quality and quantity assessment, and labeling of the sample sequence. Depending on how much sample RNA is available, amplification procedures have to be included.

For hybridization with the probe DNA on the microarray, a labeled sample nucleic acid is required. RNA, which has been extracted from the sample material, can be used either for synthesis of labeled complementary DNA (cDNA) without amplification of the sample material or for production of antisense RNA (aRNA). The process of aRNA synthesis allows high amplification of the sample material, which is of relevance if you have only low amounts of sample material, for example, after laser microdissection. Whether amplification changes the outcome of the experiment, for example, by asymmetric amplification of high- and low-abundant genes is still under discussion [5].

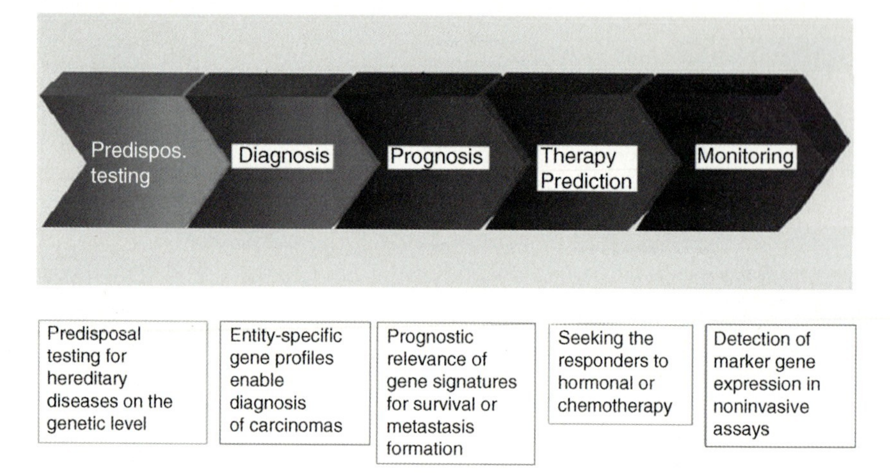

Figure 12.2 Microarray applications in cancer research.

Finally, the labeled sample sequence is hybridized on the microarray. Normalization of the data, statistical analysis, and various bioinformatics methods to dig the relevant information out of the flood of data stand at the end. A major challenge of DNA microarray technology is, of course, the massive data output (Figure 12.2). Several approaches are used to classify patients on the basis of expression profiles: Fisher's linear discriminant analysis, nearest-centroid prediction rule, and support vector machines (SVMs), among others [6, 7]. Recently, a critical review summarized some of the statistical flaws commonly found in published microarray studies [7].

To estimate the accuracy of a classification method, the standard strategy is a training–validation approach, in which a training set is used to identify the molecular signature and a validation set is used to estimate the proportion of misclassifications [8]. One approach is to randomly split the initial data set into two subsets. One part, the so-called training set, is used for identifying the gene signature that can be used to distinguish the tumor subtypes or between tumor and normal tissue. The other part of the cohort, the so-called test set, is then used to evaluate the classifier developed from the training set. The fundamental principle is that the samples used for validation must not have been used in any way before being tested. Most importantly, the outcome information of the tested samples must not have been used for identification of the gene signature that leads to the classifier.

12.2.1
How to Use Your Gene Signatures to Carry Out Cancer Risk Assessment

If you are interested in estimating the prognostic potential of only one biomarker molecule, then the usual way is to construct the receiver operating characteristic (ROC) of this biomarker according to its expression values, for instance, between normal mucosa and carcinoma tissue (see, for example, Ref. [9]). From this ROC curve, the cutoff expression value giving the highest sensitivity and specificity of the biomarker for detecting carcinomas can be calculated. This cutoff value can be used for dividing the samples or patients into groups, for instance, one group of cases with a high biomarker expression and another complementary one with a low biomarker expression. This division then is applied for evaluating the effect of the biomarker expression on the survival of patients showing a high or low expression of this biomarker. Commonly used is the Kaplan–Meier estimator (or curve) that gives an estimation of the survival probability of the two patient groups.

However, one needs another procedure if a set of genes or the so-called gene signature should be applied for risk assessment. One obvious solution would be to sum up the expression values of the different genes to get a scoring that could be used for the construction of the ROC curve. However, on the one hand, not all the genes might be overexpressed in one of the groups. On the other hand, the signal values of the genes might be very different, some of them being high-abundance genes and expression of others being very low. In order to overcome this hurdle, several methods have been used; one is the weighted survival predictor score algorithm. This was implemented to reflect the incremental statistical power of individual covariates as predictors of therapy outcome based on a multicomponent prognostic model [10]. Here, histo- or clinico-pathological variables could be used as additional covariates. In the study cited above, the final survival predictor score comprises a sum of scores for individual genes. The negative weighting values indicate that higher expression correlates with longer survival and favorable prognosis, whereas the positive score values indicate that higher expression correlates with poor outcome and shorter survival. Thus, the weighted survival predictor model was based on a cumulative score of the weighted expression values of 11 genes. For example, the following equation describes the relapse-free survival predictor score for prostate cancer patients: relapse-free survival score $= -(0.403 \times \text{gene } 1) + (1.2494 \times \text{gene } 2) - (0.3105 \times \text{gene } 3) - (0.1226 \times \text{gene } 4) + (0.0077 \times \text{gene } 5) + (0.0369 \times \text{gene } 6) - (1.7493 \times \text{gene } 7) - (1.1853 \times \text{gene } 8) + (1.5242 \times \text{gene } 9) - (0.5628 \times \text{gene } 10) - (0.4333 \times \text{gene } 11)$. Of course, overfitting might become a problem for such methods. Therefore, appropriate validation experiments are mandatory.

Another approach in use is the so-called logistic regression. This method was used, for example, by Zhang *et al.* [11] for detection of resectable pancreatic cancers using expression of biomarkers in the saliva fluid. For each validated biomarker, they constructed the ROC curve and computed the area under the curve (AUC) value by numerical integration of the ROC curve. Next, the validated salivary biomarkers were fit into logistic regression models, and a stepwise backward model selection was performed to determine final combinations of the biomarkers. Since this approach is also using arbitrarily chosen coefficients, the same holds true as stated above.

In a recent study on gene signatures for the identification of prognostic gene expression markers in early primary colorectal carcinomas, we used genome-wide gene expression profiles (Fehlker *et al.*, submitted). This study includes retrospective cases that were stratified as to contain mainly UICC II cases (according to the classification of the International Union Against Cancer). Follow-up data were obtained for at least 41 months or until recurrence of disease with a follow-up time of 41–115 months for nonrecurrent patients. The clinical endpoint examined was distant recurrence of disease. Using an algorithm based on the recursive application of SVMs, genes were identified that discriminated between patients developing later metastasis and patients with a good prognosis. Interestingly, we found a set of genes that all were downregulated in the cases with poor prognosis. This allows us by simply counting how many of the marker genes in a given sample were above median or below the median to obtain a score for each of the samples. This score then can be used for the construction of the ROC and Kaplan–Meier survival curves. The advantage of this method is that no arbitrarily chosen coefficients have to be used. However, appropriate validation experiments with independent cohorts are necessary in this case.

12.2.2
You Need a Hypothesis!

However, to find the genes by which a given subtype can be discriminated from others, for instance, carcinoma versus normal mucosa, sometimes more is needed than only bioinformatics and statistics. In a project on esophageal adenocarcinoma (EAC) cancer, we faced the problem that using bioinformatics we were not able to discriminate between cancer and precursor lesions. EAC develops as a sequel of long-standing chronic gastroesophageal reflux disease (GERD) via a number of steps described histopathologically as Barrett's esophagus (BE), low-grade intraepithelial and high-grade intraepithelial neoplasia. In BE, the lower esophageal squamous epithelium is replaced by metaplastic columnar epithelium with goblet cells (specialized intestinal metaplasia). It has been proposed that GERD patients may be categorized into three distinct groups exhibiting nonerosive reflux disease (NERD), erosive reflux disease (ERD), and BE, the last representing the most dangerous histological subgroup of GERD. Here, gene expression profiles of 19 healthy subjects (NE), 98 GERD, and 21 EAC were examined to characterize the changes in gene expression that underlie EAC progression (Figure 12.3). However, by hierarchical classification or other methods no discrimination of these subtypes was possible, although histopathology shows striking differences among these tissues.

How to find genes whose expression would allow us to discriminate these tissues? The most difficult group is the GERD group, which consists of NERD, ERD, and BE cases. Some of the ERD and NERD cases might be similar to normal squamous epithelium, gene expression of some of the BE cases might be close to that of EAC. Therefore, we decided to omit this group and to compare only NE and EAC, which are the most differing tissues in histopathological terms. Accordingly, about 1000 genes differentially expressed between healthy normal esophageal mucosa and EAC were identified by filtering by a fold change of 5 and a multiple testing-corrected p-value of 0.0001. In the next step, this gene signature was applied on the diverse GERD groups. Hereby, a clear separation between these groups could be achieved (Q. Wang and W. Kemmner, unpublished data).

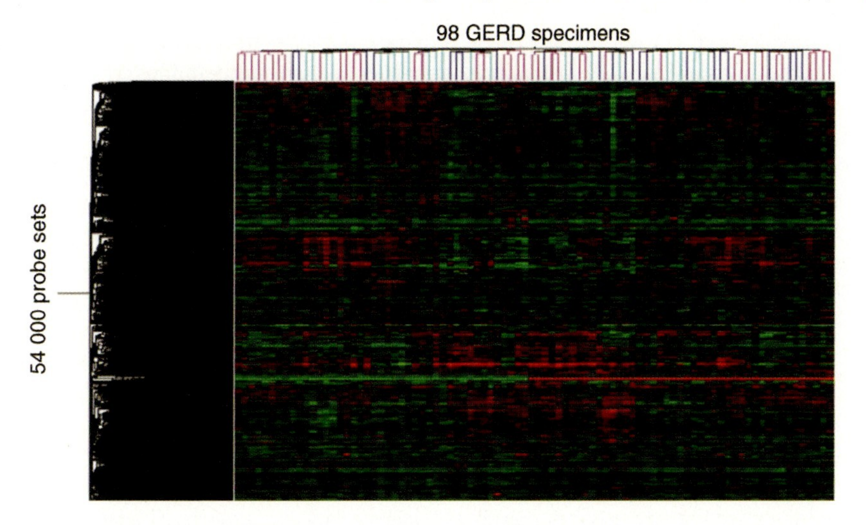

Figure 12.3 Hierarchical clustering among 98 GERD specimens. Individual tissue samples are arrayed in columns, based on gene expression profiling of about 54 000 probe sets (genes). Using ANOVA no group pattern can be detected that distinguishes NERD or ERD from BE. Picture taken from the ProGERD Study by courtesy of AstraZeneca Europe.

In summary, from the description of the lengthy and complex process of sample preparation leading to microarray analysis, it becomes clear that microarray projects should begin with an experimental design that minimizes potential biases and includes detailed protocols for the collection of high-quality samples. Moreover, successful projects depend on a close collaboration among surgeons, pathologists, molecular biologists, and statistical and bioinformatics scientists to ensure high quality of the samples, the data, and their analysis.

12.2.3 Microarray Applications in Cancer Research

Gene expression profiling has the potential to reveal underlying differences and similarities among tumors by the analysis of the whole genome. Therefore, the use of microarray technology has created optimism for the feasibility of using molecular assessments of tumors routinely in the clinical management of cancer.

12.2.4 What Are the Questions of Concern for the Clinical Management of Cancer?

Figure 12.2 gives a brief overview of the questions that might be answered with the help of gene expression. Here, we will not discuss predisposal testing for hereditary diseases (Figure 12.2), although gene expression can also be used to analyze the so-called mutational landscapes of cancers [12]. Other topics are entity-specific gene profiles that enable an accurate diagnosis of also unknown carcinomas. Meanwhile, we have learned that cancer not only is a disease but also encompasses a group of complex diseases that vary in many aspects, such as genetic alterations, cell types involved, response to treatment, and clinical prognosis. For instance, breast cancer is described as a heterogeneous disease consisting of about 16 epithelial subtypes [13]. By DNA microarray analysis, classifiers can be constructed that allow us to discriminate between such tumor subtypes. These subtypes might be different not only with regard to patient prognosis but also in their response to different treatment strategies, which is of particular relevance with regard to the advent of targeted molecular therapies.

Gene expression can also be used for deciphering genes with prognostic relevance for survival or metastasis formation. For breast cancer, it is known that patients with the same stage of that disease behave differently in terms of treatment response and survival. Unfortunately, histopathological assessment does not allow to discriminate between patients with a good prognosis and the ones who will develop metastases and therefore have a poor prognosis. In a pioneering study, van't Veer *et al.* [14] demonstrated the power of microarray analysis for cancer diagnosis. The study was based on 117 patients with breast cancer, including 78 sporadic primary invasive carcinomas with a size of less than 5 cm. All cases were early carcinomas without lymph node metastases. Sample RNA was extracted and reverse transcribed into cDNA, which after labeling was hybridized to an oligonucleotide microarray covering more than 24 000 human genes. Bioinformatics analysis was performed by using a hierarchical clustering approach based on the expression values of 5000 genes with a *p*-value of less than 0.01 and a more than twofold change in expression between the two groups. This analysis revealed the presence of two distinct groups of carcinomas. Within the one group, only 34% of the cases developed distant metastases, while more than 70% of the patients of the other group developed metastasis within the follow-up time of 5 years. Finally, the number was reduced to 78 genes, which were able to predict the clinical outcome, development of distant metastasis, with an accuracy of 83%.

This gene signature, therefore, may help to distinguish those patients who would really need adjuvant systemic therapy from those who could be spared from such treatment. Meanwhile, results of a larger study on 307 patients with a median follow-up of 13.6 years from 5 European countries showed that the 70-gene signature adds independent prognostic information to clinicopathologic risk assessment for patients with early breast cancer [15]. Besides, later findings of these researchers imply that the development of lymph node metastases and the development of distant metastases are independent properties of the tumor. It is perhaps obvious that spread of disease to the lymph nodes, which occurs through the lymphatic vessels, probably requires cancer cell properties different from those needed for spread to distant sites, and hence that these events represent two distinct pathways in the genetic evolution of cancer [16]. In this way, gene expression helps us find answers also for basic questions of cancer research.

Another question is how to find the responders to hormonal or chemotherapy. The majority of breast cancers express estrogen receptor (ER), and most patients with ER-positive tumors are treated with adjuvant tamoxifen therapy. However, approximately 40% of ER breast cancers fail to respond or eventually develop resistance to tamoxifen, leading to disease progression. Histopathological features such as tumor stage and grade fail to accurately identify individuals who are at risk for tumor recurrence. By genome-wide microarray analysis of hormone receptor-positive invasive breast tumors from 60 patients treated with adjuvant tamoxifen, Ma *et al.* [17] were able to identify 19 differentially expressed genes (p-value less than 0.001). In an attempt to refine the analysis of the tumor cells and circumvent potential variability due to stromal cell contamination, the same cohort was reanalyzed following laser-capture microdissection. Finally, the researchers identified two genes, HOXB13 and IL17BR, that have opposite patterns of expression. High expression of HOXB13 predicted recurrence and high expression of IL17BR predicted nonrecurrence. These results suggest that a simple two-gene expression ratio of HOXB13 to IL17BR accurately predicts tumor recurrence in the setting of adjuvant tamoxifen monotherapy. Hence, by using this simple assay, patients can be identified who are likely to benefit from tamoxifen therapy or who are in need to be treated by aromatase inhibitors or other chemotherapeutic agents. Unfortunately, later clinical studies with larger patient cohorts showed that the predictive ability of this assay was only modest [18]. However, discrimination between responders and nonresponders to a certain treatment providing a rational approach for tailoring treatments to subsets of patients will remain to be a major aim of gene expression [19, 20].

Detection of marker gene expression in noninvasive assays is another issue. Tumor marker molecules that can be determined in a noninvasive way from blood samples are indispensable to cancer risk assessment. However, the clinical utility of current protein biomarkers such as CEA or CA15-3 in order to monitor tumor initiation, progression, and outcome is limited because of their insufficient diagnostic sensitivity and specificity. Thus, at present, these markers are used only as an auxiliary approach to assist in clinical decision making.

Meanwhile, it has been shown that mRNA in blood, semen, urine, and saliva can provide a novel resource to supplant conventional tools for disease identification. A recent study [21] was designed to analyze the global transcriptome from serum of oral squamous cell carcinomas (OSCC). Blood samples were collected from 32 patients with primary early OSCC and matched healthy patients. Circulating RNA was isolated from serum and linearly amplified using T7 polymerase. Microarrays were then applied for profiling the transcriptome in serum from 10 cancer patients and controls. Out of 354 serum RNAs with a significant differential expression between the two groups ($p \leq 0.05$, fold change ≥ 2), five cancer-related gene transcripts were selected. In this way, an assay could be developed for OSCC detection with sensitivity (91%) and specificity (71%).

12.3 Conclusions

Overall, the use of microarrays in cancer research has become an established method for cancer risk assessment. In many cases, gene expression profiling has proven to be a better tool for classification of human cancers than classical histopathology. The relation of these biological parameters with those derived from clinical outcome allows stratifying patients at higher risk of recurrence or metastasis and also has provided researchers with target genes/pathways for therapy. Furthermore, gene expression profiling has been successfully used to develop gene signatures that allow to predict not only recurrence and metastasis but also therapy response with better accuracy than standard clinical assays. Therefore, gene expression profiling has become a powerful tool for basic, pharmacological, and clinical oncology research holding the promise that in the near future a better and more personalized treatment of cancer patients will be possible.

References

1 Yamasaki, C., Murakami, K., Fujii, Y., Sato, Y., Harada, E., Takeda, J., Taniya, T., Sakate, R., Kikugawa, S., Shimada, M., Tanino, M., Koyanagi, K.O., Barrero, R.A., Gough, C., Chun, H.W., Habara, T., Hanaoka, H., Hayakawa, Y., Hilton, P.B., Kaneko, Y., Kanno, M., Kawahara, Y., Kawamura, T., Matsuya, A., Nagata, N., Nishikata, K., Noda, A.O., Nurimoto, S., Saichi, N., Sakai, H., Sanbonmatsu, R., Shiba, R., Suzuki, M., Takabayashi, K., Takahashi, A., Tamura, T., Tanaka, M., Tanaka, S., Todokoro, F., Yamaguchi, K., Yamamoto, N., Okido, T., Mashima, J., Hashizume, A., Jin, L., Lee, K.B., Lin, Y.C., Nozaki, A., Sakai, K., Tada, M., Miyazaki, S., Makino, T., Ohyanagi, H., Osato, N., Tanaka, N., Suzuki, Y., Ikeo, K., Saitou, N., Sugawara, H., O'Donovan, C., Kulikova, T., Whitfield, E., Halligan, B., Shimoyama, M., Twigger, S., Yura, K., Kimura, K., Yasuda, T., Nishikawa, T., Akiyama, Y., Motono., C., Mukai, Y., Nagasaki, H., Suwa, M., Horton, P., Kikuno, R., Ohara, O., Lancet, D., Eveno, E., Graudens, E., Imbeaud, S., Debily, M.A., Hayashizaki, Y., Amid, C., Han, M., Osanger, A., Endo, T., Thomas, M.A., Hirakawa, M., Makalowski, W., Nakao, M., Kim, N.S., Yoo, H.S., De Souza, S.J., de Bonaldo, M.F., Niimura, Y., Kuryshev, V., Schupp, I., Wiemann, S., Bellgard, M., Shionyu, M., Jia, L., Thierry-Mieg, D., Thierry-Mieg, J., Wagner, L., Zhang, Q., Go, M., Minoshima, S., Ohtsubo, M., Hanada, K., Tonellato, P., Isogai, T., Zhang, J., Lenhard, B., Kim, S., Chen, Z., Hinz, U., Estreicher, A., Nakai, K., Makalowska, I., Hide, W., Tiffin, N., Wilming, L., Chakraborty, R., Soares, M.B., Chiusano, M.L., Suzuki, Y., Auffray, C., Yamaguchi-Kabata, Y., Itoh, T., Hishiki, T., Fukuchi, S., Nishikawa, K., Sugano, S., Nomura, N., Tateno, Y., Imanishi, T., and Gojobori, T. (2008) The H-Invitational Database (H-InvDB): a comprehensive annotation resource for human genes and transcripts. *Nucleic Acids Res.*, **36** (database issue), D793–D799.

2 Lander, E.S., Linton, L.M., Birren, B. *et al.* (2001) Initial sequencing and analysis of the human genome. *Nature*, **409** (6822), 860–921.

3 Velculescu, V.E., Zhang, L., Vogelstein, B., and Kinzler, K.W. (1995) Serial analysis of gene expression. *Science*, **270** (5235), 484–487.

4 Morozova, O. and Marra, M.A. (2008) Applications of next-generation sequencing technologies in functional genomics. *Genomics*, **92** (5), 255–264.

5 Naderi, A., Ahmed, A.A., Barbosa-Morais, N.L., Aparicio, S., Brenton, J.D., and Caldas, C. (2004) Expression microarray reproducibility is improved by optimising purification steps in RNA amplification and labelling. *BMC Genomics*, **5** (1), 9.

6 Simon, R. (2008) Interpretation of genomic data: questions and answers. *Semin. Hematol.*, **45** (3), 196–204.

7 Dupuy, A. and Simon, R.M. (2007) Critical review of published microarray studies for cancer outcome and guidelines on

statistical analysis and reporting. *J. Natl. Cancer Inst.*, **99** (2), 147–157.

8 Wessels, L.F., Reinders, M.J., Hart, A.A., Veenman, C.J., Dai, H., He, Y.D., and van't Veer, L.J. (2005) A protocol for building and evaluating predictors of disease state based on microarray data. *Bioinformatics*, **21** (19), 3755–3762.

9 Astrosini, C., Roeefzaad, C., Dai, Y.Y., Dieckgraefe, B.K., Jöns, T., and Kemmner, W. (2008) REG1A expression is a prognostic marker in colorectal cancer and associated with peritoneal carcinomatosis. *Int. J. Cancer*, **123** (2), 409–413.

10 Golub, T.R., Slonim, D.K., Tamayo, P., Huard, C., Gaasenbeek, M., Mesirov, J.P., Coller, H., Loh, M.L., Downing, J.R., Caligiuri, M.A., Bloomfield, C.D., and Lander, E.S. (1999) Molecular classification of cancer: class discovery and class prediction by gene expression monitoring. *Science*, **286** (5439), 531–537.

11 Zhang, L., Farrell, J.J., Zhou, H., Elashoff, D., Akin, D., Park, N.H., Chia, D., and Wong, D.T. (2010) Salivary transcriptomic biomarkers for detection of resectable pancreatic cancer. *Gastroenterology*, **138** (3), 949–957.

12 Wood, L.D., Parsons, D.W., Jones, S. *et al.* (2007) The genomic landscapes of human breast and colorectal cancers. *Science*, **318** (5853), 1108–1113.

13 Yerushalmi, R., Hayes, M.M., and Gelmon, K.A. (2009) Breast carcinoma – rare types: review of the literature. *Ann. Oncol.*, **20** (11), 1763–1770.

14 van't Veer, L.J., Dai, H., van de Vijver, M.J. *et al.* (2002) Gene expression profiling predicts clinical outcome of breast cancer. *Nature*, **415** (6871), 530–536.

15 Buyse, M., Loi, S., van't Veer, L., Viale, G., Delorenzi, M., Glas, A.M., d'Assignies, M.S., Bergh, J., Lidereau, R., Ellis, P., Harris, A., Bogaerts, J., Therasse, P., Floore, A., Amakrane, M., Piette, F., Rutgers, E., Sotiriou, C., Cardoso, F., Piccart, M.J., and TRANSBIG Consortium (2006) Validation and clinical utility of a 70-gene prognostic signature for women with node-negative breast cancer. *J. Natl. Cancer Inst.*, **98** (17), 1183–1192.

16 van de Vijver, M.J., He, Y.D., van't Veer, L.J. *et al.* (2002) A gene-expression signature as a predictor of survival in breast cancer. *N. Engl. J. Med.*, **347** (25), 1999–2009.

17 Ma, X.J., Wang, Z., Ryan, P.D. *et al.* (2004) A two-gene expression ratio predicts clinical outcome in breast cancer patients treated with tamoxifen. *Cancer Cell*, **5** (6), 607–616.

18 Goetz, M.P., Suman, V.J., Ingle, J.N., Nibbe, A.M., Visscher, D.W., Reynolds, C.A., Lingle, W.L., Erlander, M., Ma, X.J., Sgroi, D.C., Perez, E.A., and Couch, F.J. (2006) A two-gene expression ratio of homeobox 13 and interleukin-17B receptor for prediction of recurrence and survival in women receiving adjuvant tamoxifen. *Clin. Cancer Res.*, **12** (7), 2080–2087.

19 Korde, L.A., Lusa, L., McShane, L., Lebowitz, P.F., Lukes, L., Camphausen, K., Parker, J.S., Swain, S.M., Hunter, K., and Zujewski, J.A. (2010) Gene expression pathway analysis to predict response to neoadjuvant docetaxel and capecitabine for breast cancer. *Breast Cancer Res. Treat.*, **119** (3), 685–699.

20 Jorissen, R.N., Gibbs, P., Christie, M., Prakash, S., Lipton, L., Desai, J., Kerr, D., Aaltonen, L.A., Arango, D., Kruhøffer, M., Orntoft, T.F., Andersen, C.L., Gruidl, M., Kamath, V.P., Eschrich, S., Yeatman, T.J., and Sieber, O.M. (2009) Metastasis-associated gene expression changes predict poor outcomes in patients with Dukes stage B and C colorectal cancer. *Clin. Cancer Res.*, **15** (24), 7642–7651.

21 Li, Y., Elashoff, D., Oh, M., Sinha, U., St John, M.A., Zhou, X., Abemayor, E., and Wong, D.T. (2006) Serum circulating human mRNA profiling and its utility for oral cancer detection. *J. Clin. Oncol.*, **24** (11), 1754–1760.

13
Proteomics and Cancer Risk Assessment

Alexander Schramm

13.1
Introduction

Proteomics can be defined as functional analysis of the entire repertoire of proteins, usually performed at the cellular level. Proteome analyses turned out to be a useful tool to obtain holistic insights into the physiological status of a cell or a cellular system. For about a decade, global profiling studies such as monitoring responses to external stimuli [1, 2] or prediction of disease outcome by analyses of genome-wide expression patterns were performed mainly on the RNA level [3–5]. In recent years, proteomics also came into focus as a technological alternative for risk assessment in various diseases, including cancer. There are some obvious advantages in proteomic studies when compared to RNA-based studies: (1) posttranslational modifications, for example, phosphorylation, contributing to the activity status can be detected at the protein level only, (2) mRNA levels do not necessarily coincide with protein levels for a particular gene, and (3) feedback mechanisms within regulatory pathways can control protein activity, while the mRNA content is not affected. Taken together, protein analyses might end up in a more complete picture to distinguish between diseased and normal states and might allow the identification of robust biomarkers of disease.

As pointed out in Ref. [6], the term "proteomics" nowadays comprises several technical disciplines, including imaging by microscopy, array-based experiments, and genetic readout assays such as yeast two-hybrid screening and variations thereof [7, 8]. The section of proteomics, which will be discussed here in the context of cancer research and risk assessment, is a multistep process comprising sample preparation, separation, quantification, and protein identification. Since protein species can occur in complexes and the dynamic range of protein amount within a cell system varies, appropriate analysis methods are required. Different approaches have been developed for relative and absolute quantification of proteins based on mass spectrometry [6] and the field is still evolving. Labeling techniques as well as label-free approaches and combinations thereof [9] have been developed to facilitate the understanding of the proteome composition and protein–protein interactions.

Cancer Risk Evaluation: Methods and Trends,
Edited by Günter Obe, Burkhard Jandrig, Gary E. Marchant, Holger Schütz, and Peter M. Wiedemann.

Considerable efforts are nowadays undertaken to bring proteomics further to clinical application in the field of cancer risk assessment. For that purpose, proteomic analyses in cancer research have focused on both biomarker discovery and target identification. Still, several challenges in translating the findings of proteomics research into clinical practice exist. Lack of reproducibility, sample availability, sample selection bias, and the strict requirement for a very high level of clinical sensitivity and specificity are practical problems that need to be addressed by current proteomics approaches to assist in early detection and treatment of cancer [10]. There is clearly a need for harmonization of experimental methods, which has been recently addressed by the MIAPE and the HUPO consortia [11, 12]. Standardization of reporting for all aspects of a proteomics experiment has been recently published [13–15], providing a framework also for clinically relevant data sets.

This chapter discusses technological advances and applications of proteomics in the context of cancer risk assessment. The route from sample acquisition to potential implementation in clinical practice will be highlighted.

13.2 Sample Preparation and Storage: A Challenge in Clinical Settings

Application of proteomics in clinical settings is a demanding task in terms of logistics, since sample acquisition and processing have to be integrated into the clinical work routine. Ideally, biobanking or biorepository concepts include workflows for collecting and storing samples according to standard operating procedures (SOPs). Documentation of the entire process of sampling is also essential for the definition of inclusion and exclusion criteria for proteomic studies. These data have then to be linked to clinical information pertaining to the donor of a biospecimen. Since proteomics nowadays can deal with virtually any body fluid, from bronchoalveolar lavage to serum/plasma, every biological sample available from routine examinations as well as surgery is in principle suited for proteomic analyses.

Notably, for protein identification from blood-based samples, EDTA-stabilized plasma was recommended by the HUPO-PPP (Human Proteome Organization – Plasma Proteome Project) consortium [16, 17]. This is important when considering a setup for clinical trials using blood samples. It should also be mentioned that ethical and legal aspects, security systems, access rights, and the availability of storage facilities have to be addressed prior to setting up proteomics in the area of cancer research and risk assessment.

13.3 Caveats and Hurdles in Protein Analysis Using Cancer Specimens and Clinical Samples

Using and processing of clinical samples in cancer proteomics is complicated by the fact that in serum, for example, the 20 most abundant proteins account for 99% of all proteins [18, 19]. Since the dynamic range of protein expression is on the order of

10^8–10^{10}, the least abundant proteins in a sample containing 10–20 mg protein would be present at starting levels not exceeding picograms. This can be accounted for by immunodepletion approaches removing high-abundance proteins, which is also recommended by the HUPO-PPP for the direct identification [20]. Still, the detection limit in nonarray-based profiling methods is restricted to medium-abundance proteins, narrowing the window for identification of biologically relevant key players. In turn, array-based approaches are biased and limited to the availability of suitable antibodies.

13.4 Separation and Fractionation of Protein Mixtures as a Prerequisite to Proteomic Analyses and Protein Quantification

The classical approach in protein analysis is to separate proteins by gel electrophoresis either in one or two dimensions according to charge and molecular weight. The introduction of 2D fluorescence difference gel electrophoresis (2D-DIGE) has improved the comparability of biological samples, since this method allows the simultaneous detection of proteins from different biological sources in one gel [21–23]. Thus, it gives more reliable and accurate information on protein abundance than conventional 2D gel electrophoresis. In all of the gel-based methods, the intensity of protein stains is regarded as being proportional to the amount of a particular protein. Still, it has to be taken into account that spots on a 2D gel may contain more than one protein, thereby hampering quantification of changes in the levels of a specific protein species. In addition, gel-based methods suffer from a limited dynamic range, difficulties in the separation of hydrophobic proteins, and handling of proteins with extreme isoelectric points and molecular weights. Over the past years, at least three different strategies have been developed to overcome these limitations: (1) combining liquid chromatography and mass spectrometry (e.g., MudPIT, multidimensional protein identification) [24], (2) stable isotope labeling approaches such as isotope-coded affinity tagging (ICAT), isotope-coded protein labeling (ICPL), isobaric tags for relative and absolute quantification (iTRAQ), and others (reviewed in Ref. [25]), and (3) label-free quantitative proteomics. Although providing unprecedented flexibility and resolution in quantitative proteomics, limitations of labeling-based approaches include high costs, increased time, and a requirement for higher sample concentration, which is sometimes difficult to achieve for clinical samples. Therefore, considerable efforts are underway to achieve faster and simpler quantification results using label-free protein analyses. Label-free proteomics can also be performed quantitatively, either relative by analyzing peak intensity of liquid chromatography– mass spectrometry (LC–MS) spectra or absolute. A method to achieve absolute quantification of proteins per cell has recently been developed and termed APEX (absolute protein expression profiling) [26]. It implies that appropriate correction factors can be calculated that make the fraction of the expected number of peptides and the fraction of the observed number of peptides (e.g., in a MudPIT experiment) proportional to one another. Using APEX, it was possible to determine the abundance of 10 proteins that were spiked into a yeast cell extract [26]. Surely, optimization of separation and fractionation methods will facilitate efforts for exact and robust protein quantification needed for clinical and cancer proteomics.

13.5
Identification of Proteins by Mass Spectrometry

In the typical workflow, fractionation and separation of samples is followed by the generation of peptide mixtures, most often using trypsin. The resulting peptide fragments can then be analyzed using either matrix-assisted laser desorption ionization mass spectrometry (MALDI-MS) or electrospray ionization followed by mass spectrometry (ESI-MS). The principal difference between both methods is that ESI ionizes the analytes out of a solution, while MALDI sublimates and ionizes samples out of a dry, crystalline matrix using laser pulses. ESI can be coupled to liquid-based chromatography devices, rendering this method useful for the analysis of complex samples. MALDI in turn has its strength when dealing with less complex mixtures.

Identification of proteins can be achieved by comparing the mass/charge ratios of peptides with their respective databases. Most commonly, a combination of 2D gel electrophoresis and mass spectrometry is used for that purpose. Proteins are first separated by electrophoresis, stained, excised from the gel, digested, and then analyzed by mass spectrometry. To identify proteins based on their peptide masses, respective spectra have to be processed and compared with known or predicted masses in proteome databases such as ProteinScape (Bruker Daltonics). Scoring methods and sequence coverage in percent are used for confirming protein identity.

Alternatively, proteins can be subjected to limited purification followed by automated peptide MS/MS. Both tracks require suitable data processing, storage, and visualization infrastructure to be used as a high-throughput platform.

As there is clearly a need to directly process clinical samples for use in mass spectrometry, surface-enhanced laser desorption ionization coupled with time-of-flight mass spectrometry (SELDI-TOF-MS) has been developed. A number of studies have identified this technology as useful for biomarker discovery [27–29], but some limitations regarding peak resolution and subsequent protein identification have also been found [30]. Nevertheless, when coupled to high-resolution MS/MS, SELDI can also be used for direct protein identification. A multi-institutional effort initiated by the HUPO-PPP reported the potential of the SELDI platform for reproducible and consistent analysis of serum/plasma across multiple sites [31].

Recently, the development of ion trap spectrometers has emerged as an important tool in high-throughput protein sequencing. The combination of FTICR (Fourier transform ion cyclotron resonance) or orbitrap, which are ion traps based on magnetic and electrostatic forces, respectively, with a MS instrument allows the unprecedented accuracy and efficiency in protein identification [32, 33]. Certainly, the advances both in mass spectrometry and in the assembly of a human protein reference database [34–36] will complement each other to give precise readouts of proteomic experiments.

13.6
Array-Based Proteome Technology in Cancer Research

A complete proteomic analyses, including determination of abundance, modification, activity, localization, and interaction, is usually impracticable when using clinical samples in cancer research and cancer risk assessment. Therefore, usually only one or two of these parameters are interrogated and only a fraction of proteins are analyzed in studies. These experiments can be referred to as proteomic profiling as they are corresponding to RNA-based expression profiling. Proteomic profiling using antibody-based arrays has been introduced in the late 1990s [37]. In these experiments, the proof of principle was demonstrated that IgG subclass-specific recognition was possible with minimal cross-reactivity. Shortly thereafter, first reports on array-based detection of multiple cytokines in patient's sera were published [38, 39]. Since then, several commercial vendors have developed dedicated antibody-based arrays for the simultaneous detection of proteins in multiwell formats in biological samples, including serum. Potential applications in cancer risk assessment include is identification of biomarkers associated with response to therapy [40] or treatment toxicity [41]. These analyses are already integrated into clinical trials [42]. Other promising approaches apply proteomic profiling to the individual risk assessment in patients with malignancies (see also below). A typical setting includes a discovery phase, in which biomarkers are defined, and a validation phase, in which candidates are evaluated [43]. In principle, it is desirable to combine gene expression and proteome profiling, but this is still in the phase of basic research [44].

Profiling is not only limited to the detection of tumor-specific antigens but can also be used for the detection of tumor-specific autoantibodies by using protein arrays, which were developed first in yeast by expressing previously annotated open reading frames [45, 46]. Protein arrays can be produced recombinantly starting with cDNA libraries, for example, a fetal brain cDNA library subcloned into a bacterial expression vector [47]. Immobilized and tagged proteins can be used for a variety of other applications, including analyses of protein–protein interactions, DNA–protein interactions, and receptor–ligand interactions or allergen screening. In addition to the classical laboratory methods, protein arrays can be used to detect new antigens that elicit antibody responses in patients with tumors and other diseases such as allergy (allergen arrays) [48–50]. In cancer, antibody responses to tumor-associated antigens have been identified for several entities. These autoantibodies represent new biomarkers with potentially high levels of specificity and sensitivity, although the processes by which such self-proteins become immunogenic are not entirely understood (reviewed in Ref. [51]). One source of neoantigenicity is the aberrant localization of proteins in tumors, which has been observed for cAMP-dependent protein kinase A (PKA). The tumor-specific form of this intracellular enzyme can be secreted by cancer cells and is then referred to as ECPKA [52]. Upregulation of ECPKA in the serum correlates with the presence of ECPKA autoantibodies and this autoantibody is highly specific for the detection of cancer [53]. Autoantibodies were either found independent of the cancer type [54], exemplified by antibodies against TP53 (tumor suppressor p53) [55], or were found to be specific for a given cancer type, such as antibodies against HIP-1 (Huntingtin interacting protein 1) in prostate cancer [56].

Another important array-based technology in cancer research is represented by tissue microarrays (TMAs), which can also be subsumed to the field of applied proteomics. In contrast to other array-based methods, TMAs interrogate only one protein at a time but provide information on protein expression on a cellular level for up to 100 clinical samples at a glance. Tissue microarray technology has been recently applied for the evaluation of new biomarkers in larger cancer cohort studies (e.g., Ref.[57]). It can be anticipated that TMAs will be extensively used in the years to come for the definition of biomarkers, which are informative of treatment outcome [58].

13.7 The Present and the Future: Proteomics for Individualized Cancer Therapy

Personalized or individualized medicine is one of the promises for advanced therapy of cancer in the twenty-first century. It is based on the selection of treatments best suited for an individual patient and disease phenotype. Currently, single biomarkers are used for the detection of cancer (e.g., PSA for the diagnosis and monitoring of prostate cancer), but they often lack specificity. Proteomics potentially generate multiple biomarkers containing many different pieces of information and can therefore be regarded as one of the key technologies in the development of individualized cancer therapy. As pointed out above, access to adequately sized sample collections with corresponding reliable diagnostic data is essential for this purpose. A strategy pointing toward personalized medicine involving proteomics has been suggested and demonstrated by Marko-Varga *et al.* [59]. Their starting point was that treatment with gefitinib (Iressa®), a small-molecule inhibitor of the epidermal growth factor receptor, proved to be effective in the treatment of nonsmall cell lung cancer (NSCLC) for defined subgroups in a large phase III trial [60]. Developing interstitial lung disease (ILD) is a comorbidity in patients suffering from NSCLC. Marko-Varga *et al.* addressed the question, whether proteomics will enable prediction of occurrence of ILD in seriously ill NSCLC patients prior to treatment. Their aim was to develop a simple blood test for detection of patients treated by gefitinib and prone to ILD, but the workflow described hereafter is principally suited for all kinds of studies related to cancer risk assessment. They started to analyze blood plasma using a LC/MS–MS platform. Basically, after depletion of high-abundance proteins, all proteins were trypsinized, fractionated by high-resolution nanoflow chromatography, and subjected to mass spectrometry. The combination of modern MS/MS devices and advanced search engines like MASCOT™ [61] coupled to protein databases available at SwissProt or NCBI allowed the identification of almost all proteins in a given sample.

The challenge in defining biomarkers is therefore no longer identification of proteins, but rather to find signals against a background of high variation due to individual nonillness-related differences in plasma composition. Typically, these data analyses can be divided into three steps: data normalization, signal detection, and identification of discriminating peptides. For the latter, statistical modeling approaches are the methods of choice. These include explorative data analyses, mainly to understand the sources of variation, followed by univariate analyses and multivariate predictive modeling. Machine learning algorithms such as support vector machines, SAM or PAM, which have been successfully applied to high-dimensional genomics data [62–65], are also suitable for the analyses of proteomics data. Still, it remains to be determined for each study design individually as to which modeling approach or combination of modeling approaches will perform best for generating the most predictive and robust model. Once a model is obtained, it has to be validated either by internal or by external cross-validation to evaluate this potential biomarker tool in a way that reflects the real-life setting of the prospective clinical use. It can be expected that the combination of developments in different technological areas (cf. Chapter 12) and in data handling, storage, and analyses will further proceed in the next years. This will bring us one step closer to what is anticipated as treatment of patients with cancer based on their individual biological background and risk profile. It can be envisioned that proteomics will contribute together with other "omics" technologies into a decision-making algorithm comprising data also from imaging, pathology, or routine diagnostics (Figure 13.1) for improving the diagnoses and treatment of patients with cancer.

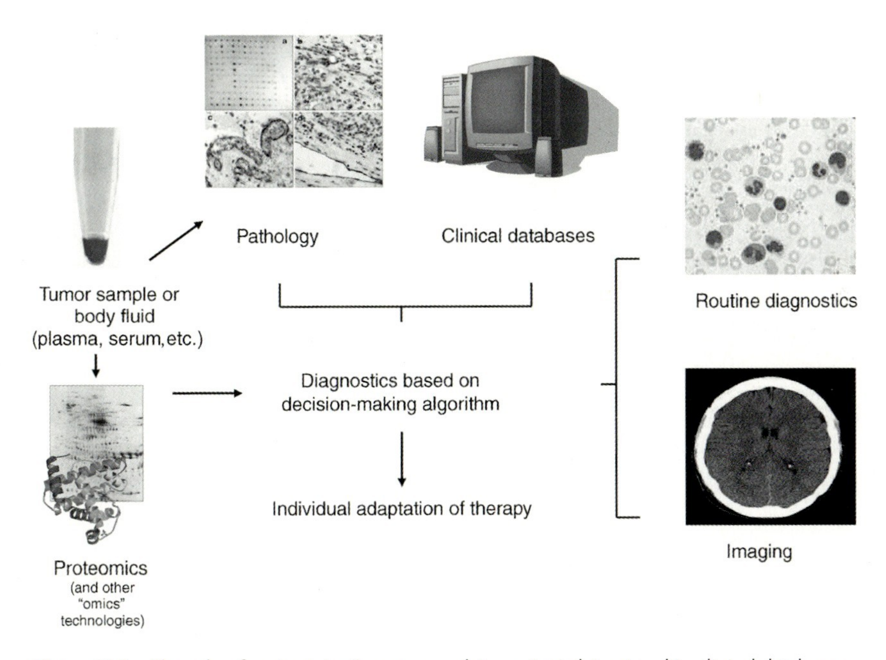

Figure 13.1 The role of proteomics in cancer risk assessment and individualized therapy. A possible scenario is highlighted in which multiple lines of evidence, including pathology data, patient data stored in clinical databases, and results from routine diagnostics and imaging devices, are collected for precise definition of diagnosis and subsequent therapy.

References

1 Iyer, V.R., Eisen, M.B., Ross, D.T., Schuler, G., Moore, T. *et al.* (1999) The transcriptional program in the response of human fibroblasts to serum. *Science*, **283**, 83–87.

2 Schulte, J.H., Schramm, A., Klein-Hitpass, L., Klenk, M., Wessels, H. *et al.* (2005) Microarray analysis reveals differential gene expression patterns and regulation of single target genes contributing to the opposing phenotype of TrkA- and TrkB-expressing neuroblastomas. *Oncogene*, **24**, 165–177.

3 Schramm, A., Schulte, J.H., Klein-Hitpass, L., Havers, W., Sieverts, H. *et al.* (2005) Prediction of clinical outcome and biological characterization of neuroblastoma by expression profiling. *Oncogene*, **24**, 7902–7912.

4 Schramm, A., Vandesompele, J., Schulte, J.H., Dreesmann, S., Kaderali, L. *et al.* (2007) Translating expression profiling into a clinically feasible test to predict neuroblastoma outcome. *Clin. Cancer Res.*, **13**, 1459–1465.

5 Vermeulen, J., De Preter, K., Laureys, G., Speleman, F., and Vandesompele, J. (2009) 59-gene prognostic signature sub-stratifies high-risk neuroblastoma patients. *Lancet Oncol.*, **10**, 1030.

6 Aebersold, R. and Mann, M. (2003) Mass spectrometry-based proteomics. *Nature*, **422**, 198–207.

7 Albers, M., Kranz, H., Kober, I., Kaiser, C., Klink, M. *et al.* (2005) Automated yeast two-hybrid screening for nuclear receptor-interacting proteins. *Mol. Cell. Proteomics*, **4**, 205–213.

8 Koegl, M. and Uetz, P. (2007) Improving yeast two-hybrid screening systems. *Brief Funct. Genomics Proteomics*, **6**, 302–312.

9 Wepf, A., Glatter, T., Schmidt, A., Aebersold, R., and Gstaiger, M. (2009) Quantitative interaction proteomics using mass spectrometry. *Nat. Methods*, **6**, 203–205.

10 Meani, F., Pecorelli, S., Liotta, L., and Petricoin, E.F. (2009) Clinical application of proteomics in ovarian cancer prevention and treatment. *Mol. Diagn. Ther.*, **13**, 297–311.

11 Taylor, C.F., Paton, N.W., Lilley, K.S., Binz, P.A., Julian, R.K., Jr. *et al.* (2007) The minimum information about a proteomics experiment (MIAPE). *Nat. Biotechnol.*, **25**, 887–893.

12 Human Proteome Organisation (2010) http://www.hupo.org (accessed August 26, 2010).

13 Binz, P.A., Barkovich, R., Beavis, R.C., Creasy, D., Horn, D.M. *et al.* (2008) Guidelines for reporting the use of mass spectrometry informatics in proteomics. *Nat. Biotechnol.*, **26**, 862.

14 Gibson, F., Anderson, L., Babnigg, G., Baker, M., Berth, M. *et al.* (2008) Guidelines for reporting the use of gel electrophoresis in proteomics. *Nat. Biotechnol.*, **26**, 863–864.

15 Taylor, C.F., Binz, P.A., Aebersold, R., Affolter, M., Barkovich, R. *et al.* (2008) Guidelines for reporting the use of mass spectrometry in proteomics. *Nat. Biotechnol.*, **26**, 860–861.

16 Omenn, G.S. and Ping, P. (2005) HUPO 3rd Annual World Congress on Proteomics, 25–27th October, 2004, Beijing, China. *Expert Rev. Proteomics*, **2**, 9–12.

17 Omenn, G.S., States, D.J., Adamski, M., Blackwell, T.W., Menon, R. *et al.* (2005) Overview of the HUPO Plasma Proteome Project: results from the pilot phase with 35 collaborating laboratories and multiple analytical groups, generating a core dataset of 3020 proteins and a publicly-available database. *Proteomics*, **5**, 3226–3245.

18 Siegmund, R., Kiehntopf, M., and Deufel, T. (2009) Evaluation of two different albumin depletion strategies for improved analysis of human CSF by SELDI-TOF-MS. *Clin. Biochem.*, **42**, 1136–1143.

19 Emanuele, V.A., 2nd and Gurbaxani, B.M. (2009) Benchmarking currently available SELDI-TOF MS preprocessing techniques. *Proteomics*, **9**, 1754–1762.

20 Roche, S., Tiers, L., Provansal, M., Piva, M.T., and Lehmann, S. (2006) Interest of major serum protein removal for surface-

enhanced laser desorption/ ionization–time of flight (SELDI-TOF) proteomic blood profiling. *Proteome Sci.*, **4**, 20.

21 Sitek, B., Apostolov, O., Stuhler, K., Pfeiffer, K., Meyer, H.E. *et al.* (2005) Identification of dynamic proteome changes upon ligand activation of Trk-receptors using two-dimensional fluorescence difference gel electrophoresis and mass spectrometry. *Mol. Cell. Proteomics*, **4**, 291–299.

22 Sitek, B., Luttges, J., Marcus, K., Kloppel, G., Schmiegel, W. *et al.* (2005) Application of fluorescence difference gel electrophoresis saturation labelling for the analysis of microdissected precursor lesions of pancreatic ductal adenocarcinoma. *Proteomics*, **5**, 2665–2679.

23 Sitek, B., Potthoff, S., Schulenborg, T., Stegbauer, J., Vinke, T. *et al.* (2006) Novel approaches to analyse glomerular proteins from smallest scale murine and human samples using DIGE saturation labelling. *Proteomics*, **6**, 4337–4345.

24 Wu, C.C., MacCoss, M.J., Howell, K.E., and Yates, J.R., 3rd (2003) A method for the comprehensive proteomic analysis of membrane proteins. *Nat. Biotechnol.*, **21**, 532–538.

25 Veenstra., T.D. (2007) Global and targeted quantitative proteomics for biomarker discovery. *J. Chromatogr. B*, **847**, 3–11.

26 Lu, P., Vogel, C., Wang, R., Yao, X., and Marcotte, E.M. (2007) Absolute protein expression profiling estimates the relative contributions of transcriptional and translational regulation. *Nat. Biotechnol.*, **25**, 117–124.

27 Braunschweig, T., Krieg, R.C., Bar-Or, R., Smeets, D., Schwamborn, K. *et al.* (2009) Protein profiling of non-malignant and malignant ascites by SELDI-TOF MS: proof of principle. *Int. J. Mol. Med.*, **23**, 3–8.

28 Hong, M., Zhang, X., Hu, Y., Wang, H., He, W. *et al.* (2009) The potential biomarkers for thromboembolism detected by SELDI-TOF-MS. *Thromb. Res.*, **123**, 556–564.

29 van Winden, A.W., Gast, M.C., Beijnen, J.H., Rutgers, E.J., Grobbee, D.E. *et al.* (2009) Validation of previously identified serum biomarkers for breast cancer with SELDI-TOF MS: a case control study. *BMC Med. Genomics*, **2**, 4.

30 Wang, Q., Shen, J., Li, Z.F., Jie, J.Z., Wang, W.Y. *et al.* (2009) Limitations in SELDI-TOF MS whole serum proteomic profiling with IMAC surface to specifically detect colorectal cancer. *BMC Cancer*, **9**, 287.

31 Rai, A.J., Stemmer, P.M., Zhang, Z., Adam, B.L., Morgan, W.T. *et al.* (2005) Analysis of Human Proteome Organization Plasma Proteome Project (HUPO PPP) reference specimens using surface enhanced laser desorption/ ionization-time of flight (SELDI-TOF) mass spectrometry: multi-institution correlation of spectra and identification of biomarkers. *Proteomics*, **5**, 3467–3474.

32 Olsen, J.V., de Godoy, L.M., Li, G., Macek, B., Mortensen, P. *et al.* (2005) Parts per million mass accuracy on an orbitrap mass spectrometer via lock mass injection into a C-trap. *Mol. Cell Proteomics*, **4**, 2010–2021.

33 Yates, J.R., Cociorva, D., Liao, L., and Zabrouskov, V. (2006) Performance of a linear ion trap–orbitrap hybrid for peptide analysis. *Anal. Chem.*, **78**, 493–500.

34 Peri, S., Navarro, J.D., Amanchy, R., Kristiansen, T.Z., Jonnalagadda, C.K. *et al.* (2003) Development of human protein reference database as an initial platform for approaching systems biology in humans. *Genome Res.*, **13**, 2363–2371.

35 Peri, S., Navarro, J.D., Kristiansen, T.Z., Amanchy, R., Surendranath, V. *et al.* (2004) Human protein reference database as a discovery resource for proteomics. *Nucleic Acids Res.*, **32**, D497–D501.

36 Prasad, T.S., Kandasamy, K., and Pandey, A. (2009) Human Protein Reference Database and Human Proteinpedia as discovery tools for systems biology. *Methods Mol. Biol.*, **577**, 67–79.

37 Silzel, J.W., Cercek, B., Dodson, C., Tsay, T., and Obremski, R.J. (1998) Mass-sensing, multianalyte microarray immunoassay with imaging detection. *Clin. Chem.*, **44**, 2036–2043.

38 Huang, R.P., Huang, R., Fan, Y., and Lin, Y. (2001) Simultaneous detection of multiple cytokines from conditioned media and patient's sera by an antibody-based protein array system. *Anal. Biochem.*, **294**, 55–62.

39 Moody, M.D., Van Arsdell, S.W., Murphy, K.P., Orencole, S.F., and Burns, C. (2001) Array-based ELISAs for high-throughput analysis of human cytokines. *Biotechniques*, **31**, 186–190, 192–194.

40 Sabatino, M., Kim-Schulze, S., Panelli, M.C., Stroncek, D., Wang, E. *et al.* (2009) Serum vascular endothelial growth factor and fibronectin predict clinical response to high-dose interleukin-2 therapy. *J. Clin. Oncol.*, **27**, 2645–2652.

41 Christensen, E., Pintilie, M., Evans, K.R., Lenarduzzi, M., Menard, C. *et al.* (2009) Longitudinal cytokine expression during IMRT for prostate cancer and acute treatment toxicity. *Clin. Cancer Res.*, **15**, 5576–5583.

42 Hanrahan, E.O., Lin, H.Y., Kim, E.S., Yan, S., Du, D.Z. *et al.* (2010) Distinct patterns of cytokine and angiogenic factor modulation and markers of benefit for vandetanib and/or chemotherapy in patients with non-small-cell lung cancer. *J. Clin. Oncol.*, **28**, 193–201.

43 Findeisen, P., Zapatka, M., Peccerella, T., Matzk, H., Neumaier, M. *et al.* (2009) Serum amyloid A as a prognostic marker in melanoma identified by proteomic profiling. *J. Clin. Oncol.*, **27**, 2199–2208.

44 Chen, Q.R., Song, Y.K., Yu, L.R., Wei, J.S., Chung, J.Y. *et al.* (2010) Global genomic and proteomic analysis identifies biological pathways related to high-risk neuroblastoma. *J. Proteome Res.*, **9**, 373–382.

45 Zhu, H., Bilgin, M., Bangham, R., Hall, D., Casamayor, A. *et al.* (2001) Global analysis of protein activities using proteome chips. *Science*, **293**, 2101–2105.

46 Zhu, H., Klemic, J.F., Chang, S., Bertone, P., Casamayor, A. *et al.* (2000) Analysis of yeast protein kinases using protein chips. *Nat. Genet.*, **26**, 283–289.

47 Bussow, K., Cahill, D., Nietfeld, W., Bancroft, D., Scherzinger, E. *et al.* (1998) A method for global protein expression and antibody screening on high-density filters of an arrayed cDNA library. *Nucleic Acids Res.*, **26**, 5007–5008.

48 Deinhofer, K., Sevcik, H., Balic, N., Harwanegg, C., Hiller, R. *et al.* (2004) Microarrayed allergens for IgE profiling. *Methods*, **32**, 249–254.

49 Hiller, R., Laffer, S., Harwanegg, C., Huber, M., Schmidt, W.M. *et al.* (2002) Microarrayed allergen molecules: diagnostic gatekeepers for allergy treatment. *FASEB J.*, **16**, 414–416.

50 Jahn-Schmid, B., Harwanegg, C., Hiller, R., Bohle, B., Ebner, C. *et al.* (2003) Allergen microarray: comparison of microarray using recombinant allergens with conventional diagnostic methods to detect allergen-specific serum immunoglobulin E. *Clin. Exp. Allergy*, **33**, 1443–1449.

51 Zinkernagel, R.M. (2000) What is missing in immunology to understand immunity? *Nat. Immunol.*, **1**, 181–185.

52 Cvijic, M.E., Kita, T., Shih, W., DiPaola, R.S., and Chin, K.V. (2000) Extracellular catalytic subunit activity of the cAMP-dependent protein kinase in prostate cancer. *Clin. Cancer Res.*, **6**, 2309–2317.

53 Nesterova, M.V., Johnson, N., Cheadle, C., Bates, S.E., Mani, S. *et al.* (2006) Autoantibody cancer biomarker: extracellular protein kinase A. *Cancer Res.*, **66**, 8971–8974.

54 Kijanka, G. and Murphy, D. (2009) Protein arrays as tools for serum autoantibody marker discovery in cancer. *J. Proteomics*, **72**, 936–944.

55 Soussi, T. (2000) p53 Antibodies in the sera of patients with various types of cancer: a review. *Cancer Res.*, **60**, 1777–1788.

56 Bradley, S.V., Oravecz-Wilson, K.I., Bougeard, G., Mizukami, I., Li, L. *et al.* (2005) Serum antibodies to huntingtin interacting protein-1: a new blood test for prostate cancer. *Cancer Res.*, **65**, 4126–4133.

57 Chia, S., Norris, B., Speers, C., Cheang, M., Gilks, B. *et al.* (2008) Human epidermal growth factor receptor 2 overexpression as a prognostic factor in a large tissue microarray series of node-negative breast cancers. *J. Clin. Oncol.*, **26**, 5697–5704.

58 Vergis, R., Corbishley, C.M., Norman, A.R., Bartlett, J., Jhavar, S. *et al.* (2008) Intrinsic markers of tumour hypoxia and angiogenesis in localised prostate cancer

and outcome of radical treatment: a retrospective analysis of two randomised radiotherapy trials and one surgical cohort study. *Lancet Oncol.*, **9**, 342–351.

59 Marko-Varga, G., Ogiwara, A., Nishimura, T., Kawamura, T., Fujii, K. *et al.* (2007) Personalized medicine and proteomics: lessons from non-small cell lung cancer. *J. Proteome Res.*, **6**, 2925–2935.

60 Thatcher, N., Chang, A., Parikh, P., Rodrigues Pereira, J., Ciuleanu, T. *et al.* (2005) Gefitinib plus best supportive care in previously treated patients with refractory advanced non-small-cell lung cancer: results from a randomised, placebo-controlled, multicentre study (Iressa Survival Evaluation in Lung Cancer). *Lancet*, **366**, 1527–1537.

61 Perkins, D.N., Pappin, D.J., Creasy, D.M., and Cottrell, J.S. (1999) Probability-based protein identification by searching sequence databases using mass spectrometry data. *Electrophoresis*, **20**, 3551–3567.

62 Tusher, V.G., Tibshirani, R., and Chu, G. (2001) Significance analysis of microarrays applied to the ionizing radiation response. *Proc. Natl. Acad. Sci. USA*, **98**, 5116–5121.

63 De Preter, K., De Brouwer, S., Van Maerken, T., Pattyn, F., Schramm, A. *et al.* (2009) Meta-mining of neuroblastoma and neuroblast gene expression profiles reveals candidate therapeutic compounds. *Clin. Cancer Res.*, **15**, 3690–3696.

64 Schramm, A., Mierswa, I., Kaderali, L., Morik, K., Eggert, A. *et al.* (2009) Reanalysis of neuroblastoma expression profiling data using improved methodology and extended follow-up increases validity of outcome prediction. *Cancer Lett.*, **282**, 55–62.

65 Tibshirani, R., Hastie, T., Narasimhan, B., and Chu, G. (2002) Diagnosis of multiple cancer types by shrunken centroids of gene expression. *Proc. Natl. Acad. Sci. USA*, **99**, 6567–6572.

Part Six
Current Use of Omics Studies for Cancer Risk Assessment

Cancer Risk Evaluation: Methods and Trends,
Edited by Günter Obe, Burkhard Jandrig, Gary E. Marchant, Holger Schütz, and Peter M. Wiedemann.

14
Omics in Cancer Risk Assessment: Pathways to Disease

Christopher J. Portier and Reuben Thomas

14.1
Introduction

Cancer risk assessment has predominantly relied upon the use of chronic exposure animal studies both to identify potential human carcinogens and to quantify the risks from these carcinogens. This creates three classic problems in risk assessment: studies to assess the carcinogenicity of a new agent take a considerable amount of time and expenses (timeliness), results must be extrapolated from a laboratory species (usually rats and/or mice) to humans (species extrapolation), and doses that are large enough to produce a detectable risk in a reasonable size laboratory study must be extrapolated to the usually much lower doses to which humans are exposed (low-dose extrapolation) [1, 2]. Even in cases where there is good human data available to identify and quantify human carcinogenic risks, the data are generally from an occupational setting and require some degree of extrapolation to lower exposures in order to address risks to the general public. Since some of the earliest uses of information for guiding decisions about cancer risks, these three problems have been known to exist and have been the subject of lively debate.

Historically, there have been many solutions to all three problems [1, 2]. Since the earliest risk assessments, safety factors and uncertainty factors have been used to address low-dose extrapolation and species extrapolation. In general, a dose is defined where there is expected to be little or no risk, and depending on the type and quality of data being used, safety factors (usually in multiples of 10) are applied to produce a reasonable margin of safety. It is not unusual to have a total safety factor adjustment range from 10-fold to 3000-fold [3] below the presumed no-risk dose or level of exposure.

Species extrapolation has seen several modifications over the years. The basic modification that most scientist work on for currently advancing risk assessment is not new and is based on a simple parallelogram (Figure 14.1) [2]. The original proposal suggested that if you had three of the four sides of a parallelogram, one of the two parallel lines referred to mechanistic assays in animals or animal cells versus *in vivo* assays in animals and the other line referred to human-based mechanistic data versus human response data. Using the parallelogram approach, if you had an

Cancer Risk Evaluation: Methods and Trends,
Edited by Günter Obe, Burkhard Jandrig, Gary E. Marchant, Holger Schütz, and Peter M. Wiedemann.

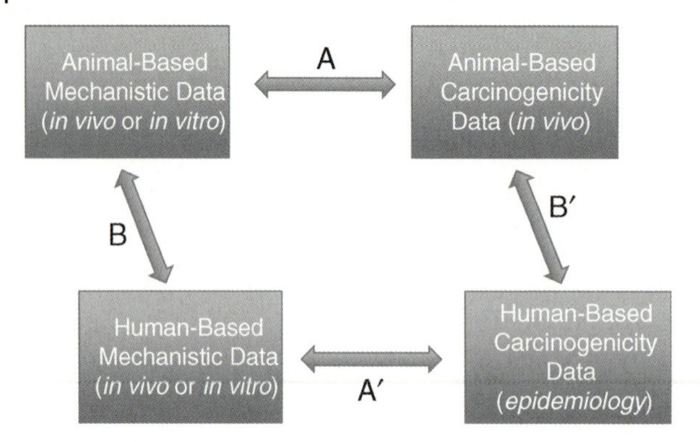

Figure 14.1 Parallelogram linking mechanistic data to cancer data for risk assessment.

animal 2-year chronic bioassay, a set of *in vitro* assays describing the mechanisms that are important to that carcinogenic finding, and did those same assays in relevant human cells, you could accurately predict the factors (relationships A and B in Figure 14.1) that could be used to predict the human response from either animal data (substitute relationship B for B′) or human mechanistic data (substitute relationship A for A′). Thus, with three types of data (human mechanism, animal mechanism, and animal toxicity), one can overcome the extrapolation issues but not the basic problem of resources in time and money to do the long-term animal toxicity studies. This basic concept is still used by the risk assessment community [4, 5].

Low-dose extrapolation went through a series of remarkable changes and attempts at improvement over the past 40 years; many of these changes tied to either statistical concepts that would improve the handling of these types of data or attempts to combine mechanistic data from multiple sources into a single grand model that best explains or predicts the potential carcinogenicity of a compound [1]. In some cases, the improvements addressed both. A recent criticism [6] of these approaches for risk assessment cites a number of statistical and scientific issues that remain unresolved and that, at least some of, the decision tree-based approaches we are using to guide the type of cancer risk assessment we plan to do, may be driven more by dogmatic beliefs than solid scientific evidence that allows one to avoid assumptions that lead to extrapolations.

What is needed for risk assessment to both improve and remain closer to the available science is more rigorous statistical handling of the data, objective use of knowledge to improve extrapolations, and approaches more closely tied to mechanisms, but not necessarily through mechanistic biomathematical models. With the onset of high-content microarray chips for genomics, genetics, proteomics, and metabolomics ("omics" data), there is a greater ability to use a large body of evidence to address mechanistic changes in cellular function that are likely to be tied to the eventual carcinogenic findings [5]. Once these relationships have been identified, then more targeted high-throughput approaches could be used to develop a smaller set of rapidly conducted assays to determine the toxic potential of thousands of chemical or physical agents.

14.2
"Omics" Data in Cancer Risk Assessment

Most approaches that have been used to apply "omics" data to cancer risk assessment have relied upon the concept of a cancer fingerprint being identified by the results of an "omics" experiment [5, 7–12]. Several groups have used small sets of data as case studies to determine if these approaches should work. For example, using data on mouse liver carcinogens and toxicogenomics in mouse liver cells, several groups have shown some degree of concordance between sets of genes and toxicity [13–15]. One major problem with these approaches is that for different chemicals and even for different doses of the same chemical you do not necessarily get the same genes responding [16–20]. To overcome this problem, scientists have begun to use ontology classes [21] for genes describing processes a set of genes are part of and using the disturbance of the processes to portray the linkage between genes and disease (see Figure 14.2). In this case, one would need to know linkage D and any of C, E, or F to be able to make a prediction. Methods have been developed to formally test for such linkages and assign a p-value for interpretation [22–24]. With reasonable mappings between human and animal genes and processes, these types of gene set enrichment analysis can play a major role in identifying chemicals that are likely to have the same mechanisms of action and possibly the same carcinogenic response.

A different approach was proposed by a third group [25] and is illustrated in Figure 14.3. They argued that gene sets, while informative, were suboptimal at using the available information on signaling and metabolic pathways within the cell that are fundamental to the molecular biology of a functioning, healthy cellular environment. They proposed a scheme where the linkage maps describing these pathways can be used as roadmaps to provide a more accurate linkage between genetic changes and cellular function known as the structurally enhanced pathway enrichment analysis (SEPEA).

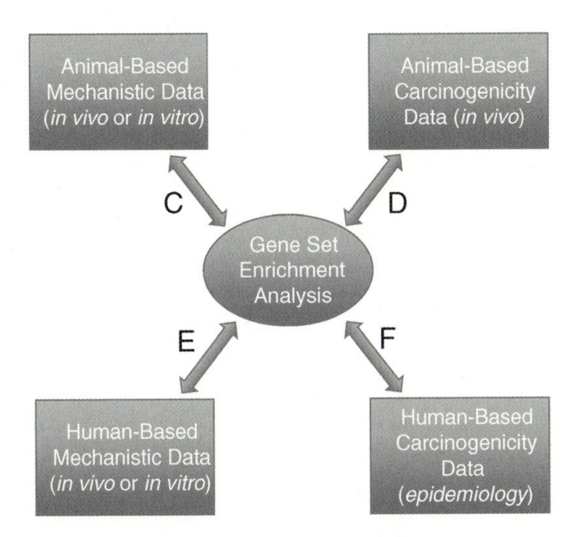

Figure 14.2 Linkage between genes and disease.

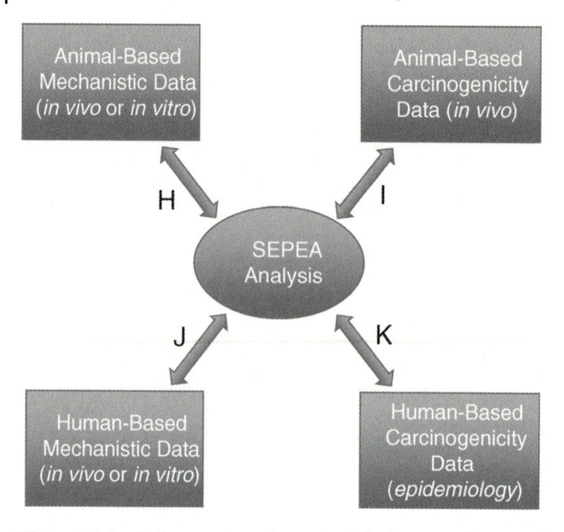

Figure 14.3 Pathways most important for human disease phenotypes using SEPEA.

They argued for a weighting scheme for alterations of genes that are in a particular pathway based upon their locations in the pathway and their "closeness." Their "heavy ends rule" provides greater weight for genes that are at the start or end of a particular pathway than those in the middle. The logic for this weighting scheme is based upon two concepts. First, many different pathways share the same genes and a pathway that is not affected by exposure may appear activated simply because it happens to be cross-linked with several other pathways that are truly activated. Second, the ends of pathways do not generally share common genes, and if you see a pathway altered at its beginning or end, there is a greater probability that this pathway has been altered. They incorporated a second weighting scheme, the "sequential best rule," argued that if a gene is altered in a specific pathway, then if its closest neighbors are also altered, you have a greater probability that this pathway is truly altered, and thus assigned greater weight to genes in a pathway when they were closely linked in the pathway mapping.

In a parallel paper [26], they were able to demonstrate both the utility and the strength of the approach they proposed. They assumed that, at least for chronic diseases such as cancer, it is the consistent perturbation of the pathways that matters and not temporal changes that are not sustained. Given this, they then argued that the best data for looking at chronic perturbation of genes in humans were those relating human genetic polymorphisms (germ line deviations in genes) to human diseases since these changes are present from birth and effectively do not change throughout life. Thus, if six different genes with specific polymorphisms have been shown to be linked to childhood leukemia and all six are in a specific signaling or metabolic pathway, then there is a higher than average likelihood that chemically induced changes in this pathway would increase the risk of childhood leukemia.

The Genetic Association Database (GAD) [27] contains about 30 000 records on human gene–phenotype relationships. Manual grouping of the phenotypes led to about 9000 unique associations between approximately 2000 genes and 200 phenotypes.

The Kyoto Encyclopedia of Genes and Genomes (KEGG) pathway database [28] describes approximately 200 signaling and metabolic pathways. Using the SEPEA algorithm to link these two databases resulted in a mapping showing what pathways were most important for any particular human disease phenotype (relationship K in Figure 14.3) using human genetic polymorphisms linked to the pathways (relationship J in Figure 14.3). To demonstrate the utility of this approach for prediction of toxicity, they used "omics" data from the Comparative Toxicogenomics Database (CTD) [29]. The CTD is a database of interactions between environmental factors and genes/proteins in diverse organisms that has been curated from the published literature using both algorithm-based methods and manual curation. Using the genomic/proteomic changes from the CTD, they linked chemicals to pathways in the KEGG database and since these pathways were already linked to disease using the GAD, they were able to make predictions of which chemicals were likely to be associated with which diseases (combining relationships H and K in Figure 14.3). The resulting gene–environment–disease interactome can be used to group chemicals by common affected pathways, group diseases by common affected pathways, and predict which chemicals are likely to affect which diseases. The interactome of chemicals from the CTD significantly associated with breast cancer is given in Figure 14.4. The distances in between the linked chemicals and the distances in the chemical to breast cancer linkages reflect the significance of the association.

These types of fingerprints, while helpful for the identification of a hazard, still have limitations with regard to the quantitation of risks from exposures to the chemicals. In order to overcome these types of limitations, the somewhat qualitative linkage provided by the gene set enrichment and SEPEA analyses need to be turned into a

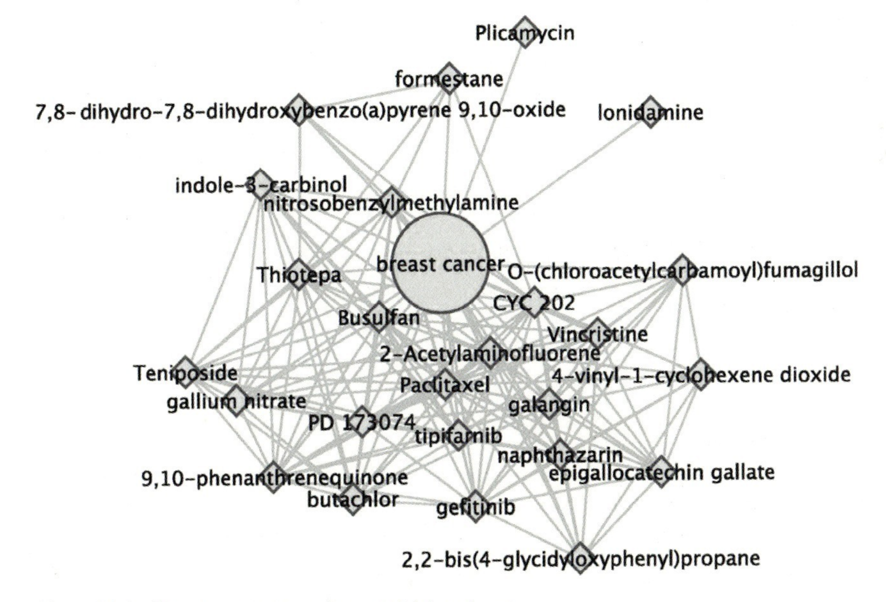

Figure 14.4 Breast cancer to environment interactome.

quantitative relationship that can be compared across several diverse sets of genes with modified message or protein in different dose groups for different chemicals.

Yu *et al.* [30] used gene ontology (GO) classification to develop a score for evaluating dose–response curves from toxicogenomic data. Their approach uses normalized gene expression data to identify significantly altered GO categories. Once this has been done, they assemble the significantly altered genes that were part of the altered GO category and calculate the average fold change for this set of genes (direction of the change does not matter in their analysis). The resulting average fold change (with associated standard error) can then be modeled using standard regression methods or tailored software to produce a dose–response curve for that GO classification. From this, it is possible to derive benchmark doses, no observed adverse effect levels, and other measures commonly used in risk assessment.

Thomas *et al.* [31] proposed a similar methodology, but instead of first evaluating GO classes and then modeling dose response, they identify gene with significant changes individually, model the gene expression for each gene separately, and calculate a benchmark dose from each of these models. They then find the GO classes for the genes and as the benchmark dose for a particular GO classification, they propose the use of the average benchmark dose for the gene in the GO class with confidence bounds derived from the averaging.

Thomas *et al.* [32] in continuation of their work using the SEPEA algorithm and the framework shown in Figure 14.3 developed a prediction model for the likelihood of a significant increase in liver tumors at a particular dose using 26 two-year chronic exposure carcinogenicity studies (10 positive, 2 equivocal, and 14 negative) from National Toxicology Program (NTP) and toxicogenomics data from the livers of B6C3F1 mice exposed for 90 days to the same 26 chemicals. They did a good job on positive predictivity (9 of the 10 positives were predicted correctly with the 10th having a different tumor pathology from the other 9). The false positive rate was high (6 of the 14 negatives were predicted as positive although there are reasons why 2 may need to be redone). One equivocal was predicted positive and the other negative. More importantly, for five of the compounds for which there was dose response for both the toxicogenomics data and the carcinogenicity data, the predictions were accurate for both sensitivity and specificity. They also evaluated cross-species extrapolation for their model by taking the genes associated with human liver cancer and running them through the mouse model. The result was a significant prediction for liver cancer. When 24 other human diseases were evaluated, they were negative with the exception of colorectal cancer and prostate cancer.

14.3 High-Throughput Screening

High-throughput screening (HTS) in toxicology refers to a method of experimentation using robotics, bioinformatics, and sensitive detectors, usually in specially designed cells or through immunohistochemistry, to screen a large number of chemical agents for specific activities. Through the use of HTS, researchers can rapidly identify compounds that target specific metabolic pathways, biochemical pathways, or cellular processes (e.g., apoptosis) to determine if these are targets for the potential toxicity of the chemicals. Large-scale screening of chemicals for these types of biological activity began in the drug industry in the 1980s and has continued to this day. HTS was first seriously introduced into toxicology testing by the NTP as a result of the 2004 Roadmap [33] to achieve the Vision for Toxicology in the twenty-first century [34]. Since then, through collaborations with the US Environmental Protection Agency (EPA), the National Chemical Genomics Research Center, and others, there is a concerted effort to evaluate HTS data as predictive tools for risk assessment known as the Tox21 Initiative [35].

The current HTS design for Tox21 uses a 1536 well plate. Each assay consists of two control plates followed by 15 plates with increasing (doubling) concentration of chemicals on each plate, followed by two more control plates. On each plate, 1353 different compounds are tested with 1298 using only one well and 55 using two wells. There are a variety of bioinformatic tools that can be used to evaluate these types of data [36–38] and several authors have already shown some promise in the use of these data for qualitative and quantitative risk assessment [39–42].

"Omics" data, as discussed above, can be used to identify pathways for screening and even help to identify where one might like to place molecular probes on these target pathways for maximum efficiency. For example, one of the pathways that was significantly linked to breast cancer in the analysis by Gohlke *et al.* [26] is the focal adhesion pathway. Examining the structure of this pathway, one can identify points where it appears that a large part of the signaling will be channeled (e.g., JUN and ERK) and focus an HTS assay on these cellular proteins. Changes in these proteins can also have an impact on mRNA levels and these could also be targeted. Another method of identifying targets is to overexpress target receptors or proteins in the network and measure activity in all the components of the pathway using genomics or proteomics. In Figure 14.5, data from Creighton *et al.* [43] were used to illustrate this approach. In their experiment, Creighton *et al.* used modified MCF-7 human cell lines that overexpressed Raf, c-erb-2, MEK, and EGFR and used microarray chips to look at changes in gene expression in other genes due to overexpression of these four genes. The results are shown in Figure 14.5 where color coding has been used to indicate the degree of change in each gene in the focal adhesion pathway. Darker nodes indicate genes with a very strong change and it is these points that would serve as good targets for identifying activity in this pathway (e.g., RhoB, CAPN1, and CCND3). Since many pathways share common genes, it will take more than a single target to be able to identify specific pathways that have been activated. However, once this is accomplished, with a model based upon the human genetics, genomics, and disease data as shown above, HTS offers a means to screen many chemicals for particular pathways in a way that is impractical for microarrays.

HTS data can also be combined with other historical prediction tools to enhance our ability to make reasonable predictions of potential long-term toxicity. At present, the most cited tool for predicting biological activity of a chemical is quantitative structure–activity relationships (QSARs). QSAR uses attributes of a chemical (e.g., solubility in oil, surface charge, and molecular width at the

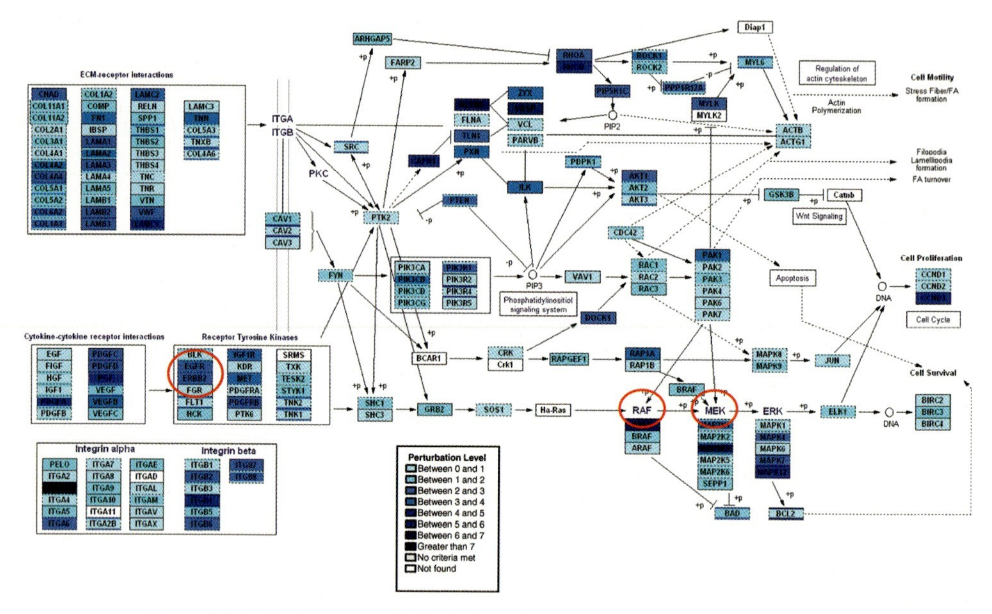

Figure 14.5 Finding targets for high-throughput screening using selective gene overexpression in the focal adhesion signaling pathway from KEGG (from Ref. [43]).

maximum diameter) to characterize the overall structure and activity of a chemical, then, using regression tools, links this with the biological activity of a compound to build a prediction model for that activity. Linking these characteristics with the HTS activity profiles may greatly enhance the predictivity of QSAR models and lead to better decisions for further testing and risk assessment [41].

At present, HTS assays allow one to potentially screen millions of compounds in a single day at a relatively small cost per data point. On the other hand, for the same amount of resources, one could only screen a few hundred compounds using microarray technology and this would take multiple days. In the future, multiplexed assays, potentially as dense as the microarray analyses that are being done, may be possible as HTS assays and the "omics" and HTS screening will likely become synonymous.

14.4 Discussion

"Omics" and HTS hold great promise for dramatically improving the scientific quality and speed of toxicological screening. But, at this point, it is only a promise. The use of these technologies for risk assessment is clear in some ways and unknown in others.

Most regulatory authorities have established guidance on how to use mechanistic data in risk assessment. By their very nature, "omics" and HTS provide information on the mechanisms through which a chemical interacts with cellular function and, in doing so, provide the potential to explain why certain chemicals cause disease in laboratory animals and humans. Hence, the initial use of these types of data for risk assessment has focused on mechanisms, looking at concordance across species and, in some cases, extrapolation to lower exposure levels. They have proven themselves to be useful for these types of assessments and, this alone, warrants the large investment in resources being directed to these activities by the NTP, the EPA, and others.

In its original vision [34] and roadmap [33], the NTP saw "omics" and HTS data as a means to set priorities for further testing rather than a direct replacement for the current screening tools used in regulatory toxicology. The idea was that by screening a large number of compounds with known and unknown toxicity, one would be able to find activity patterns in chemicals that would allow for targeting experimental studies at a hypothesized disease for a given compound and by comparing the activity patterns of compounds, group them into those with similar patterns. The groups of chemicals could then be tested in *in vivo* assays using invertebrates, such as *Caenorhabditis elegans* [44, 45], or nonmammalian vertebrates, such as zebrafish [46] to confirm common activity in a complete living organism [47]. Once a group is firmly established, complete toxicological testing could be done on a subset of the chemicals and the toxicity of the remaining compounds be inferred from the "omics," HTS, and short-term *in vivo* data. While the NTP would then test as many compounds in traditional screens as they currently do, this testing would be more effective and result in the regulatory review of a much larger number of chemicals overall. Once this approach has been well established and verified to work for several classes of compounds, the NTP envisioned a gradual evolution to eliminating the traditional toxicological screens altogether in favor of these other shorter and quicker assays.

Not all toxicologists hold the same view as that envisioned by the NTP. After the National Research Council report on Toxicity Testing in the Twenty-first Century [48], many scientists believe that it is possible to directly skip the priority setting and group validation proposed by the NTP and go directly to the development of predictive models using "omics" and HTS data. As noted, some attempts have already been made to build these models and demonstrate their utility. Many argue that since "omics" and HTS can be done on human cells and cell lines, you avoid the species extrapolation and the serious question of whether effects seen in animals are likely to occur in humans. We believe the extrapolation from cellular responses to human disease may prove to be just as daunting as species extrapolation and may not fully satisfy regulatory needs.

One thing is for certain: before "omics" and HTS data will receive regulatory acceptance in risk assessment, sufficient data on enough chemicals using a large array of targets are needed and this is now happening. There is no doubt that these new technologies will play a major role in shaping risk assessment into the twenty-first century; the only remaining question concerns the magnitude of that impact.

References

1 National Research Council (U.S.), Committee on Improving Risk Analysis Approaches Used by the US EPA, Board on Environmental Studies and Toxicology, and Division on Earth and Life Studies (2009) *Science and Decisions: Advancing Risk Assessment*, National Academies Press, Washington, DC.
2 National Research Council (U.S.), Committee on the Institutional Means for Assessment of Risks to Public Health (1983) *Risk Assessment in the Federal Government: Managing the Process*, National Academy Press, Washington, DC.
3 Stedeford, T., Zhao, Q.J., Dourson, M.L., Banasik, M., and Hsu, C.H. (2007) The application of non-default uncertainty factors in the U.S. EPA's Integrated Risk Information System (IRIS). Part I: UF(L), UF(S), and "other uncertainty factors". *J. Environ. Sci. Health C*, **25**, 245–279.
4 Kienhuis, A.S., van de Poll, M.C., Dejong, C.H., Gottschalk, R., van Herwijnen, M., Boorsma, A., Kleinjans, J.C., Stierum, R.H., and van Delft, J.H. (2009) A toxicogenomics-based parallelogram approach to evaluate the relevance of coumarin-induced responses in primary human hepatocytes *in vitro* for humans *in vivo*. *Toxicol. In Vitro*, **23**, 1163–1169.
5 National Research Council (U.S.), Committee on Applications of Toxicogenomic Technologies to Predictive Toxicology and National Academies Press (U.S.) (2007) *Applications of Toxicogenomic Technologies to Predictive Toxicology and Risk Assessment*, National Academies Press, Washington, DC.
6 Crump, K.S., Chen, C., Chiu, W.A., Louis, T.A., Portier, C.J., Subramaniam, R.P., and White, P.D. (2010) What role for biologically based dose-response models in estimating low-dose risk? *Environ. Health Perspect.*, **118**, 585–588.
7 Foster, W.R., Chen, S.J., He, A., Truong, A., Bhaskaran, V., Nelson, D.M., Dambach, D.M., Lehman-McKeeman, L.D., and Car, B.D. (2007) A retrospective analysis of toxicogenomics in the safety assessment of drug candidates. *Toxicol. Pathol.*, **35**, 621–635.
8 Mattes, W.B. (2006) Cross-species comparative toxicogenomics as an aid to safety assessment. *Expert Opin. Drug Metab. Toxicol.*, **2**, 859–874.
9 Sone, H., Okura, M., Zaha, H., Fujibuchi, W., Taniguchi, T., Akanuma, H., Nagano, R., Ohsako, S., and Yonemoto, J. (2010) Profiles of Chemical Effects on Cells (pCEC): a toxicogenomics database with a toxicoinformatics system for risk evaluation and toxicity prediction of environmental chemicals. *J. Toxicol. Sci.*, **35**, 115–123.
10 Thybaud, V., Le Fevre, A.C., and Boitier, E. (2007) Application of toxicogenomics to genetic toxicology risk assessment. *Environ. Mol. Mutagen.*, **48**, 369–379.
11 Waters, M.D., Jackson, M., and Lea, I. (2010) Characterizing and predicting carcinogenicity and mode of action using conventional and toxicogenomics methods. *Mutat. Res.*, **705**, 184–200.
12 Zhou, T., Chou, J., Watkins, P.B., and Kaufmann, W.K. (2009) Toxicogenomics: transcription profiling for toxicology assessment. *EXS*, **99**, 325–366.
13 Heinloth, A.N., Irwin, R.D., Boorman, G.A., Nettesheim, P., Fannin, R.D., Sieber, S.O., Snell, M.L., Tucker, C.J., Li, L., Travlos, G.S., Vansant, G., Blackshear, P.E., Tennant, R.W., Cunningham, M.L., and Paules, R.S. (2004) Gene expression profiling of rat livers reveals indicators of potential adverse effects. *Toxicol. Sci.*, **80**, 193–202.
14 Tang, W. (2007) Drug metabolite profiling and elucidation of drug-induced hepatotoxicity. *Expert Opin. Drug Metab. Toxicol.*, **3**, 407–420.
15 Ulrich, R.G., Rockett, J.C., Gibson, G.G., and Pettit, S.D. (2004) Overview of an interlaboratory collaboration on evaluating the effects of model hepatotoxicants on hepatic gene expression. *Environ. Health Perspect.*, **112**, 423–427.
16 Toyoshiba, H., Sone, H., Yamanaka, T., Parham, F.M., Irwin, R.D., Boorman, G.A., and Portier, C.J. (2006) Gene

interaction network analysis suggests differences between high and low doses of acetaminophen. *Toxicol. Appl. Pharmacol.*, **215**, 306–316.

17 Auerbach, S.S., Shah, R.R., Mav, D., Smith, C.S., Walker, N.J., Vallant, M.K., Boorman, G.A., and Irwin, R.D. (2010) Predicting the hepatocarcinogenic potential of alkenylbenzene flavoring agents using toxicogenomics and machine learning. *Toxicol. Appl. Pharmacol.*, **243**, 300–314.

18 Thomas, R.S., Bao, W., Chu, T.M., Bessarabova, M., Nikolskaya, T., Nikolsky, Y., Andersen, M.E., and Wolfinger, R.D. (2009) Use of short-term transcriptional profiles to assess the long-term cancer-related safety of environmental and industrial chemicals. *Toxicol. Sci.*, **112**, 311–321.

19 Zhang, L.P., McHale, C.M., Rothman, N., Li, G.L., Ji, Z.Y., Vermeulen, R., Hubbard, A.E., Ren, X.F., Shen, M., Rappaport, S.M., North, M., Skibola, C.F., Yin, S.N., Vulpe, C., Chanock, S.J., Smith, M.T., and Lan, Q. (2010) Systems biology of human benzene exposure. *Chem. Biol. Interact.*, **184**, 86–93.

20 Judson, R.S., Houck, K.A., Kavlock, R.J., Knudsen, T.B., Martin, M.T., Mortensen, H.M., Reif, D.M., Rotroff, D.M., Shah, I., Richard, A.M., and Dix, D.J. (2010) *In vitro* screening of environmental chemicals for targeted testing prioritization: the ToxCast Project. *Environ. Health Perspect.*, **118**, 485–492.

21 Ashburner, M., Ball, C.A., Blake, J.A., Botstein, D., Butler, H., Cherry, J.M., Davis, A.P., Dolinski, K., Dwight, S.S., Eppig, J.T., Harris, M.A., Hill, D.P., Issel-Tarver, L., Kasarskis, A., Lewis, S., Matese, J.C., Richardson, J.E., Ringwald, M., Rubin, G.M., and Sherlock, G. (2000) Gene ontology: tool for the unification of biology. The Gene Ontology Consortium. *Nat. Genet.*, **25**, 25–29.

22 Efron, B. and Tibshiran, R. (2007) On testing the significance of sets of genes. *Ann. Appl. Stat.*, **1**, 107–129.

23 Subramanian, A., Tamayo, P., Mootha, V.K., Mukherjee, S., Ebert, B.L., Gillette, M.A., Paulovich, A., Pomeroy, S.L., Golub, T.R., Lander, E.S., and Mesirov, J.P. (2005) Gene set enrichment analysis: a knowledge-based approach for interpreting genome-wide expression profiles. *Proc. Natl. Acad. Sci. USA*, **102**, 15545–15550.

24 Tian, L., Greenberg, S.A., Kong, S.W., Altschuler, J., Kohane, I.S., and Park, P.J. (2005) Discovering statistically significant pathways in expression profiling studies. *Proc. Natl. Acad. Sci. USA*, **102**, 13544–13549.

25 Thomas, R., Gohlke, J.M., Stopper, G.F., Parham, F.M., and Portier, C.J. (2009) Choosing the right path: enhancement of biologically relevant sets of genes or proteins using pathway structure. *Genome Biol.*, **10**, R44.

26 Gohlke, J.M., Thomas, R., Zhang, Y., Rosenstein, M.C., Davis, A.P., Murphy, C., Becker, K.G., Mattingly, C.J., and Portier, C.J. (2009) Genetic and environmental pathways to complex diseases. *BMC Syst. Biol.*, **3**, 46.

27 Becker, K.G., Barnes, K.C., Bright, T.J., and Wang, S.A. (2004) The genetic association database. *Nat. Genet.*, **36**, 431–432.

28 Kanehisa, M. and Goto, S. (2000) KEGG: Kyoto Encyclopedia of Genes and Genomes. *Nucleic Acids Res.*, **28**, 27–30.

29 Mattingly, C.J., Colby, G.T. Forrest, J.N., and Boyer, J.L. (2003) The Comparative Toxicogenomics Database (CTD). *Environ. Health Perspect.*, **111** 793–795.

30 Yu, X.Z., Griffith, W.C. Hanspers, K., Dillman, J.F. Ong, H., Vredevoogd, M.A., and Faustman, E.M. (2006) A system-based approach to interpret dose- and time-dependent microarray data: quantitative integration of gene ontology analysis for risk assessment. *Toxicol. Sci.*, **92**, 560–577.

31 Thomas, R.S., Allen, B.C., Nong, A., Yang, L., Bermudez, E., Clewell, H.J., and Andersen, M.E. (2007) A method to integrate benchmark dose estimates with genomic data to assess the functional effects of chemical exposure. *Toxicol. Sci.*, **98**, 240–248.

32 Thomas, R., Thomas, R., Auerbach, S.S., and Portier, C.J. (2010) Biological networks for predicting chemical

hepatocarcinogenicity using gene expression data from treated mice and its relevance across human and rat species. *Environ. Health Perspect.*, submitted.

33 National Toxicology Program (2004) A National Toxicology Program for the 21st Century, National Institute of Environmental Health Sciences, Research Triangle Park, NC, http://ntp.niehs.nih.gov/files/NTPrdmp.pdf (accessed July 13, 2009).

34 National Toxicology Program (2003) Toxicology in the 21st Century: The Role of the National Toxicology Program, National Institute of Environmental Health Sciences, Research Triangle Park, NC, http://ntp.niehs.nih.gov/ntp/main_pages/NTPVision.pdf (accessed July 13, 2009).

35 Collins, F.S., Gray, G.M., and Bucher, J.R. (2008) Toxicology: transforming environmental health protection. *Science*, **319**, 906–907.

36 Inglese, J., Auld, D.S., Jadhav, A., Johnson, R.L., Simeonov, A., Yasgar, A., Zheng, W., and Austin, C.P. (2006) Quantitative high-throughput screening: a titration-based approach that efficiently identifies biological activities in large chemical libraries. *Proc. Natl. Acad. Sci. USA*, **103**, 11473–11478.

37 Parham, F., Austin, C., Southall, N., Huang, R., Tice, R., and Portier, C. (2009) Dose–response modeling of high-throughput screening data. *J. Biomol. Screen.*, **14**, 1216–1227.

38 Smith, M.V., Boyd, W.A., Kissling, G.E., Rice, J.R., Snyder, D.W., Portier, C.J., and Freedman, J.H. (2009) A discrete time model for the analysis of medium-throughput *C. elegans* growth data. *PLoS One*, **4**, e7018.

39 Boyd, W.A., Smith, M.V., Kissling, G.E., Rice, J.R., Snyder, D.W., Portier, C.J., and Freedman, J.H. (2009) Application of a mathematical model to describe the effects of chlorpyrifos on *Caenorhabditis elegans* development. *PLoS One*, **4**, e7024.

40 Xia, M., Huang, R., Witt, K.L., Southall, N., Fostel, J., Cho, M.H., Jadhav, A., Smith, C.S., Inglese, J., Portier, C.J., Tice, R.R., and Austin, C.P. (2008) Compound cytotoxicity profiling using quantitative high-throughput screening. *Environ. Health Perspect.*, **116**, 284–291.

41 Zhu, H., Rusyn, I., Richard, A., and Tropsha, A. (2008) Use of cell viability assay data improves the prediction accuracy of conventional quantitative structure–activity relationship models of animal carcinogenicity. *Environ. Health Perspect.*, **116**, 506–513.

42 Zhu, H., Ye, L., Richard, A., Golbraikh, A., Wright, F.A., Rusyn, I., and Tropsha, A. (2009) A novel two-step hierarchical quantitative structure–activity relationship modeling work flow for predicting acute toxicity of chemicals in rodents. *Environ. Health Perspect.*, **117**, 1257–1264.

43 Creighton, C.J., Hilger, A.M., Murthy, S., Rae, J.M., Chinnaiyan, A.M., and El-Ashry, D. (2006) Activation of mitogen-activated protein kinase in estrogen receptor alpha-positive breast cancer cells *in vitro* induces an *in vivo* molecular phenotype of estrogen receptor alpha-negative human breast tumors. *Cancer Res.*, **66**, 3903–3911.

44 Boyd, W.A., McBride, S.J., Rice, J.R., Snyder, D.W., and Freedman, J.H. (2010) A high-throughput method for assessing chemical toxicity using a *Caenorhabditis elegans* reproduction assay. *Toxicol. Appl. Pharmacol.*, **245**, 153–159.

45 Boyd, W.A., Smith, M.V., Kissling, G.E., and Freedman, J.H. (2009) Medium- and high-throughput screening of neurotoxicants using *C. elegans*. *Neurotoxicol. Teratol.*, **32**, 68–73.

46 Kaufman, C.K., White, R.M., and Zon, L. (2009) Chemical genetic screening in the zebrafish embryo. *Nat. Protoc.*, **4**, 1422–1432.

47 Portier, C.J. and Schwartz, D.A. (2005) The NIEHS and the national toxicology program: an integrated scientific vision. *Environ. Health Perspect.*, **113**, A440–A540.

48 National Research Council (U.S.), Committee on Toxicity Testing and Assessment of Environmental Agents . (2007) *Toxicity Testing in the 21st Century: A Vision and a Strategy*, National Academies Press, Washington, DC.

15
What Have "Omics" Taught Us about the Health Risks Associated with Exposure to Low Doses of Ionizing Radiation

William F. Morgan and Marianne B. Sowa

15.1
Introduction

There are numerous agencies and regulatory bodies worldwide whose goal is to protect mankind from the potentially deleterious health effects of exposure to ionizing radiation. These deleterious health effects include leukemia and solid tumors, as well as noncancer effects such as hypertension, stroke, and other cardiovascular complications. While the carcinogenic effects of exposure to high doses of radiation (>200 mSv) are relatively well established, the dilemma for radiation protection is what is the scientific basis for radiation standards to protect mankind from exposures to low levels of ionizing radiation (<100 mSv). This is because there are considerable uncertainties in the epidemiological data. Based primarily on the data from the Japanese A-bomb survivors, there appears to be a linear dose response for cancer risk at doses between ~200 mSv and ~2 Sv [1].

At doses below 100 mSv, there are significant controversies as to potential human responses. These include the induction of nontargeted effects [2, 3], which suggest that the tissue at risk is greater than the volume actually irradiated or, alternatively, could indicate that the irradiated host is "adapted" to better respond to the radiation insult and more rapidly regain adaptive homeostasis. Adaptive responses suggest that in a protracted exposure situation man can "adapt" to subsequent insults and is better equipped to deal with both induced DNA damage [4] and individual genetic susceptibilities and sensitivities, lifestyle factors, and host environment. Such risk(s) at low radiation doses are also modulated by age at exposure, sex, total radiation dose, dose rate, radiation quality, dose distribution, mode of exposure (internal versus external exposure), and time since exposure. Obviously, evaluating human risk(s) associated with exposure to low doses of ionizing radiation is complicated and fraught with complications for a given individual.

Here, we will evaluate the contribution of advances in high-throughput "omics" technologies to examine what they have taught us about health risk(s) to humans associated with exposure to ionizing radiation. In this evaluation, we will

Cancer Risk Evaluation: Methods and Trends,
Edited by Günter Obe, Burkhard Jandrig, Gary E. Marchant, Holger Schütz, and Peter M. Wiedemann.

"cherry-pick" published papers to illustrate points rather than attempt to provide a comprehensive review of this rapidly expanding field.

15.2
Pre-"Omics"

It has long been recognized that cells respond to ionizing radiation through an intricate network of interacting signaling cascades to regulate diverse cellular responses and functions. It is not our goal to present an exhaustive review of the vast *in vitro* biochemical, cellular, or *in vivo* animal literature to substantiate this claim. Instead, we will briefly highlight work from two of the many research groups that in our opinion spearheaded the effort to move radiation effects into the molecular era. The Fornace Laboratory used subtractive hybridization and identified the growth arrest DNA damage (GADD) genes. They concluded that these genes may represent part of a novel regulatory pathway involved in the negative control of mammalian cell growth after DNA damage [5]. Such studies ultimately resulted in the seminal paper by Kastan *et al.* [6], who in collaboration with the Fornace Laboratory demonstrated that following DNA damage primary murine fibroblasts became deficient in a G1 checkpoint activated by ionizing radiation when both wild-type p53 alleles were disrupted. In addition, cells from patients with the radiosensitive, cancer-prone disease ataxia telangiectasia lacked the radiation-induced increase in p53 protein levels seen in normal cells. Parallel and complementary studies from Boothman and colleagues investigated the biochemical events involved in potentially lethal DNA damage repair and identified a pleiotropic protein expression response that is activated after X-irradiation of confluence-arrested human malignant melanoma (U1-Mel) cells. They found that the majority of induced proteins synthesized by X-irradiated cells were of low molecular weights and that their expression characteristics suggested their role in either gross chromosomal repair or X-ray-induced adaptive responses [7–10].

15.3
Functional Genomics

Subsequent sequencing-based techniques including differential display and serial analysis of gene expression accelerated gene discovery. Amundson *et al.* [11] used cDNA microarray hybridization to discover potential biomarkers of radiation exposure and identified genes expressed at increased levels in human peripheral blood lymphocytes after *ex vivo* irradiation. While a number of candidate genes were identified, for example, DDB2, CDKN1A (also known as CIP1/WAF1), and XPC [11], their response varied as a function of time after exposure indicating that such techniques provided a temporal "snapshot" of the dynamic process of cellular response. Furthermore, Amundson *et al.* [11] screened a panel of 12 human cancer cell lines and clearly demonstrated significant variability in radiation-induced gene expression between the different cell lines. Extending this study, Amundson *et al.* [12] interrogated the 60 cell lines of the National Cancer Institute Anticancer Drug Screen (NCI-60) that constitute the most extensively characterized *in vitro* cancer cell model to develop a database of responses of cancer cells to ionizing radiation. They compared clonogenic survival and apoptosis with gene expression response by microarray analysis. Twenty-two genes were differentially regulated in cells with low survival after 2 Gy of γ-rays, while 14 genes identified lines more sensitive to a very high dose of 8 Gy. Unlike reported basal gene expression patterns, changes in the expression of radiation responsive genes showed little tissue-of-origin effect, except for differentiating the lymphoblastoid cell lines from other cell types. Basal expression patterns appeared to discriminate between radiosensitive and radioresistant lines. The most striking patterns in the radiation data were a set of genes upregulated preferentially in the p53 wild-type lines and a set of cell cycle regulatory genes downregulated across the entire NCI-60 panel [12]. The response of those genes to γ-rays was apparently unaffected by the genetic differences across this diverse cell set.

15.4 Gene Expression Profiling for Nontargeted Effects Induced by Exposure to Ionizing Radiation

Nontargeted effects associated with exposure to ionizing radiation are those effects elicited by irradiated cells that occur in cells that were not subject to energy deposition events, that is, effects occurring in nonirradiated cells that were stimulated by signals from irradiated cells [2, 3]. Such signals may be transmitted via cell-to-cell gap junction communication or by secreted or shed factors from irradiated cells [13].

Azzam *et al.* [14] used gene expression profiling to examine cDNA from confluent human diploid fibroblast cultures exposed to very low fluences of α-particles. In such instances not all cells in the population would be "hit" by the high linear energy radiation and under such conditions a pronounced bystander effect had previously been observed [15]. Their data, supported by Northern and Western analyses, indicated that radiation induced upregulation of connexin 43 expression. This phenomenon was observed in a variety of irradiated cell types and was consistent with previous observations that connexin 43-mediated gap junction intercellular communication is involved in the bystander response [15]. Using the Columbia University charged particle mircobeam in conjunction with a strip dish design, Zhou *et al.* [16] showed that the cyclooxygenase-2 (COX-2) signaling cascade plays an essential role in the bystander process. Treatment of bystander cells with NS-398, which suppresses COX-2 activity, significantly reduced the bystander effect. Because a critical event in COX-2 signaling is the activation of the mitogen-activated protein kinase pathways, the finding that inhibition of the extracellular signal-related kinase phosphorylation suppressed the bystander response further indicated a role for mitogen-activated protein kinase signaling cascades in the bystander process. These results provide evidence that the COX-2-related pathway, which is essential in mediating cellular inflammatory response, may be a critical signaling link for the bystander phenomenon [17].

Bystander effects after low linear energy transfer irradiation have also been demonstrated following transfer of medium from irradiated cells to nonirradiated cells that can lead to reduced cell survival in the nontargeted bystander cells [18]. Chaudhry *et al.* [19] took medium from irradiated normal human diploid lung fibroblasts and examined gene expression alterations in bystander cells. The microarray data revealed that the radiation-induced gene expression profile in irradiated cells was different from nonirradiated bystander cells suggesting that the pathways leading to bystander effects were different from the directly targeted (irradiated) cells. Not surprisingly, genes known to be responsive to ionizing radiation were observed in irradiated cells. Several genes were upregulated but no genes were significantly downregulated in these cells. A number of genes belonging to extracellular signaling, growth factors, and several receptors were identified in bystander cells. Interestingly, 15 genes involved in the cell-to-cell communication processes were found to be upregulated, a result that supported the observations of Azzam and colleagues described above. The induction of receptors and the cell communication processes in bystander cells receiving media from irradiated cells supports the active involvement of these processes in inducing the bystander effect.

Another nontargeted effect of ionizing radiation is genomic instability. This can occur in the progeny of an irradiated cell, multiple cell generations after the initial exposure [2, 3]. Snyder and Morgan [20] examined differential gene expression patterns associated with the instability phenotype using microarray analysis. They used the GM10115 human–hamster hybrid-derived chromosomally stable and radiation-induced unstable clones and identified a number of candidate genes that were all underexpressed relative to the chromosomally stable reference sample. Among this set of differentially expressed genes identified were two candidates with a relationship to the ubiquitin/proteasome pathway. While follow-up gene expression analyses have confirmed the underexpression of these two genes in some of the chromosomally unstable clones, functional studies were unable to demonstrate a link to instability [21].

15.5
Gene Expression Profiling for Adaptive Responses Induced by Exposure to Ionizing Radiation

Adaptive responses are cell and tissue responses that appear to make cells more refractory to exposures to ionizing radiations. These can occur after stimulation with a very low "adapting" or "priming" dose of radiation and make cells more "resistant" to a higher challenge dose of radiation (reviewed in Ref. [4]). Alternatively, adaptive responses can be observed when cells are exposed to a very low dose of radiation and a reduction in the spontaneous level of background effects occurs [22, 23]. It is important to remember here that the background frequency of a particular endpoint has to be high enough in order to observe a significant reduction after radiation exposure – thus, a very limited number of endpoints are available for such analysis. Coleman and colleagues [24] investigated the cytogenetic adaptive response in human cell lines exposed to 5 cGy priming dose followed by a subsequent 2 Gy challenge dose. Their goal was to determine how the priming dose influenced gene transcript expression in adapting and nonadapting cell lines. Transcript profiles were evaluated using oligonucleotide arrays and RNA obtained 4 h after the challenge dose. A set of 145 genes affected by the priming dose fell into two categories. First, a set of common genes that were modulated by the 5 cGy priming dose, irrespective of whether the cells subsequently adapted, and second, genes with differential transcription in accordance with whether or not the cell lines adapted. The common priming dose response genes showed upregulation for protein synthesis genes and downregulation of metabolic and signal transduction. The genes associated with subsequent adaptive and nonadaptive outcomes involved DNA repair, stress response, cell cycle control, and apoptosis.

15.6 *In Vivo* Gene Profiling after Irradiation

In vitro data demonstrating that the dose, dose rate, radiation quality, and elapsed time since ionizing radiation exposure result in variations in cellular responses of stress genes suggest that gene expression signatures may be informative markers of radiation exposure. Changes in gene expression in human cell lines occur after as little as 0.02 Gy and in peripheral blood lymphocytes after as little as 0.2 Gy [11]. Diverse genes are also elevated *in vivo* in mice 24 h after 0.2 Gy irradiation, but how these relate to either the *in vitro* situation in human cell lines or the *in vivo* situation in the event of a radiation accident or radiotherapy is far from clear [25].

There have been a number of reports on gene expression profiling in tissue samples from humans, primarily radiation therapy patients, exposed to radiation. Goldberg and colleagues [26, 27] used a patient-based tissue acquisition protocol and performed a series of genomic analyses on the temporal dynamics over a 24 h period to determine the radiation response after a single exposure of 100 mGy. RNA from each patient tissue sample was hybridized to an Affymetrix Human Genome U133 Plus 2.0 array. They found 19 gene groups and 7 gene pathways that were radiation responsive. Of these, nine gene groups showed significant transient transcriptional changes in the human tissue samples, which returned to baseline by 24 h postexposure. These investigators concluded that doses of ionizing radiation on full-thickness human skin produced a definable temporal response. Genes involved in DNA and tissue remodeling, cell cycle transition, and inflammation showed statistically significant changes in expression, despite variability between patients. Goldberg *et al.* [27] evaluated transcriptomic responses to a single exposure in the normal skin of men undergoing therapeutic radiation for prostate cancer and demonstrated a dose–response pattern in gene expression in a number of pathways and gene groups that were biologically plausible responses to low doses of ionizing radiation. These included the Akt/phosphoinositide-3-kinase pathway, the growth factor pathway, the stress/apoptosis pathway, and the pathway initiated by transforming growth factor-beta signaling, whereas gene groups with altered expression included the keratins, the zinc finger proteins, and signaling molecules in the mitogen-activated protein kinase gene group. Interestingly, these investigators showed that although there was significant individual variability in radiation response that makes the detection of effects difficult, it was still feasible when analyzed according to gene group and pathway. The genes and pathways involved indicated that the skin tissue (1) does detect the injury, (2) initiates a stress/inflammatory response, (3) undergoes DNA remodeling, and (4) initiates a "prosurvival," or as described above, an adaptive response. These observations have important implications for risk assessment both in radiation therapy and in the event of a radiological terrorist event [28].

Clearly, ionizing radiation exposure to biological systems can result in transcriptional responses that vary as a function of time, dose, and dose rate, across the individual organism. This creates major difficulties in detecting these responses, especially when the data for any one individual are limited and when the number of genes, probes, or probe sets is large. Rocke *et al.* [29] have developed a procedure that allows sensitive detection of transcriptional responses that differ between individuals in type or in timing following exposure to ionizing radiation. This consists of four steps: the first is to identify a group of genes, probes, or probe sets that detect genes belonging to a molecular class or to a common pathway. The second is to conduct a statistical test of the hypothesis that the gene is differentially expressed for each individual and for each gene in the set. The third is to examine the collection of these statistics to determine if there is a detectable signal. The final step is to assess the significance of this by resampling to avoid correlation biases (http://www.idav.ucdavis.edu/~dmrocke/software).

15.7
Radiation-Induced Oscillatory Signaling

The dynamics of radiation-induced signaling are relatively poorly understood. This is in part due to the presence of regulatory mechanisms that can modulate the response on the timescale of many experimental measurements. In human breast cancer cells, exposure to γ-radiation produced damped oscillations in the levels of p53 expressions and its negative regulator Mdm2, occurring with a periodicity of 4–7 h [30]. A single cell can show repeated oscillations in both p53 and Mdm2 levels; however, the dose delivered does not determine the number of pulses, but rather the probability of a given cell oscillating increases as dose increases. The biological significance of such oscillatory signaling is not fully understood [31]. It may represent a means of sensing the environment to turn on repair mechanisms or to signal for cell survival or death [30]. Negative feedback mechanisms such as these may provide a way of returning to homeostasis after a damaging insult such as exposure to radiation [32]. Oscillations in the p53 pathway have also been observed *in vivo* for the mouse intestine and spleen, occurring with a similar period as seen *in vitro* (~5 h) [33].

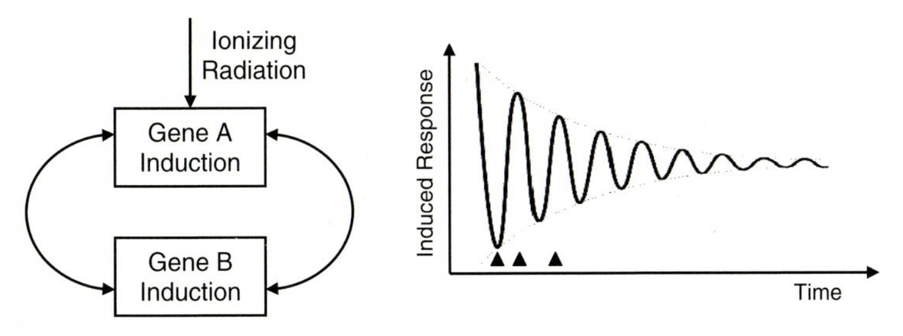

Figure 15.1 Ionizing radiation induces (represses) a given gene A. This in turn induces gene B that then feeds back to induce gene A causing "pulses" (oscillations) in a given response. The filled triangles represent sample points.

Oscillations occurring on a much shorter timescale (periodicity of ~15 min) have been observed for the EGF-induced regulation of ERK signaling in human mammary epithelial cells [34]. Such rapid oscillations like those seen for ERK are difficult to detect as they are more rapid than the typical sampling period of most experiments. Measurements made as a function of time may provide differential responses depending on where they fall on the oscillation pattern (Figure 15.1). In addition, oscillatory patterns for ERK are highly context dependent and do not occur for all cells in a population [34]. They are inhibited by cell contact and, in 3D mammary culture, low LET radiation exposures did not produce nuclear translocation of ERK (Sowa, unpublished results). Using a transformed mouse epidermal cell model to study malignant transformation in skin carcinogenesis, Weber *et al.* [35] found that cells also demonstrate oscillatory translocation of ERK to the nucleus and these oscillations are regulated by basic fibroblast growth factor in the premalignant but not malignant phenotype.

15.8 Proteomic Profiling after Exposure to Ionizing Radiation

Both the dynamic changes in the proteome in terms of the abundance of cellular and secreted proteins and posttranslational protein modifications that include methylation, acetylation, and phosphorylation, among others, in response to radiation are important for understanding cell signaling pathways. Lin *et al.* [36] irradiated C57BL/6 mice with 20 Gy and 24 or 48 h later liver cell homogenates were subfractionated into nuclei, mitochondria, and cytosol, for analysis by MALDI-TOF mass spectrometry (MS). Thirty-seven proteins at 24 h and 29 proteins at 48 h were matched with known proteins after database searching in nuclei, mitochondria, and cytosol, respectively, among which nine proteins exhibited changes at both time points. Not surprisingly, most of these were involved in antioxidant response, energy metabolism, molecular chaperones, and inflammatory responses. Reverse transcriptase polymerase chain reaction (RT-PCR) and Western blotting further validated these proteins. This particular study is described to illustrate the temporal nature of the radiation response. Obviously, if there are gene expression oscillations/pulses following radiation exposure as discussed earlier, it is not unreasonable to expect protein levels to fluctuate or oscillate as a function of time after irradiation as autocrine and paracrine signaling pathways feed back upon the induced protein response [32]. Not surprisingly, long lists of proteins modified by irradiation have been profiled, and vary as a function of cell line/organism studied, time after irradiation, dose, and dose rate [37–40].

Clearly, concerns about radiological terrorism have led to a number of studies in an effort to discover proteomic biomarkers of radiation exposure. For example, Menard *et al.* [41] studied radiotherapy patients before and during radiation therapy. High-resolution surface-enhanced laser desorption and ionization time-of-flight (SELDI-TOF) MS was used to generate high-throughput proteomic profiles of unfractionated serum samples using an immobilized metal ion-affinity chromatography nickel-affinity chip surface. MS-based protein identification was then done on pooled sera to identify specific protein fragments that were altered after radiation exposure. Computer-based analyses of the SELDI protein spectra could distinguish unexposed from radiation-exposed patient samples with 91–100% sensitivity and 97–100% specificity using various classifier models. Given the present interest in radiation-induced protein biomarkers, it is reasonable to expect an explosion of information in this area in the future. Human biofluids, including blood and urine as well as relatively noninvasive methodologies for tissue samples, for example, hair pulp samples and fingernails, will likely be developed and interrogated (see Ref. [42]).

15.9
Metabolomic Profiling after Exposure to Ionizing Radiation

Metabolomics can be thought of as the study of small molecule profiles that are the end product of multiple cellular processes. This is a relatively new "omic" platform and provides information of the physiology of the cell, tissue, or organ. In a series of publications, Idle and coworkers [42, 43, 44] used metobolomic profiling to investigate potential biomarkers for γ-radiation exposure. Mice were γ-irradiated at doses of 0, 3, and 8 Gy and urine samples collected 24 h after exposure were analyzed by ultra-performance liquid chromatography time-of-flight mass spectrometry (UPLC-TOF MS). They analyzed the top 22 ions for each dose and found that *N*-hexanoylglycine and β-thymidine were urinary biomarkers of exposure to 3 and 8 Gy, that 3-hydroxy-2-methylbenzoic acid 3-*O*-sulfate was elevated in urine of mice exposed to 3 Gy but not 8 Gy, and that taurine was elevated after 8 Gy but not 3 Gy [43]. These studies were extended to profile the hydrophilic metabolome of TK6 cells as a function of time and radiation dose. Using ultra-performance liquid chromatography coupled with electrospray time-of-flight mass spectrometry, Patterson *et al.* [42] found dose- and time-dependent clustering of the irradiated cells and significant depletion of a number of metabolites associated with oxidative stress and DNA repair pathways as early as 1 h after irradiation.

15.10
Conclusions

In this brief snapshot of some of the "omics" approaches to investigating the effects of ionizing radiation, it is clear that significant progress has been made. However, one of the challenges for understanding tissue, organ, and even organismal responses to ionizing radiation is integrating all these "omics" platforms. This requires a "systems level" approach to combining these technologies in a living organism and interrogating them as a function of dose, dose rate, and time after exposure [45, 46]. This is a major challenge and will require both long-term funding commitments and collaboration of a diverse team of skilled investigators and is likely to be a long complex process as befits the plethora of studies to date [47].

Acknowledgment

This research was supported by Battelle Memorial Institute, Pacific Northwest Division, under Contract No. DE-AC05-76RL0 1830 with the US Department of Energy, Office of Biological and Environmental Research Low Dose Science Program.

References

1 NAS/NRC (2005) *Health Effects from Exposure to Low Levels of Ionizing Radiation, BEIR VII*, The National Academies Press, Washington DC.
2 Morgan, W.F. (2003) Non-targeted and delayed effects of exposure to ionizing radiation: II. Radiation-induced genomic instability and bystander effects *in vivo*, clastogenic factors and transgenerational effects. *Radiat. Res.*, **159**, 581–596.
3 Morgan, W.F. (2003) Non-targeted and delayed effects of exposure to ionizing radiation: I. Radiation-induced genomic instability and bystander effects *in vitro*. *Radiat. Res.*, **159**, 567–580.
4 Tapio, S. and Jacob, V. (2007) Radioadaptive response revisited. *Radiat. Environ. Biophys.*, **46**, 1–12.
5 Fornace, A.J., Jr., Nebert, D.W., Hollander, M.C., Luethy, J.D., Papathanasiou, M., Fargnoli, J., and Holbrook, N.J. (1989) Mammalian genes coordinately regulated by growth arrest signals and DNA-damaging agents. *Mol. Cell. Biol.*, **9**, 4196–4203.
6 Kastan, M.B., Zhan, Q., el-Deiry, W.S., Carrier, F., Jacks, T., Walsh, W.V., Plunkett, B.S., Vogelstein, B., and Fornace, A.J., Jr. (1992) A mammalian cell cycle checkpoint pathway utilizing p53 and GADD45 is defective in ataxia-telangiectasia. *Cell*, **71**, 587–597.
7 Boothman, D.A., Bouvard, I., and Hughes, E.N. (1989) Identification and characterization of X-ray-induced proteins in human cells. *Cancer Res.*, **49**, 2871–2878.
8 Boothman, D.A., Majmudar, G., and Johnson, T. (1994) Immediate X-ray-inducible responses from mammalian cells. *Radiat Res.*, **138**, S44–S46.
9 Boothman, D.A., Meyers, M., Fukunaga, N., and Lee, S.W. (1993) Isolation of X-ray-inducible transcripts from radioresistant human melanoma cells. *Proc. Natl. Acad. Sci. USA*, **90**, 7200–7204.
10 Leskov, K.S., Criswell, T., Antonio, S., Li, J., Yang, C.R., Kinsella, T.J., and Boothman, D.A. (2001) When

X-ray-inducible proteins meet DNA double strand break repair. *Semin. Radiat. Oncol.*, **11**, 352–372.

11 Amundson, S.A., Do, K.T., Shahab, S., Bittner, M., Meltzer, P., Trent, J., and Fornace, A.J., Jr. (2000) Identification of potential mRNA biomarkers in peripheral blood lymphocytes for human exposure to ionizing radiation. *Radiat. Res.*, **154**, 342–346.

12 Amundson, S.A., Do, K.T., Vinikoor, L.C., Lee, R.A., Koch-Paiz, C.A., Ahn, J., Reimers, M., Chen, Y., Scudiero, D.A., Weinstein, J.N., Trent, J.M., Bittner, M.L., Meltzer, P.S., and Fornace, A.J., Jr. (2008) Integrating global gene expression and radiation survival parameters across the 60 cell lines of the National Cancer Institute Anticancer Drug Screen. *Cancer Res.*, **68**, 415–424.

13 Morgan, W.F. and Sowa, M.B. (2009) Non-targeted effects of ionizing radiation: implications for risk assessment and the radiation dose response profile. *Health Phys.*, **97**, 426–432.

14 Azzam, E.I., de Toledo, S.M., and Little, J.B. (2003) Expression of Connexin43 is highly sensitive to ionizing radiation and other environmental stresses. *Cancer Res.*, **63**, 7128–7135.

15 Azzam, E.I., de Toledo, S.M., and Little, J.B. (2001) Direct evidence for the participation of gap junction-mediated intercellular communication in the transmission of damage signals from alpha-particle irradiated to nonirradiated cells. *Proc. Natl. Acad. Sci. USA*, **98**, 473–478.

16 Zhou, H., Ivanov, V.N., Gillespie, J., Geard, C.R., Amundson, S.A., Brenner, D.J., Yu, Z., Lieberman, H.B., and Hei, T.K. (2005) Mechanism of radiation-induced bystander effect: role of the cyclooxygenase-2 signaling pathway. *Proc. Natl. Acad. Sci. USA*, **102**, 14641–14646.

17 Hei, T.K., Zhou, H., Ivanov, V.N., Hong, M., Lieberman, H.B., Brenner, D.J., Amundson, S.A., and Geard, C.R. (2008) Mechanism of radiation-induced bystander effects: a unifying model. *J. Pharm. Pharmacol.*, **60**, 943–950.

18 Mothersill, C. and Seymour, C. (1997) Medium from irradiated human epithelial cells but not human fibroblasts reduces the clonogenic survival of unirradiated cells. *Int. J. Radiat. Biol.*, **71**, 421–427.

19 Chaudhry, M.A. (2006) Bystander effect: biological endpoints and microarray analysis. *Mutat. Res.*, **597**, 98–112.

20 Snyder, A.R. and Morgan, W.F. (2004) Radiation-induced chromosomal instability and gene expression profiling: searching for clues to initiation and perpetuation. *Mutat. Res.*, **568**, 89–96.

21 Snyder, A.R. and Morgan, W.F. (2005) Lack of consensus gene expression changes associated with radiation-induced chromosomal instability. *DNA Repair (Amst.)*, **4**, 958–970.

22 Azzam, E.I., de Toledo, S.M., Raaphorst, G.P., and Mitchel, R.E. (1996) Low-dose ionizing radiation decreases the frequency of neoplastic transformation to a level below the spontaneous rate in C3H 10T1/2 cells. *Radiat. Res.*, **146**, 369–373.

23 Redpath, J.L., Short, S.C., Woodcock, M., and Johnston, P.J. (2003) Low-dose reduction in transformation frequency compared to unirradiated controls: the role of hyper-radiosensitivity to cell death. *Radiat. Res.*, **159**, 433–436.

24 Coleman, M.A., Yin, E., Peterson, L.E., Nelson, D., Sorensen, K., Tucker, J.D., and Wyrobek, A.J. (2005) Low-dose irradiation alters the transcript profiles of human lymphoblastoid cells including genes associated with cytogenetic radioadaptive response. *Radiat. Res.*, **164**, 369–382.

25 Amundson, S.A. and Fornace, A.J., Jr. (2001) Gene expression profiles for monitoring radiation exposure. *Radiat. Prot. Dosimetry*, **97**, 11–16.

26 Berglund, S.R., Rocke, D.M., Dai, J., Schwietert, C.W., Santana, A., Stern, R.L., Lehmann, J., Hartmann Siantar, C.L., and Goldberg, Z. (2008) Transient genome-wide transcriptional response to low-dose ionizing radiation *in vivo* in humans. *Int. J. Radiat. Oncol. Biol. Phys.*, **70**, 229–234.

27 Goldberg, Z., Rocke, D.M., Schwietert, C., Berglund, S.R., Santana, A., Jones, A.,

Lehmann, J., Stern, R., Lu, R., and Hartmann Siantar, C. (2006) Human *in vivo* dose-response to controlled, low-dose low linear energy transfer ionizing radiation exposure. *Clin. Cancer Res.*, **12**, 3723–3729.

28 Lehmann, J., Stern, R.L., Daly, T.P., Rocke, D.M., Schwietert, C.W., Jones, G.E., Arnold, M.L., Siantar, C.L., and Goldberg, Z. (2006) Dosimetry for quantitative analysis of the effects of low-dose ionizing radiation in radiation therapy patients. *Radiat. Res.*, **165**, 240–247.

29 Rocke, D.M., Goldberg, Z., Schweitert, C., and Santana, A. (2005) A method for detection of differential gene expression in the presence of inter-individual variability in response. *Bioinformatics*, **21**, 3990–3992.

30 Wee, K.B., Surana, U., and Agunda, B.D. (2009) Oscillations of the p53-Akt network: implications on cell survival and death. *PLoS One*, **4**, e4407.

31 Tyson, J.J. (2006) Another turn for p53. *Mol. Syst. Biol.*, **10**, 1–3.

32 Lev Bar-Or, R., Maya, R., Segel, L.A., Alon, U., Levine, A.J., and Oren, M. (2000) Generation of oscillations by the p53-Mdm2 feedback loop: a theoretical and experimental study. *Proc. Natl. Acad. Sci. USA*, **97**, 11250–11255.

33 Hamstra, D.A., Bhojani, M.S., Griffin, L.B., Laxman, B., Ross, B.D., and Rehemtulla, A. (2006) Real-time evaluation of p53 oscillatory behavior *in vivo* using bioluminescent imaging. *Cancer Res.*, **66**, 7482–7489.

34 Shankaran, H., Ippolito, D.L., Chrisler, W.B., Resat, H., Bollinger, N., Opresko, L.K., and Wiley, H.S. (2009) Rapid and sustained nuclear–cytoplasmic ERK oscillations induced by epidermal growth factor. *Mol. Syst. Biol.*, **5**, 1–13.

35 Weber, T.J., Shankaran, H., Wiley, H.S., Opresko, L.K., and Chrisler, W.B., and Quesenberry, R.D. (2009) Basic fibroblast growth factor regulates persistent ERK oscillations in premalignant but not malignant JB6 cells. *J. Invest. Dermatol.*, **130** (5), 1444–1456.

36 Lin, R.X., Zhao, H.B., Li, C.R., Sun, Y.N., Qian, X.H., and Wang, S.Q. (2009) Proteomic analysis of ionizing radiation-induced proteins at the subcellular level, *J. Proteome Res.*, **8**, 390–399.

37 Berglund, S.R., Santana, A.R., Li, D., Rice, R.H., Rocke, D.M., and Goldberg, Z. (2009) Proteomic analysis of low dose arsenic and ionizing radiation exposure on keratinocytes. *Proteomics*, **9**, 1925–1938.

38 Jenkins, L.M., Mazur, S.J., Rossi, M., Gaidarenko, O., Xu, Y., and Appella, E. (2008) Quantitative proteomics analysis of the effects of ionizing radiation in wild type and p53 K317R knock-in mouse thymocytes. *Mol. Cell. Proteomics*, **7**, 716–727.

39 Lee, Y.S., Chang, H.W., Jeong, J.E., Lee, S.W., and Kim, S.Y. (2008) Proteomic analysis of two head and neck cancer cell lines presenting different radiation sensitivity. *Acta Otolaryngol.*, **128**, 86–92.

40 Park, E.C., Yoon, J.B., Seong, J.S., Choi, K.S., Kong, E.S., Kim, Y.J., Park, Y.M., and Park, E.M. (2006) Effect of ionizing radiation on rat tissue: proteomic and biochemical analysis. *Prep. Biochem. Biotechnol.*, **36**, 19–35.

41 Menard, C., Johann, D., Lowenthal, M., Muanza, T., Sproull, M., Ross, S., Gulley, J., Petricoin, E., Coleman, C.N., Whiteley, G., Liotta, L., and Camphausen, K. (2006) Discovering clinical biomarkers of ionizing radiation exposure with serum proteomic analysis, *Cancer Res.*, **66**, 1844–1850.

42 Patterson, A.D., Li, H., Eichler, G.S., Krausz, K.W., Weinstein, J.N., Fornace, A.J., Jr., Gonzalez, F.J., and Idle, J.R. (2008) UPLC-ESI-TOFMS-based metabolomics and gene expression dynamics inspector self-organizing metabolomic maps as tools for understanding the cellular response to ionizing radiation. *Anal. Chem.*, **80**, 665–674.

43 Tyburski, J.B., Patterson, A.D., Krausz, K.W., Slavik, J., Fornace, A.J., Jr., Gonzalez, F.J., and Idle, J.R. (2008) Radiation metabolomics. 1. Identification of minimally invasive urine biomarkers for gamma-radiation exposure in mice. *Radiat. Res.*, **170**, 1–14.

44 Tyburski, J.B., Patterson, A.D., Krausz, K.W., Slavik, J., Fornace, A.J., Jr., Gonzalez, F.J., and Idle, J.R. (2009) Radiation metabolomics. 2. Dose- and time-dependent urinary excretion of deaminated purines and pyrimidines after sublethal gamma-radiation exposure in mice. *Radiat. Res.*, **172**, 42–57.

45 Barcellos-Hoff, M.H. (2008) Cancer as an emergent phenomenon in systems radiation biology. *Radiat. Environ. Biophys.*, **47**, 33–38.

46 Barcellos-Hoff, M.H. and Costes, S.V. (2006) A systems biology approach to multicellular and multi-generational radiation responses. *Mutat Res.*, **597**, 32–38.

47 Morgan, W.F. (2010) Integrated systems biology as an approach to understand the effects of exposure to ionizing radiation. *Radiat. Protect. Dosimetry*, in press.

16
Transcriptomics Approach in RF EMF Research

Meike Mevissen

16.1
Introduction

The abundance of specific mRNA transcripts in a biological sample is a reflection of the expression levels of the corresponding genes [1].

Gene expression profiling is the identification and characterization of the mixture of mRNA that is present in a specific sample. An important application of gene expression profiling is to associate differences in mRNA mixtures originating from different groups of individuals with phenotypic differences between the groups [2, 3].

In contrast to genotyping, gene expression profiling allows characterization of the level of gene expression. Both, the presence of specific forms of mRNA and the levels in which these forms occur are parameters that provide information on gene expression [4].

In contrast to the genome, the transcroptome is highly variable over time, between cell types, and will change in response to environmental changes. A gene expression profile provides a quantitative overview of the mRNA transcripts that were present in a sample at the time of collection. Therefore, gene expression profiling can be used to determine which genes are differently expressed as a result of changes in environmental conditions. A typical gene expression profiling study includes a group of individuals with similar phenotype (e.g., exposure level, disease status) and compares the gene expression profile of this group with the profile of a reference group matched on selected factors such as age and sex to the group of interest. Studies of this type usually report a set of genes that are differentially expressed between the groups.

Microarrays represent an innovative and comprehensive technology that allows researchers to assess the expression level of thousands of genes in a high-throughput fashion. The development of high-quality, commercially available gene arrays has allowed this technology to become a standard tool in molecular toxicology.

In spite of years of research, there is still ongoing discussion whether RF EMFs (radio frequency electromagnetic fields) (100 kHz–300 GHz) can induce health effects, especially in long-time users and children. Studies evaluating changes in

Cancer Risk Evaluation: Methods and Trends,
Edited by Günter Obe, Burkhard Jandrig, Gary E. Marchant, Holger Schütz, and Peter M. Wiedemann.

transcription profiles are possibly able to speed up the identification of genes responding to RF EMF.

16.2 Transcriptomics in RF EMF Research

To date, 26 studies examining transcriptomic changes have been published using a variety of species, tissues, and cell types. The following overview of the available transcriptomics studies in RF EMF research is presented in the context of tissue/cell type. Table 16.1 provides a summary of the transcriptomics studies after RF EMF exposure.

GenChip technology was recently used to assess the effect of RF EMF on gene expression after *in vivo* exposure in mouse cerebellum [5]. Interestingly, the short-term RF EMF exposure (2 h) caused expression changes in a smaller number of genes compared to the 6 h exposure. The functional classification of the affected genes revealed that these genes are associated primarily with apoptosis and cell cycle regulation. However, the differentially expressed genes were not confirmed by other methods [5].

Gene expression was also studied in whole mouse brains by microarrays containing over 22 600 probes. RF EMF exposure (1805 MHz; SAR: 0.56 W/kg maximal value in the brain and 0.2 W/kg average value in the brain) of mice for 1 h did not indicate consistent gene expression modulation in whole mouse brain [6]. Overall, none of the changes in gene expression (>1.5- or >2-fold) passed stringent statistical filters, suggesting the absence of gene expression variation across samples. The results provide no clear evidence that RF EMF might induce major gene expression changes in the whole mouse brain after 2 h of exposure [6]. In a recent study, significant alterations in a gene expression were seen after short-term exposure of Fischer 344 rats to RF EMF (1800 MHz; SAR of 30 mW/kg average value in the brain). The Gene Ontology analysis revealed significantly altered genes belonging to Gene Ontology categories associated with membrane activity and function and signal transduction in both cortex and hippocampus. The results were not confirmed by RT PCR or other methods [7].

In human neuroblastoma cells (SK-N-SH), RF EMF exposure (900 MHz; SAR (average special absorption rate): 0.2 W/kg) for 2 h did not result in different gene expression when sham-exposed cells were compared to RF EMF-exposed cells [8]. Even though six genes, including genes important for cell cycle regulation and regulation of the circadian rhythm, were found to be downregulated after a 2 h exposure and a 2 h recovery period, these findings could not be verified by RT PCR. Hsp70 protein levels were found unchanged, and similarly, the gene microarray indicated no change in the gene expression levels of several large heat shock genes [8]. RF EMF-exposed human microglia cells and human neuroblastoma or glioblastoma cells did not show different gene expression when compared to sham controls [9–14]. Zhao *et al.* [15] identified a significant change in gene expression of 34 genes after 24 h of intermittent exposure in rat neurons at an SAR of 2 W/kg. These genes are

Table 16.1 RF EMF transcriptomics studies.

Reference	Organism/tissue/cell	Exposure			Assay: no. of experiments	Results: no. of screened genes	Function
		RF EMF (MHz)	SAR (W/kg)	Duration			
Harvey and French [23]	Human mast cells (HMC-1)	864 CW	7.3	3×20 min/day, 7 days	2	3	Apoptosis, stress response
Pacini *et al.* [20]	Human skin fibroblasts	902 GSM	0.6	1 h	1	14	Stress response, cell cycle, nuclear receptors, cytoskeleton, apoptosis, extracellular matrix
Port [27]	Human. leukemia cells (HL-60)	400	50 kV/m	6 min	2	0	None
Leszczynski *et al.* [18]	Human endothelial cells (EA.hy926)	900 GSM	2.5	1 h	10	49	Apoptosis
Lee *et al.* [26]	Human leukemia cells (HL-60)	2450	10	2 h	1	221	Apoptosis, cell cycle, metabolism, RNA processing and translation, glia function, BBB function
				6 h		759	
Balyaev [5]	Rat (Fischer), cerebellum	915 GSM	0.4	2 h	3	12[a)]	Apoptosis, metabolism, cell cycle, blood coagulation, polysaccharide biosynthesis, RNA processing

(*Continued*)

Table 16.1 (*Continued*)

Reference	Organism/tissue/cell	Exposure			Assay: no. of experiments	Results: no. of screened genes	Function
		RF EMF (MHz)	SAR (W/kg)	Duration			
Gruisik *et al.* [8]	Human neuroblastoma cells (SK-N-SH)	900 GSM CW	0.2	1, 2 h	1	6[a)]	Cell cycle (CCPG1)
Hirose *et al.* [12]	Human glioblastoma cells (A172)	2143 CW	0.08	24, 48 h	2	0	
		2143 CDMA	0.08 0.25 0.8	24, 48 h			
	Human fibroblasts (IMR-90)	2142 CW	0.8	24, 48 h	2	0	
		2143 CDMA	0.08	28 h			
Huang *et al.* [24]	Human T-lymphoma cells (Jurkat)	1762 CDMA	10	7 days, 1 h/day	5	68	Apoptosis, metabolism
Quotob *et al.* [14]	Human glioblastoma cells (U87MG)	1900 GSM	0.1, 1, 10	4 h	5	0	
Whitehead *et al.* [22]	Mouse fibroblasts (C3H 10T1/2)	836 FDMA	5	24 h	3	0	
		848 CDMA	5	24 h	3	0	

Zeng *et al.* [17]	Human breast cancer cells (MCF-7)	1800 PW	2.0, 3.5	24 h, 5 min on/10 min off	2	5[a)]	
Remondini *et al.* [9]	Human neuroblastoma cells (NB69)	1800 GSM	2.0	24 h, 5 min on/10 min off	3	0	
	Human endothelial cell (EA.hy926)	1800 GSM	2.5	1 h	2	0	
		900 GSM	2.5	1 h	4	32	Metabolism, stress response, differentiation
	Human T-lymphocytes	1800 GSM	1.4	44 h, 10 min on/20 min off	3	0	
	Human leukemia cells (HL-60)	1800 GSM	1.0	24 h, 5 min on/5 min off	6	12	Stress response, metabolism
				24 h	6	0	
	Human lymphoma monocytes (U937)	900 GSM	2.5	1 h	5	34	Metabolism, differentiation, signaling
	Human microglia cells (CHME5)	900 GSM	2.0	1 h	5	0	
Nylund and Leszczynski [19]	Human endothelial cells						
	EA.hy926	900 GSM	2.8	1 h	3	1	Not specified
	EA.hy926v1	900 GSM	2.8	1 h	10	13	Various
Chauhan *et al.* [10]	Human glioblastoma cells (U87MG)	1900 GSM	0.1, 1.0, 10	24 h	5	0	

(Continued)

Table 16.1 (*Continued*)

Reference	Organism/tissue/cell	Exposure			Assay: no. of experiments	Results: no. of screened genes	Function
		RF EMF (MHz)	SAR (W/kg)	Duration			
	Human monocyte cells (MM6)		1.0, 10	6 h (5 min on/10 min off)	5	0	
Hirose *et al.* [12]	Human glioblastoma cells (A172)	2143 CDMA	0.08	2, 24, 48 h	3	0	
			0.250	2, 24, 48 h		0	
			0.8	2, 24, 48 h		0	
		2143 CW	0.08	24, 48 h		0	
	Human fibroblasts (IMR90)	2143 CDMA	0.08	2, 28 h	3	0	
			0.8	2, 28 h		0	
		2143 CW	0.08	28 h		0	
Zhao *et al.* [15]	Rat (Sprague Dawley), neurons	1800 GSM	2.0	24 h, 5 min on/10 min off	1	34	Cytoskeleton, signaling, metabolism, ion channels, G-protein-coupled receptors, and others
Zhao *et al.* [16]	Mouse (ICR), neurons	1900 GSM	n.d.	2 h	1	9	Apoptosis
	Mouse (ICR), astrocytes	1900 GSM		2 h	1		Apoptosis
Huang *et al.* [28]	Mouse auditory hair cells (HEI-OC1)	1763 GSM	20	24, 48 h	3	18	No consistent group of functional categories

Huang *et al.* [25]	Human T-lymphoma cells (Jurkat)	1763 GSM	2, 10 10	1, 2, 3 days, 1 h/day 1, 4, 24 h	5	10	Chemotaxis of immune cells
Engelmann *et al.* [29]	Plant, *Arabidopsis thaliana*	1900 GSM CW	2 (peak) 0.75 (average)	24 h	4	10	Dark-inducible genes (three), function of seven genes is unknown
Millenbaugh *et al.* [21]	Rat (Sprague Dawley), skin	35 000	75 mW/cm^2 0.56 max. (brain) 0.2 average (brain)	1 h	1	56[b),c)]	Transcription, protein folding, oxidative stress, immune response, tissue matrix turnover, extracellular matrix structure, chemokine activity[c)]
Nittby *et al.* [7]	Rat (Fischer 344), brain	1800 GSM	0.03 (brain)	6 h	1	45[a)]	Membrane function, receptor activity and function, signaling
Dawe *et al.* [30]	Nematodes, *C. elegans*	1000 CW	0.009–0.03	1.5, 2.5, 6 h	5	27[d)]	Signaling, metabolism, embryonic development, differentiation, transcription, membrane organization, and so on
Paparini *et al.* [6]	Mouse (Balb/cJ), brain	1805 GSM	0.2–1.1	1 h	1	73[a)]	Various
Sekijima *et al.* [13]	Human glioblastoma cells (A172)	2143 CW	0.08	2–96 h	2	20[e)] (4[f)])	Metabolism, signaling (in H4 and A172)

(*Continued*)

Table 16.1 *(Continued)*

Reference	Organism/ tissue/cell	Exposure			Assay: no. of experiments	Results: no. of screened genes	Function
		RF EMF (MHz)	SAR (W/kg)	Duration			
	Human neuroglioma cells (H4)		0.25				
	Human fibroblasts (IMR90)		0.8				
		21 453 CDMA	0.08	2–96 h	2		
			0.25				
			0.8				

BBB: blood–brain barrier; CW: continuous wave minutes; CDMA: code division multiple access; FDMA: frequency division multiple access; n.d.: not determined.

a) Not confirmed by RT PCR.
b) Genes at 24 h, but not 6 h, after exposure were verified by RT PCR.
c) Effects were temperature related.
d) Changes >1.4-fold and occurred mostly in only three or four out of five runs, none of the gene expression changes were significant if a correction for multiple testings was performed.
e) Not repeated in two independent experiments.
f) Function of differently expressed genes compared to heat shock at 0.8 W/kg, 21 453 CDMA.

associated with multiple cellular functions, including cytoskeleton, signal transduction pathway, metabolism, G-protein-coupled receptors, ion channels, and so on, after functional classification.

In primary mouse neurons and astrocytes, genes related to early apoptosis were significantly upregulated after 2 h of RF EMF exposure (1800 MHz) [16]. In this study, increased transcription of the proapoptotic Bax gene was seen in astrocytes but not in neurons, suggesting that the responses of neurons and glia to RF EMF exposure differ. In contrast, no significant difference was obtained in p53-related genes in human neuroblastoma cells and fibroblasts exposed to 2.1425 GHz (CW and CDMA) [11].

Zeng *et al.* [17] reported an upregulation of genes involved in the regulation of muscle contraction, transcription, cell–cell signaling, DNA topological changes, and cell adhesive interactions in human breast cancer cells exposed for 24 h to RF EMF (1.8 GHz, SAR: 3.5 W/kg). No changes in gene expression were obtained at an SAR of 2.0 W/kg. However, these genes showing high fold changes in a single pairwise comparison at an SAR of 3.5 W/kg could not be confirmed by RT-PCR analyses. In a follow-up experiment, the authors performed a systematic proteome analysis of the same cells under the same RF EMF conditions. Based on this experiment, the authors concluded that the observed effects in gene expression might have occurred by chance.

Exposure of two human endothelial cell lines at 900 MHz (2.5 and 2.8 W/kg) for 1 h showed that RF EMF altered expression levels of multiple genes [18, 19], including downregulated genes involved in forming the Fas/TNFα apoptotic pathway [18]. In one of these studies, a comparison of the genes affected by RF EMF revealed that both cells lines (EA.hy926 and EA.hy926v1) responded differently to the same RF EMF exposure [19]. The results were confirmed by RT PCR only in one of the two studies [19]. The same endothelial cell line (EA.hy926) responded differently to the GSM signal at 900 and at 1800 MHz. Thirty-two genes were up- or downregulated at 900 MHz, including genes related to metabolism, stress response, signaling, and differentiation, where no effect on gene expression was seen at 1800 MHz [9]. The data suggest that cell responses might be genome dependent.

It has been reported that at the level of gene expression in human fibroblasts, a GSM signal is the source of significant upregulation of genes involved in stress response, cell cycle regulation, nuclear receptors, the cytoskeleton, and extracellular matrix [20]. Data from microarrays and RT PCR demonstrated that exposure to 35 GHz millimeter waves caused thermally related stress and injury in the skin of rats while triggering repair processes involving inflammation and tissue matrix recovery [21]. RF EMF exposure of mouse or human fibroblasts did not result in a significant change in gene expression when compared to controls [12, 13, 22]. The experiments were repeated three times and a sham–sham analysis was performed in order to evaluate the false positives. The authors concluded that the number of gene expression changes induced by RF EMF was not greater than the number of false positives expected based on the sham–sham comparison. Changes to the cell plasma membrane permeability in response to RF EMF have been reported. GSM signals induced deep infolding of the plasmalemma [20]. A significant decrease in the expression of the apoptosis suppressor gene, DAD-1, has been demonstrated by Harvey and French [23] in human mast cells, whereas no change in gene expression was seen in human monocytes [10]. In the study performed by Harvey and French [23], no quantitative confirmation experiments were performed.

In an effort to characterize the cellular responses to RF EMF radiation (1763 MHz; SAR 2 and 10 W/kg), human T-lymphoblastoma cells were exposed for 1–3 days for 1 h/day and continuous exposure for 1, 4, and 24 h. The same author reported a change in gene expression in this cell line that is related to apoptosis and metabolism [24]. Ten genes were shown to be up- or downregulated [25]. The downregulation of two cytokine receptor genes suggests that RF EMF exposure might influence the chemotaxis of various immune cells. However, no gene changed more than twofold in their expression levels. The changes in gene expression reported were >1.3-fold. The level of CXCR3, the CXC motif chemokine receptor 3, a receptor for interferon-inducible protein and interferon gamma-induced monokine-2, was increased 1.8-fold. Binding of CXCR3 to these ligands induces chemotaxis of activated lymphocytes. The authors concluded that down-regulation of two cytokine receptor genes suggests that RF EMF can influence chemotaxis of various immune cells [25]. In this study, RF EMF exposure did not alter changes in global gene expression. No significant change in gene expression was seen in human T lymphocytes after RF EMF exposure for 44 h at 1800 MHz (SAR: 1.4) [9]. Interestingly, the short-term irradiation caused expression changes in a smaller number of genes and the effect was transient. In contrast, the same study demonstrated 34- and 12-fold up- or downregulated genes in human lymphoblastoma cells and leukemia cells, respectively. Interestingly, the leukemia cells responded to 1800 MHz intermittent exposure, but not to continuous RF EMF exposure. Analysis of the affected gene families does not point toward stress response, but primarily to genes related to metabolism [9]. Apoptosis-associated genes were among the upregulated ones, whereas the cell cycle genes were downregulated in human leukemia cells after 6 h of RF EMF exposure at 2.45 GHz [26]. In contrast, exposure of human leukemia cells to RF EMFs 25 times higher than the ICNIRP reference levels for occupational exposure failed to induce any changes in gene expression [27]. However, the exposure duration was only 6 min and the authors do not mention an SAR value.

Gene expression was investigated in mouse auditory hair cells, an organ that is close to cell phone exposure. Auditory hair cells are important for hearing and vestibular function in the ear. Huang *et al.* [28] reported changes of 18 annotated genes and 11 ESTs after 24 and 48 h of RF EMF exposure (1763 MHz; 20 W/kg). The authors concluded that the changes in gene expression did not group into any functional categories, such as cell cycle and apoptosis.

Only few genes displayed an altered transcription level after 24 h of exposure to high-frequency electromagnetic fields of the plant *Arabidopsis thaliana*. The data set clustered most closely to experiments in which plants were exposed to different light conditions. The authors concluded that it is unlikely that these small changes in the activity of very few genes will have pronounced effects on the physiology of plant cells [29].

Continuous RF EMF exposure of nematodes (*Caenorhabditis elegans*) at 3 mW/kg (max) at 1 GHz resulted in gene expression changes of 27 genes associated with differentiation, signaling, and embryonic development (see Table 16.1) when a cutoff value of 1.2 was used. However, the authors concluded that there was no evidence for major changes in gene expression based on the fact that the changes observed were quantitatively small and can be considered as false discoveries under the null hypothesis. None of the changes was significant after adjusting the *P*-values for instances of multiple testings [30].

16.3 Discussion

Despite extensive research, the evidence that RF EMFs are detrimental to health remains controversial. Although the majority of the research has provided no evidence that RF EMF pose any health risk, sufficient suggestive evidence exists such that concerns remain that these fields may induce stress responses and alter cellular functions. Stress response could initiate the activation of signal transduction cascades that can possibly lead to the development of cancer. There is some evidence suggesting that RF EMF results in gene expression related to cell cycle regulation, apoptosis, metabolism, and differentiation. However, even when biological effects were reported, the mechanisms behind their occurrence are still unknown.

A broad spectrum of data is available that has focused on the effects of RF EMF on the expression and the individual endpoints such as genes and proteins involved in stress response and/or apoptosis. In contrast to the approach investigating the effect of RF EMF on a limited number of genes, high-throughput techniques are useful in localizing specific patterns of effects.

First, findings on changes in genes associated with stress response and apoptosis are discussed, including hypothesis-based studies with defined endpoints. Second, general issues relevant for transcriptomics studies are addressed.

Stress response: Some hypothesis-driven studies focusing on single endpoints after RF EMF exposure reported activation of stress response genes (e.g., heat shock genes) and proteins in a variety of organisms, tissues, and cells [31–43], while others reported no effect on genes associated with stress response [44–48]. Immediately after the exposure of rats to a 915 MHz GSM signal (SAR 7.5 W/kg), an increase in Hsp70 mRNA was detected in the cerebellum and hippocampus, but 24 h after exposure, there was no alteration of Hsp70 in these brain areas [49]. The exposure of rats twice a day for 3 h per day up to 18 weeks to 1.9 GHz electromagnetic field showed an upregulation of injury-associated genes in the brain, including genes responsible for neurite outgrowth, nerve sprouting, nerve regeneration, and wound healing [50].

A few studies tried to pinpoint signal transduction pathway involved in stress response. These investigations demonstrated that RF EMF activates the p38 MAPK (mitogen-activated protein kinase) stress response pathway [35, 51]. Further evidence of MAPK involved in the effect of RF EMF was reported by Friedmann *et al.* [52], demonstrating that long-term exposure of cells to mobile phone irradiation results in the activation of p38 and the ERK (extracellular signal-regulated kinase) MAPKs. These authors presented a detailed molecular mechanism by which RF EMF induces the activation of the ERK cascade and thereby induces transcription and other cellular processes [52]. The effect of RF EMF on a variety of proteins involved in transcription and protein stability can be mediated by the MAPK cascades, which serve as central signaling pathways and govern essentially all stimulated cellular processes. In the brain, MAPK operates through constitutive phosphorylation activity to regulate microtubule function in neurons [53, 54]. An increase in reactive oxygen species (ROS) being part of the molecular mechanism was presented by Friedman *et al.* [52]. The brain is particularly vulnerable to ROS due to its high metabolic rate. In a study using microarrays, a downregulation of the gene Sod2, its product Sod2 is a ROS scavenger, was seen in rat neurons after 24 h of intermittent RF EMF exposure at an average SAR of 2.0 W/kg, which is defined as safety limit for mobile phone emission radiation by ICNIRP [15].

On the other hand, there is only little evidence from transcriptomics studies that RF EM causes an up- or downregulation of heat shock genes [9, 20, 23, 26], while only four studies report changes of genes associated with stress response [9, 20, 21, 23].

Apoptosis and cell cycle control: While seven transcriptomics studies demonstrated gene expression changes of apoptosis or cell cycle-related genes, several studies investigating the expression of single or multiple genes or proteins using a hypothesis-based approach gave evidence that RF EMF exposure resulted in changes in cell cycle control and apoptosis [43, 55–59].

Apoptosis is a cellular suicide mechanism that occurs in mammalian cells not only during normal development but also as a response to injury, DNA damage, or disease. If RF EMF is not directly carcinogenic, mobile phone radiation could act as repetitive stressor leading to expression of heat shock proteins in cells, which in turn affects their normal regulation. Heat shock proteins have been reported to inhibit apoptosis via direct interaction with key regulatory proteins, such as Apaf-1, thereby preventing events downstream of Apaf-1 activation [60]. Therefore, they play an important role in the maintenance and survival potential of cells acting as antiapoptotic proteins, a function that appeared to be independent of their chaperoning activity. Egr-I, a gene essential for proliferation, differentiation, and apoptosis was shown to be upregulated after a short-time RF EMF exposure at 900 MHz at 1 W/kg [59]. RF EMF did not result in different expression levels of p53-related genes in human fibroblasts and neuroblastoma cells when compared to sham controls [11].

In contrast, Czyz *et al.* [38] demonstrated that GSM 217 Hz (SAR: 1.5 and 2 W/kg) signals induced a significant upregulation of heat shock protein mRNA levels in embryonic stem cells that differentiated *in vitro*, but not in wild-type cells. Zhao *et al.* [16] reported an upregulation of genes related to apoptosis after intermittent exposure to RF EMF showing that neurons appear to be more sensitive to this effect than astrocytes. An increased expression of apoptosis-associated genes was reported by Lee *et al.* [26]. A cutoff value of 4 was used in this study. Based on their finding on downregulation of genes associated with cell cycle regulation and the upregulation of genes related to apoptosis, the authors suggest that the cell response to RF EMF may cause a delayed cell division providing opportunities for repairing on the one hand and programmed cell death on the other hand.

There is evidence that the same genes are differentially affected by the RF EMF exposure depending on the cell line, suggesting that the cell response to RF EMF might be epigenome and/or genome dependent. It is likely that different types of cells derived from different tissues and/or different species might respond differently to RF EMF and that there might be a variable threshold sensitivity depending on the cell type. Moreover, the variability in response to RF EMF of even closely related cell lines derived from the same species and tissue [19] suggests that the response depends on their origin and metabolic status at the time of RF EMF exposure. It is well known that genotypically different cells might respond differently to the same stimulus, for example, ionizing radiation [61, 62], showing that the cell genome/epigenome and the cellular response are closely related. Furthermore, immortalized cell lines can be standardized, but they are genetically different from primary cells, which can lead to changes in cell proliferation, cell differentiation, and apoptosis.

In addition, the molecular response of a normal cell to injury is largely governed by interactions with neighboring cells as well as immune cells [63]. A key finding from single cell gene expression profiling in the CNS and other tissues is the degree of variation between neighboring, apparently identical cells [64]. On the other hand, a specific gene pattern may be induced in one type of cell, but may be masked if a heterogeneous tissue is analyzed. Subcellular mRNA localization and local translation enable the generation of protein gradients that are essential for cell polarity [65]. In addition, the presence of noncoding RNAs in the cellular transcriptome is of increasing interest. MicroRNAs (miRNAs) are endogenous noncoding RNAs that posttranscriptionally repress protein coding genes by base pairing to 3′-UTR [63]. Thus, the biological context is crucial to gaining a complete picture of the molecular response of each cell type to RF EMF. The variations in culture conditions, for example, medium, density, and passage, may contribute to observed differences. In addition, differences in cell size, shape, composition of cell membranes, organelle distribution, coupling between adjacent cells, and stage of cell cycle might also contribute to different outcomes. In some experiments, thermal effects could not be excluded [42].

Changes in gene expression are observed within specific time windows with and without recovery time that might depend on various variables, including exposure time, exposure pattern, and recovery time after exposure, type of biological system (organism, tissue, and cell), and experimental techniques. As already mentioned, the same endothelial cell line responded to 1800 MHz intermittent exposure, but not to continuous exposure [9].

The importance of the genetic background was shown by comparison of different cell lines [38].

Contrary to the transformed cells *in vitro*, the changes in mammalian tissues, especially in the brain, are subject of feedback mechanisms. Discrepancies may be ascribed to tissue- or cell-specific compensatory mechanisms responsible for masking the effect. Finally, the parameters of RF EMF exposure (field intensity, frequency modulation, polarization, and dose in SAR) differ between studies. Taking in mind the importance of these physical and biological variables for the effect of RF EMF, only results from studies with identical parameters of exposure should be compared.

In gene expression profiling, one has to consider that it is an assessment of variability in composition and abundance of the transcriptome [3]. The temporal variance is high and the expression profile depends on the tissue or cell type.

In studies where gene expression was studied in the entire brain after RF EMF *in vivo* exposure of animals, specific response patterns that are associated with specific brain regions might have been masked. Therefore, the lack of gene expression differences between sham- and RF EMF-exposed animals does not rule out the possibility of effects within specific brain regions or cells in the brain region.

Some findings indicate that the effects of RF EMF on gene expression are transient [49]. Investigations at various exposure durations and recovery times after the exposure are necessary to address this question. In addition, an up- or down-regulation of one or more genes at a time can result in the expression of other genes, while the change in the previous gene might be absent later on.

Although hypothesis-free studies might contribute considerably to the elucidation of highly complex biological processes, the interpretation of generated data by transcriptomics requires a different approach from the interpretation of data generated by hypothesis-based studies. The increased number of possible endpoints in a study is accompanied by higher probability of detecting statistically significant false positive results [66].

Appropriate statistical approaches and standardized protocols must be applied to correctly model the parameters and interpret biological variability [67]. The different cutoffs in the reported studies, varying from 1.2- to 4-fold can at least partly explain the different outcome. Studies can suffer from bias of a pooled variance approach, thereby minimizing the generation of false positives or negatives that can occur when a common variance is assumed [68]. Furthermore, data integration is necessary to manage the overwhelming quantity of data, to cross-validate the noise in data sets, and to gain broad interdisciplinary views of large data sets [69]. The replication of the initial findings in follow-up studies remains the strongest safeguard against false positive studies. Quantitative methods such as RT-PCR require template cDNA that is synthesized from normalized RNA preparations. This condition is crucial to ensure that the relative amount of a given cDNA is proportional to the relative amount of its RNA template.

Hypothesis-driven research and discovery-driven research are complementary and synergistic [67]. Studies that incorporate thousands of biological endpoints should therefore be seen as discovery studies that can aid the generation of new hypotheses [3]. Comparisons across studies should carefully consider differences in experimental design, such as animal or cell model, exposure conditions, and experimental protocols.

The analytical reproducibility of transcriptomic approaches also needs to be considered.

In most of the RF EMF transcriptomics studies, the fold changes in gene expression are modest (less than twofold). The differences in coverage of genes represented on the different types of array, their reproducibility, and confirmation by independent techniques differ across the reports. In addition, the dynamic nature of each of these systems may militate against long-term exposure assessment, unless some of the changes prove stable over time [70].

In summary, although there are a number of studies that investigated the effects of RF EMF on gene expression using transcriptomics, the data are insufficient to develop any generalized conclusions as to the mechanisms involved and the biological impact of RF EMF exposure on living organisms. One has to keep in mind that the published studies differ in scientific quality, focus, and exposure and many are based on a single experiment. A better understanding of the mechanism by which RF EMF exert their possible biological effects is necessary for determining the potential health risks. In the future, the use of analytical tools/databases that recently became available will allow the integrated analyses of biological functions and changes in biological functions as a result of environmental factors (see Chapter 14).

References

1 Manning, A.T., Garvin, J.T., Shahbazi, R.I., Miller, N., McNeill, R.E., and Kerin, M.J. (2007) Molecular profiling techniques and bioinformatics in cancer research. *Eur. J. Surg. Oncol.*, **33**, 255–265.

2 Nachtomy, O., Shavit, A., and Yakhini, Z. (2007) Gene expression and the concept of the phenotype. *Stud. Hist. Philos. Biol. Biomed. Sci.*, **38**, 238–254.

3 Vlaanderen, J., Moore, L.E., Smith, M.T., Lan, Q., Zhang, L., Skibola, C.F., Rothman, N., and Vermeulen, R. (2010) Application of OMICS technologies in occupational and environmental health research; current status and projections. *Occup. Environ. Med.*, **67**, 136–143.

4 Celis, J.E., Kruhoffer, M., Gromova, I., Frederiksen, C., Ostergaard, M., Thykjaer, T., Gromov, P., Yu, J., Palsdottir, H., Magnusson, N., and Orntoft, T.F. (2000) Gene expression profiling: monitoring transcription and translation products using DNA microarrays and proteomics. *FEBS Lett.*, **480**, 2–16.

5 Belyaev, I.Y., Koch, C.B., Terenius, O., Roxström-Lindquist, K., Malmgren, L.O.G., Sommer, W.H., Salford, L.G., and Persson, B.R.R. (2006) Exposure of rat brain to 915MHz GSM microwaves induces changes in gene expression but not double stranded DNA breaks or effects on chromatin conformation. *Bioelectromagnetics*, **27**, 295–306.

6 Paparini, A., Rossi, P., Gianfranceschi, G., Brugaletta, V., Falsaperla, R., De Luca, P., and Romano Spica, V. (2008) No evidence of major transcriptional changes in the brain of mice exposed to 1800MHz GSM signal. *Bioelectromagnetics*, **29**, 312–323.

7 Nittby, H., Widegren, B., Krogh, M., Grafström, G., Berlin, H., Rehn, G., Eberhardt, J.L., Malmgren, L., Persson, B.R.R., and Salford, L.G. (2008) Exposure to radiation from global system for mobile communications at 1,800MHz significantly changes gene expression in rat hippocampus and cortex. *Environmentalist*, **28**, 458–465.

8 Gurisik, E., Warton, K., Martin, D.K., and Valenzuela, S.M. (2006) An *in vitro* study of the effects of exposure to a GSM signal in two human cell lines: monocytic U937 and neuroblastoma SK-N-SH. *Cell Biol. Int.*, **30**, 793–799.

9 Remondini, D., Nylund, R., Reivinen, J., De Gannes, F.P., Veyret, B., Lagroye, I., Haro, E., Trillo, M.A., Capri, M., Franceschi, C., Schlatterer, K., Gminski, R., Fitzner, R., Tauber, R., Schuderer, J., Kuster, N., Leszczynski, D., Bersani, F., and Maercker, C. (2006) Gene expression changes in human cells after exposure to mobile phone microwaves. *Proteomics*, **6**, 4745–4754.

10 Chauhan, V., Qutob, S.S., Lui, S., Mariampillai, A., Bellier, P.V., Yauk, C.L., Douglas, G.R., Williams, A., and McNamee, J.P. (2007) Analysis of gene expression in two human-derived cell lines exposed *in vitro* to a 1.9GHz pulse-modulated radiofrequency field. *Proteomics*, **7**, 3896–3905.

11 Hirose, H., Sakuma, N., Kaji, N., Suhara, T., Sekijima, M., Nojima, T., and Miyakoshi, J. (2006) Phosphorylation and gene expression of p53 are not affected in human cells exposed to 2.1425GHz band CW or W-CDMA modulated radiation allocated to mobile radio base stations. *Bioelectromagnetics*, **27**, 494–504.

12 Hirose, H., Sakuma, N., Kaji, N., Nakayama, K., Inoue, K., Sekijima, M., Nojima, T., and Miyakoshi, J. (2007) Mobile phone base station-emitted radiation does not induce phosphorylation of Hsp27. *Bioelectromagnetics*, **28**, 99–108.

13 Sekijima, M., Takeda, H., Yasunaga, K., Sakuma, N., Hirose, H., Nojima, T., and Miyakoshi, J. (2010) 2-GHz band CW and W-CDMA modulated radiofrequency fields have no significant effect on cell proliferation and gene expression profile in human cells. *J. Radiat. Res. (Tokyo)*, **51**, 277–284.

14 Qutob, S.S., Chauhan, V., Bellier, P.V., Yauk, C.L., Douglas, G.R., Berndt, L., Williams, A., Gajda, G.B., Lemay, E., Thansandote, A., and McNamee, J.P. (2006) Microarray gene expression profiling of a human glioblastoma cell line exposed *in vitro* to a 1.9GHz pulse-modulated radiofrequency field. *Radiat. Res.*, **165**, 636–644.

15 Zhao, R., Zhang, S., Xu, Z., Ju, L., Lu, D., and Yao, G. (2007) Studying gene expression profile of rat neuron exposed to 1800MHz radiofrequency electromagnetic fields with cDNA microassay. *Toxicology*, **235**, 167–175.

16 Zhao, T.Y., Zou, S.P., and Knapp, P.E. (2007) Exposure to cell phone radiation up-regulates apoptosis genes in primary cultures of neurons and astrocytes. *Neurosci. Lett.*, **412**, 34–38.

17 Zeng, Q., Chen, G., Weng, Y., Wang, L., Chiang, H., Lu, D., and Xu, Z. (2006) Effects of global system for mobile communications 1800MHz radiofrequency electromagnetic fields on gene and protein expression in MCF-7 cells. *Proteomics*, **6**, 4732–4738.

18 Leszczynski, D., Nylund, R., Joenvaara, S., and Reivinen, J. (2004) Applicability of discovery science approach to determine biological effects of mobile phone radiation. *Proteomics*, **4**, 426–431.

19 Nylund, R. and Leszczynski, D. (2006) Mobile phone radiation causes changes in gene and protein expression in human endothelial cell lines and the response seems to be genome- and proteome-dependent. *Proteomics*, **6**, 4769–4780.

20 Pacini, S., Ruggiero, M., Sardi, I., Aterini, S., Gulisano, F., and Gulisano, M. (2002) Exposure to global system for mobile communication (GSM) cellular phone radiofrequency alters gene expression, proliferation, and morphology of human skin fibroblasts. *Oncol. Res.*, **13**, 19–24.

21 Millenbaugh, N.J., Roth, C., Sypniewska, R., Chan, V., Eggers, J.S., Kiel, J.L., Blystone, R.V., and Mason, P.A. (2008) Gene expression changes in the skin of rats induced by prolonged 35GHz millimeter-wave exposure. *Radiat. Res.*, **169**, 288–300.

22 Whitehead, T.D., Moros, E.G., Brownstein, B.H., and Roti Roti, J.L. (2006) Gene expression does not change significantly in C3H 10T1/2 cells after exposure to 847.74 CDMA or 835.62 FDMA radiofrequency radiation. *Radiat. Res.*, **165**, 626–635.

23 Harvey, C. and French, P.W. (1999) Effects on protein kinase C and gene expression in a human mast cell line, HMC-1, following microwave exposure. *Cell Biol. Int.*, **23**, 739–748.

24 Huang, T., Lee, M.S., Bae, Y., Park, H., and Seo, J. (2006) Prediction of exposure to 1763MHz radiofrequency radiation using support vector machine algorithm in Jurkat cell model system. *Genomics Informatics*, **4**, 71–76.

25 Huang, T.Q., Lee, M.S., Oh, E., Zhang, B.T., Seo, J.S., and Park, W.Y. (2008) Molecular responses of Jurkat T-cells to 1763MHz radiofrequency radiation. *Int. J. Radiat. Biol.*, **84**, 734–741.

26 Lee, S., Johnson, D., Dunbar, K., Dong, H., Ge, X., Kim, Y.C., Wing, C., Jayathilaka, N., Emmanuel, N., Zhou, C.Q., Gerber, H.L., Tseng, C.C., and Wang, S.M. (2005) 2.45GHz radiofrequency fields alter gene expression in cultured human cells. *FEBS Lett.*, **579**, 4829–4836.

27 Port, M., Abend, M., Römer, B., and Van Beuningen, D. (2003) Influence of high-frequency electromagnetic fields on different modes of cell death and gene expression. *Int. J. Radiat. Biol.*, **79**, 701–708.

28 Huang, T.Q., Lee, M.S., Oh, E.H., Kalinec, F., Zhang, B.T., Seo, J.S., and Park, W.Y. (2008) Characterization of biological effect of 1763MHz radiofrequency exposure on auditory hair cells. *Int. J. Radiat. Biol.*, **84**, 909–915.

29 Engelmann, J.C., Deeken, R., Mueller, T., Nimtz, G., Roelfsema, M.R.G., and Hedrich, R. (2008) Is gene activity in plant cells affected by UMTS-irradiation? A whole genome approach. *Comput. Biol. Chem. Adv. Appl.*, **1**, 71–83.

30 Dawe, A.S., Nylund, R., Leszczynskj, D., Kuster, N., Reader, T., and De Pomerai, D.I. (2008) Continuous wave and simulated GSM exposure at 1.8W/kg and 1.8GHz do not induce hsp16-1 heat-shock gene expression in *Caenorhabditis elegans. Bioelectromagnetics*, **29**, 92–99.

31 Lixia, S., Yao, K., Kaijun, W., Deqiang, L., Huajun, H., Xiangwei, G., Baohong, W., Wei, Z., Jianling, L., and Wei, W. (2006) Effects of 1.8GHz radiofrequency field on DNA damage and expression of heat shock protein 70 in human lens epithelial cells. *Mutat. Res.*, **602**, 135–142.

32 Tonomura, H., Takahashi, K.A., Mazda, O., Arai, Y., Shin-Ya, M., Inoue, A., Honjo, K., Hojo, T., Imanishi, J., and Kubo, T. (2008) Effects of heat stimulation via microwave applicator on cartilage matrix gene and HSP70 expression in the rabbit knee joint. *J. Orthop. Res.*, **26**, 34–41.

33 Weisbrot, D., Lin, H., Ye, L., Blank, M., and Goodman, R. (2003) Effects of mobile phone radiation on reproduction and development in *Drosophila melanogaster. J. Cell. Biochem.*, **89**, 48–55.

34 Lin, H., Opler, M., Head, M., Blank, M., and Goodman, R. (1997) Electromagnetic field exposure induces rapid, transitory heat shock factor activation in human cells. *J. Cell. Biochem.*, **66**, 482–488.

35 Leszczynski, D., Joenväärä, S., Reivinen, J., and Kuokka, R. (2002) Non-thermal activation of the hsp27/p38MAPK stress pathway by mobile phone radiation in human endothelial cells: molecular mechanism for cancer- and blood–brain barrier-related effects. *Differentiation*, **70**, 120–129.

36 Kwee, S., Raskmark, P., and Velizarov, S. (2001) Changes in cellular proteins due to environmental non-ionizing radiation. I. Heat-shock proteins. *Electromagn. Biol. Med.*, **20**, 141–152.

37 Lee, K.S., Choi, J.S., Hong, S.Y., Son, T.H., and Yu, K. (2008) Mobile phone electromagnetic radiation activates MAPK signaling and regulates viability in *Drosophila. Bioelectromagnetics*, **29**, 371–379.

38 Czyz, J., Guan, K., Zeng, Q., Nikolova, T., Meister, A., Schönborn, F., Schuderer, J., Kuster, N., and Wobus, A.M. (2004) High frequency electromagnetic fields (GSM signals) affect gene expression levels in tumor suppressor p53-deficient embryonic stem cells. *Bioelectromagnetics*, **25**, 296–307.

39 Wang, J., Koyama, S., Komatsubara, Y., Suzuki, Y., Taki, M., and Miyakoshi, J. (2006) Effects of a 2450MHz high-frequency electromagnetic field with a wide range of SARs on the induction of heat-shock proteins in A172 cells. *Bioelectromagnetics*, **27**, 479–486.

40 Tian, F., Nakahara, T., Wake, K., Taki, M., and Miyakoshi, J. (2002) Exposure to 2.45GHz electromagnetic fields induces hsp70 at a high SAR of more than 20W/kg but not at 5W/kg in human glioma MO54 cells. *Int. J. Radiat. Biol.*, **78**, 433–440.

41 Yu, Y., Yao, K., Wu, W., Wang, K., Chen, G., and Lu, D. (2008) Effects of exposure to 1.8GHz radiofrequency field on the expression of Hsps and phosphorylation of MAPKs in human lens epithelial cells. *Cell Res.*, **18**, 1233–1235.

42 Daniells, C., Duce, I., Thomas, D., Sewell, P., Tattersall, J., and De Pomerai, D. (1998)

Transgenic nematodes as biomonitors of microwave-induced stress. *Mutat. Res.*, **399**, 55–64.

43 Caraglia, M., Marra, M., Mancinelli, F., D'Ambrosio, G., Massa, R., Giordano, A., Budillon, A., Abbruzzese, A., and Bismuto, E. (2005) Electromagnetic fields at mobile phone frequency induce apoptosis and inactivation of the multi-chaperone complex in human epidermoid cancer cells. *J. Cell. Physiol.*, **204**, 539–548.

44 Chauhan, V., Mariampillai, A., Kutzner, B.C., Wilkins, R.C., Ferrarotto, C., Bellier, P.V., Marro, L., Gajda, G.B., Lemay, E., Thansandote, A., and McNamee, J.P. (2007) Evaluating the biological effects of intermittent 1.9GHz pulse-modulated radiofrequency fields in a series of human-derived cell lines. *Radiat. Res.*, **167**, 87–93.

45 Chauhan, V., Mariampillai, A., Gajda, G., Thansandote, A., and McNamee, J. (2006) Analysis of proto-oncogene and heat-shock protein gene expression in human derived cell-lines exposed *in vitro* to an intermittent 1.9GHz pulse-modulated radiofrequency field. *Int. J. Radiat. Biol.*, **82**, 347–354.

46 Miyakoshi, J., Takemasa, K., Takashima, Y., Ding, G.R., Hirose, H., and Koyama, S. (2005) Effects of exposure to a 1950MHz radio frequency field on expression of Hsp70 and Hsp27 in human glioma cells. *Bioelectromagnetics*, **26**, 251–257.

47 Valbonesi, P., Franzellitti, S., Piano, A., Contin, A., Biondi, C., and Fabbri, E. (2008) Evaluation of HSP70 expression and DNA damage in cells of a human trophoblast cell line exposed to 1.8GHz amplitude-modulated radiofrequency fields. *Radiat. Res.*, **169**, 270–279.

48 Zhadobov, M., Sauleau, R., Le Coq, L., Debure, L., Thouroude, D., Michel, D., and Le Dréan, Y. (2007) Low-power millimeter wave radiations do not alter stress-sensitive gene expression of chaperone proteins. *Bioelectromagnetics*, **28**, 188–196.

49 Fritze, K., Sommer, C., Schmitz, B., Mies, G., Hossmann, K.A., Kiessling, M., and Wiessner, C. (1997) Effect of global system for mobile communication (GSM) microwave exposure on blood–brain barrier permeability in rat. *Acta Neuropathol.*, **94**, 465–470.

50 Yan, J.G., Agresti, M., Zhang, L.L., Yan, Y., and Matloub, H.S. (2008) Upregulation of specific mRNA levels in rat brain after cell phone exposure. *Electromagn. Biol. Med.*, **27**, 147–154.

51 Inoue, S., Motoda, H., Koike, Y., Kawamura, K., Hiragami, F., and Kano, Y. (2008) Microwave irradiation induces neurite outgrowth in PC12m3 cells via the p38 mitogen-activated protein kinase pathway. *Neurosci. Lett.*, **432**, 35–39.

52 Friedman, J., Kraus, S., Hauptman, Y., Schiff, Y., and Seger, R. (2007) Mechanism of short-term ERK activation by electromagnetic fields at mobile phone frequencies. *Biochem. J.*, **405**, 559–568.

53 Morishima-Kawashima, M. and Kosik, K.S. (1996) The pool of map kinase associated with microtubules is small but constitutively active. *Mol. Biol. Cell*, **7**, 893–905.

54 Hoshi, M., Ohta, K., Gotoh, Y., Mori, A., Murofushi, H., Sakai, H., and Nishida, E. (1992) Mitogen-activated-protein-kinase-catalyzed phosphorylation of microtubule-associated proteins, microtubule-associated protein 2 and microtubule-associated protein 4, induces an alteration in their function. *Eur. J. Biochem.*, **203**, 43–52.

55 Marinelli, F., La Sala, D., Cicciotti, G., Cattini, L., Trimarchi, C., Putti, S., Zamparelli, A., Giuliani, L., Tomassetti, G., and Cinti, C. (2004) Exposure to 900MHz electromagnetic field induces an unbalance between pro-apoptotic and pro-survival signals in T-lymphoblastoid leukemia CCRF-CEM cells. *J. Cell. Physiol.*, **198**, 324–332.

56 Nikolova, T., Czyz, J., Rolletschek, A., Blyszczuk, P., Fuchs, J., Jovtchev, G., Schuderer, J., Kuster, N., and Wobus, A.M. (2005) Electromagnetic fields affect transcript levels of apoptosis-related genes in embryonic stem cell-derived neural progenitor cells. *FASEB J.*, **19**, 1686–1688.

57 Natarajan, M., Nayak, B.K., Galindo, C., Mathur, S.P., Roldan, F.N., and

Meltz, M.L. (2006) Nuclear translocation and DNA-binding activity of NF-κB after exposure of human monocytes to pulsed ultra-wideband electromagnetic fields (1kV/cm) fails to transactivate κB-dependent gene expression. *Radiat. Res.*, **165**, 645–654.

58 Palumbo, R., Brescia, F., Capasso, D., Sannino, A., Sarti, M., Capri, M., Grassilli, E., and Scarfi, M.R. (2008) Exposure to 900MHz radiofrequency radiation induces caspase 3 activation in proliferating human lymphocytes. *Radiat. Res.*, **170**, 327–334.

59 Buttiglione, M., Roca, L., Montemurno, E., Vitiello, F., Capozzi, V., and Cibelli, G. (2007) Radiofrequency radiation (900MHz) induces Egr-1 gene expression and affects cell-cycle control in human neuroblastoma cells. *J. Cell. Physiol.*, **213**, 759–767.

60 Kalmar, B. and Greensmith, L. (2009) Induction of heat shock proteins for protection against oxidative stress. *Adv. Drug Deliv. Rev.*, **61**, 310–318.

61 Popanda, O., Marquardt, J.U., Chang-Claude, J., and Schmezer, P. (2009) Genetic variation in normal tissue toxicity induced by ionizing radiation. *Mutat. Res.*, **667**, 58–69.

62 Averbeck, D. (2010) Non-targeted effects as a paradigm breaking evidence. *Mutat. Res.*, **687**, 7–12.

63 Munro, K.M. and Perreau, V.M. (2009) Current and future applications of transcriptomics for discovery in CNS disease and injury. *Neurosignals*, **17**, 311–327.

64 Kamme, F., Salunga, R., Yu, J., Tran, D.T., Zhu, J., Luo, L., Bittner, A., Guo, H.Q., Miller, N., Wan, J., and Erlander, M. (2003) Single-cell microarray analysis in hippocampus CA1: demonstration and validation of cellular heterogeneity. *J. Neurosci.*, **23**, 3607–3615.

65 Du, T.G., Schmid, M., and Jansen, R.P. (2007) Why cells move messages: the biological functions of mRNA localization. *Semin. Cell Dev. Biol.*, **18**, 171–177.

66 Wacholder, S., Chanock, S., Garcia-Closas, M., El Ghormli, L., and Rothman, N. (2004) Assessing the probability that a positive report is false: an approach for molecular epidemiology studies. *J. Natl. Cancer Inst.*, **96**, 434–442.

67 Blankenburg, M., Haberland, L., Elvers, H.D., Tannert, C., and Jandrig, B. (2009) High-throughput omics technologies: potential tools for the investigation of influences of EMF on biological systems. *Curr. Genomics*, **10**, 86–92.

68 Cui, X., Xu, J., Asghar, R., Condamine, P., Svensson, J.T., Wanamaker, S., Stein, N., Roose, M., and Close, T.J. (2005) Detecting single-feature polymorphisms using oligonucleotide arrays and robustified projection pursuit. *Bioinformatics*, **21**, 3852–3858.

69 Detours, V., Dumont, J.E., Bersini, H., and Maenhaut, C. (2003) Integration and cross-validation of high-throughput gene expression data: comparing heterogeneous data sets. *FEBS Lett.*, **546**, 98–102.

70 Wild, C.P. (2009) Environmental exposure measurement in cancer epidemiology. *Mutagenesis*, **24**, 117–125.

17
Proteomics Approach in Mobile Phone Radiation Research

Dariusz Leszczynski

Elucidation of the biological and health effects of mobile phone radiation (radio frequency-modulated electromagnetic fields) has been done for decades, but the reliable answers concerning potential health hazard are still missing.

There is ongoing discussion whether the users of mobile phones should be concerned about

i) the health safety of the radiation emitted by these devices,
ii) whether the safety standards are adequate, and
iii) whether continuation of research in this area is scientifically justified.

The International Commission on Non-Ionizing Radiation Protection (ICNIRP), the International Committee on Electromagnetic Safety (ICES), and the World Health Organization Electromagnetic Fields Project (WHO EMF Project) are assuring users that there is no proven health risk and that the present safety standards on radiation emitted by mobile phones protect all users [1–3].

When evaluating scientific evidence we need to not only determine what we know but also consider what kind of important information about the biological mechanisms, and their possible biological and health effects, was not yet studied. Only by combining the information of what we know and of what we did not examine yet, we can reliably evaluate the adequacy of the current knowledge and based on it health safety standards.

In case of mobile phone radiation, as with any other environmental factor, whether naturally occurring or man-made, are needed several types of scientific evidence such as

i) the possible mechanism how the effect is induced in living organism,
ii) *in vitro* laboratory studies that confirm the existence of a biophysical and biochemical mechanism of the effect,
iii) animal studies,
iv) human volunteer studies, and
v) epidemiological evidence of the effect on human population.

Cancer Risk Evaluation: Methods and Trends,
Edited by Günter Obe, Burkhard Jandrig, Gary E. Marchant, Holger Schütz, and Peter M. Wiedemann.

Each type of the evidence is of different significance and value for the estimation and proof of human health effect. The most important is epidemiological evidence, followed by the human volunteer studies and animal experiments. *In vitro* evidence does not directly inform about the possible health impact, but it provides information about the possible mechanism of the effect on cellular level. Knowing the mechanism of the effect increases the reliability of the evidence gathered in epidemiological, human volunteer, animal, and *in vitro* studies. In the ideal situation, all above-listed types of scientific evidence would point into the same direction.

At this point, the only firmly established mechanism of the mobile phone radiation effects is by increasing temperature of the irradiated living matter (thermal effect). However, there is ongoing discussion whether there are nonthermal, or athermal, effects where mobile phone radiation energy does not significantly increase the temperature of the living matter, but still is able to induce some biological effects. This discussion is fueled by new studies in which authors observe biological effects of mobile phone radiation induced at the exposure levels that do not increase temperature of the biological system by more than 0.1–0.3 °C.

In 2001, it was proposed that proteomics [4], and other high-throughput screening techniques, could be used to identify molecular targets of mobile phone radiation and by this way generate biological endpoints for further *in vitro*, animal, and human volunteer studies. It has also been suggested that this new high-throughput screening approach will help in revealing biological mechanisms of the mobile phone radiation interacting with living matter, some of which might be unpredictable using the presently available knowledge. This high-throughput screening approach seems to be particularly suited for elucidation of mobile phone radiation health hazard issue because it might reveal effects that are not possible to predict, based on the present very limited knowledge about the biological effects of mobile phone radiation. Subsequently, the usefulness of proteomics and transcriptomics in search for molecular targets of mobile phone radiation has been demonstrated in a five-step feasibility study [5].

Proteomics is the science that examines proteins present in the organism on large scale. Proteomics leads to a much better understanding of the functional processes ongoing in cells than does genomics or transcriptomics because proteins are the molecules that regulate physiological processes of cells and tissues. For this reason, use of proteomics approach to study the effects of mobile phone radiation might provide information as to potential biological and health effects. As shown in Table 17.1, only nine proteomics studies [6–14] were conducted where effects of mobile phone radiation on living matter were examined. Eight out of them are *in vitro* studies with human cells, either primary cultures or cell lines grown in laboratory, and one is an *in vivo* study with human volunteers.

In six out of the nine proteomics studies, the authors concluded that mobile phone radiation has an effect on the proteome whereas in three studies the authors did not see any effect. However, because of the variety of limiting factors, associated with the study design and/or methods used, the conclusions of all studies should be looked at with caution. As of now, it appears that all the studies can be considered as kind of

Table 17.1 To date executed proteomics studies examining effects of mobile phone radiation.

Biological model	Type of study (*in vitro* /*in vivo*)	Radiation frequency	Exposure conditions	Sampling time	Proteome analysis methods	Number of 2DE gel replicates (sham and exposed)	Protein spot identification	Validation of identified proteins	Effect yes/no	References
Human endothelial cell line EA.hy926	*In vitro*	900 MHz GSM	2.4 SAR; 1 h	Immediately and 1, 5, and 7 h after exposure[a)]	2DE, silver stain, PDQuest software	4 and 4	Yes	Yes	Yes	Leszczynski *et al.* [6]
Human endothelial cell line EA.hy926	*In vitro*	900 MHz GSM	2.4 SAR; 1 h	Immediately after exposure	2DE, silver stain, PDQuest software	10 and 10	Yes	Yes (vimentin)	Yes	Nylund and Leszczynski [7]
Human endothelial cell line EA.hy926	*In vitro*	900 MHz GSM	2.4 SAR; 1 h	Immediately after exposure	2DE, silver stain, PDQuest software	10 and10	No	No	Yes	Nylund and Leszczynski [8]
Human breast cancer MCF-7 cells	*In vitro*	1800 MHz GSM	3.5 SAR; 1, 6, 12, and 24 h continuous or 5 min on/10 min off	Immediately after exposure	2DE, silver stain, PDQuest software	3 and 3	No	No	No	Zeng *et al.* [9]

(*Continued*)

Table 17.1 (*Continued*)

Biological model	Type of study (*in vitro* /*in vivo*)	Radiation frequency	Exposure conditions	Sampling time	Proteome analysis methods	Number of 2DE gel replicates (sham and exposed)	Protein spot identification	Validation of identified proteins	Effect yes/no	References
Human lens epithelial cells	*In vitro*	1800 MHz GSM	1, 2, 3.5 SAR; 2 h	Immediately after exposure	2DE, silver stain, PDQuest software	3 and 3	Yes	No	Yes	Li *et al.* [10]
Human skin	*In vivo*	900 MHz GSM	1.3 SAR; 1 h	Immediately after exposure	2DE, silver stain, PDQuest software	1 and 1/person; 10 persons comparison	No	No	Yes	Karinen *et al.* [11]
Human endothelial cell line EA. hy926	*In vitro*	1800 MHz GSM	2 SAR; 1 h	Immediately after exposure	2DE-DIGE, PDQuest software	10 and 10	Yes	Yes (selected proteins)	No	Nylund *et al.* [12]
Jurkat cells, human fibroblasts	*In vitro*	1800 MHz GSM	2 SAR; 8 h; 5 min on/10 min off	Immediately after exposure	2DE, fluorescent dye, Progenesis software	3 and 3	Yes	No	Yes	Gerner *et al.* [13]
Human breast cancer MCF-7 cells	*In vitro*	849 MHz CDMA	2 and 10 SAR; 1 h/day for 3 days	24 h after exposure	2DE, silver stain, Progenesis software	3 and 3	Yes	Yes (selected proteins)	No	Kim *et al.* [14]

a) 2DE was analyzed immediately after exposure. Protein expression changes were validated immediately after exposure and 1, 5, and 7 h after exposure.

"feasibility studies" paving the way for further, more thorough and better designed, proteomics studies.

In all studies, a variety of mobile phone signals and exposure conditions were used (Table 17.1). Some of the exposure conditions caused changes in protein expression and some not. However, because of the very limited number of studies, it is not possible to determine whether any of the observed effects can be in any particular way correlated with certain exposure conditions.

As with all proteomics studies, the posttreatment sampling time is crucial and a "wrong" sampling time may lead to no-effect outcome of the study. In the mobile phone radiation proteomics studies conducted so far, the scale of sampling times used is very limited. In the majority of studies, the samples were harvested immediately after exposure to mobile phone radiation and in one study 24 h after exposure. In only one study, several different time points were used when cells were collected for the validation experiments but not for the 2DE (two-dimensional electrophoresis) experiments [6]. Such limited scale of time points may produce false negative results because the time point used might not be optimal for observing mobile phone radiation-induced protein expression changes in a given cellular system. Another limitation, which might have prevented detection of changes in expression of low abundant proteins, is that in all studies nonfractionated/nonenriched protein samples were used containing all proteins of the cellular lysate. Therefore, the high-abundance structural proteins dominated in the 2DE gels whereas low-abundance proteins, some of them seemed to respond to mobile phone radiation in some of the studies, might have been either difficult to detect or not detected at all. Therefore, the sample protein composition might have also led to false negative results.

In all of the studies, proteome changes were determined by separating proteins with 2DE. The protein spot detection was done in most of the studies by silver staining [6–11, 14]. The silver staining procedure for protein visualization, though a well-established and accepted method, might not be the best option for differential proteomics. This is because of the poor reproducibility of this procedure and the reduced linear dynamic protein concentration range that makes it difficult to draw clear conclusions about the changes in protein expression. Further limitation of the reliability of some studies is the low number of gel replicates. The silver-based approach requires high number (10 or more) of gel replicates [7, 8]. However, in three of the studies the authors rely on only three gel replicates [9, 10, 14], which is not sufficient to draw any reliable conclusions and which may lead to false negative results.

The use of the relatively novel DIGE system for 2DE (2D fluorescence difference gel electrophoresis), considered to be the gold standard, might have been limited in the mobile phone radiation proteomics studies by the availability of specialized hardware in the laboratories performing the studies. There is only a single study [12] where changes in protein expression were examined using 2DE-DIGE system and with a sufficiently high number of 10 replicate gels. This study has shown that the exposure of cells at SAR of 2.0 W/kg for 1 h using 1800 MHz GSM signal does not cause changes in protein expression. Because of the techniques used, this is the most reliable differential proteome analysis to date. However, there are other limitations of this study that need to be addressed, in further research, before accepting this negative result. Namely, as the authors have stated in their article [12], there appears to be a discrepancy between the responses of EA.hy926 cells to 1800 MHz GSM radiation and 900 MHz GSM radiation [7, 8]. These might be caused either by the different exposure frequencies or by the technical differences between the exposure setups, or by both. The major difference, besides the frequency, between the 900 MHz GSM and 1800 MHz GSM exposure chambers, appears to be the distribution of radiation field within the cell culture dish. In the 900 MHz GSM setup, there was a nonuniform SAR distribution with a high SAR observed in the center of the dish, with the effect that the cells growing in the center of the culture dish were exposed to much higher SAR ($\geq$5.0 W/kg) compared to the average SAR for the whole cell culture dish (2.4 W/kg). In contrast, the 1800 MHz GSM setup has uniform SAR distribution and the cells throughout the cell culture dish were exposed to the same level of radiation (2.0 W/kg). The difference in proteome response, whether it is SAR or frequency dependent, needs to be resolved before accepting the results of this study [12] as reliable. However, it is necessary to remember that in spite of some dosimetric uncertainties about the SAR field distribution, the observed effects were of nonthermal nature even in the higher SAR areas (SAR $\geq$ 5.0 W/kg) as the temperature of the cell cultures remained at 37 ± 0.1–$0.3\,^{\circ}$C throughout the experiment.

The same caution, concerning the SAR level that might be able to induce biological effects, should also be applied to studies where the same 900 MHz GSM exposure chamber, producing nonuniform SAR field distribution, was used [6–8]. Though the average SAR of the exposures was 2.4 W/kg, there were areas within the cell culture that received higher exposures ($\geq$5.0 W/kg) that could affect the cell response. Thus, it is still uncertain whether the effects observed in studies [6–8] were caused by the average SAR of 2.4 W/kg or by local SAR of $\geq$5.0 W/kg.

An interesting hypothesis has been presented in the study by Gerner *et al.* [13]. The authors suggest that the protein expression changes might be not the best endpoint to show effects of mobile phone radiation. They have demonstrated that in cells exposed to mobile phone radiation changes in the *de novo* synthesis of some proteins occur, without statistically significant changes in the overall protein expression. However, their conclusions are weakened by the fact that there were only three gel replicates performed in the study, which is not sufficient for a reliable determination of changes in protein expression. However, the observations made by Gerner *et al.* [13] and Leszczynski *et al.* [6] suggest that examining protein expression alone might not be sufficient to detect effects of mobile phone radiation and the experiments that examine both *de novo* protein synthesis and posttranslational modifications of proteins (e.g., phosphorylation) are necessary additions that may help determine the effects of mobile phone radiation on the cell proteome.

There is only a single *in vivo* human volunteer study [11] that has examined effects of mobile phone radiation on the proteome of human skin. As the authors stated, this was only a pilot study aimed at showing the feasibility of the proteomics approach to study effects of mobile phone radiation on humans. Because of the variability in proteomes between individual volunteers, the authors used samples of skin from the same person as unexposed sham controls. In this way, they were able to perform a pilot study using only 10 volunteers. However, use of 2DE and silver staining method and the low number of volunteers diminish the reliability of the changes found in the expression of a few proteins. This study should be repeated with a larger number of volunteers and using the 2DE-DIGE method to get more reliable information whether mobile phone radiation affects the proteome of human tissues. Interestingly, in spite of the continuing discussion whether mobile phone radiation can cause any health effect on people, it is still not known whether the human body responds to mobile phone radiation at the molecular level.

The numbers of proteins shown to be affected by mobile phone radiation, in all proteomics studies showing an effect, are not only low but also lower than the number of expected false positives. This argument is often used to automatically state that there is no effect of mobile phone radiation on the cell proteome. However, the low number of responding proteins, even when it is below the expected rate of false positives, does not mean automatically that every protein appearing as affected by mobile phone radiation is a sure false positive finding. The calculation of the number of expected false positives shows only the probability that the affected proteins might be false positives, but it does not prove that all of them are indeed false positives. The proof can be obtained only by a biological experiment. This has been demonstrated in a study by Nylund and Leszczynski [7] where the number of detected statistically significantly affected proteins was lower than the expected number of false positives. However, further Western blot analysis of the changes in the expression of one of the affected proteins, vimentin, has shown that this protein did respond to mobile phone radiation. Therefore, even though the number of statistically significantly affected proteins observed in this study [7] was lower than the expected number of false positives, among these "false positive" affected proteins, one protein was found that was indeed affected, which was confirmed using a non-2DE approach. Therefore, one should be cautious before automatically disregarding the changes in a small number of proteins of the whole proteome. Furthermore, hypothetically speaking, if more than 5% of the needed protein spots would be identified as affected proteins, these would include both real positives and false positives. However, this would be classified as a positive finding, even though among the identified proteins would be some false positive findings. Thus, when the statistical analysis is done, it is also necessary to look at the affected proteins themselves and reconfirm whether the change is real or false. This may apply especially to weak stimuli, such as mobile phone radiation, that are likely to exert weak effects and affect expression of only a small number of proteins.

Finally, it is necessary to keep in mind that a screening study using proteomics, or other high-throughput screening approach, is just the beginning of the process of finding out the biological effects of mobile phone radiation. The validity of expression changes detected in identified protein targets needs to be confirmed by other, non-high-throughput screening methods. Such validation experiments were performed only in four of the nine published proteomics studies. In two of them [6, 7], the authors were able to confirm that targets identified by proteomics were correct, whereas in two other studies, they were not [12, 14]. However, as seen from the study by Nylund *et al.* [12], the success of the validation experiments depends on the commercial availability of antibodies directed against the identified protein targets. Once the affected proteins are confirmed, this information can be used for determining which physiological functions of the cell might be altered by the observed change in protein expression and whether the observed change is of sufficient magnitude to change cell physiology. Only when the altered protein expression is able to alter cell physiology can we suspect that the effect might have some potential to cause biological or health-related effects, but if the cell physiology will not be affected then there will not be any risk to health.

In summary, because of the very small number of proteomics studies, with the variety of shortcomings caused by study design and by the availability of methods and reagents, the information provided by them is very limited, at best. The most convincing evidence of the effect of mobile phone radiation on cell proteome comes from studies by Leszczynski *et al.* [6] and by Nylund and Leszczynski [7], where the subsequent positive validation of the effects on protein expression seems to confirm that the effects observed in proteomics experiments might be indeed real, and from the study by Gerner *et al.* [13], where the authors demonstrated the significance of following *de novo* synthesis of proteins after exposure to mobile phone radiation.

In the future, proteomics studies should be continued [15] in order to get sufficient amount of information that will permit drawing valid conclusions about the impact of mobile phone radiation on cell proteome. Physiology of the cell or tissue will be affected only when the information encoded in genes will be translated into proteins regulating cellular functions. That is why examining proteome gives the closest estimate of the functional status of the cell. However, it is necessary to remember certain limitations of the available proteomics methods, at its present stage of technological development. Two of the important ones are (i) problems with extraction of the full proteome, many proteins are still insoluble for our extraction methods, and (ii) the most interesting proteins that regulate functions of cells are present in cells in very minute amounts, unlike the structural proteins, and thus a variety of enrichment steps need to be employed in order to study them.

References

1 ICNIRP (1998) Guidelines for limiting exposure to time-varying electric, magnetic and electromagnetic fields (up to 300GHz). *Health Phys.*, **74**, 494–522.

2 WHO Framework for Developing Health-Based EMF Standards, http://www.who.int/peh-emf/standards/EMF_standards_framework%5b1%5d.pdf (accessed December 31, 2010).

3 ICNIRP (1996) Statement: health issues related to the use of hand-held radiotelephones and base transmitters. *Health Phys.*, **70**, 578–593.

4 Leszczynski, D. and Joenväärä, S. (2001) Proteomics: new way to determine possible biological effects of mobile phone radiation. *Nat. Genet.*, **27** (Suppl.), 67.

5 Leszczynski, D., Nylund, R., Joenväärä, S., and Reivinen, J. (2004) Applicability of discovery science approach to determine biological effects of mobile phone radiation. *Proteomics*, **4**, 426–431.

6 Leszczynski, D., Joenväärä, S., Reivinen, J., and Kuokka, R. (2002) Non-thermal activation of hsp27/p38MAPK stress pathway by mobile phone radiation in human endothelial cells: molecular mechanism for cancer- and blood–brain barrier-related effects. *Differentiation*, **70**, 120–129.

7 Nylund, R. and Leszczynski, D. (2004) Proteomics analysis of human endothelial cell line EA.hy926 after exposure to GSM 900 radiation. *Proteomics*, **4**, 1359–1365.

8 Nylund, R. and Leszczynski, D. (2006) Mobile phone radiation causes changes in gene and protein expression in human endothelial cell lines and the response seems to be genome- and proteome-dependent. *Proteomics*, **6**, 4769–4780.

9 Zeng, Q., Chen, G., Weng, Y., Wang, L., Chiang, H., Lu, D., and Xu, Z. (2006) Effects of global system for mobile communications 1800MHz radiofrequency electromagnetic fields on gene and protein expression in MCF-7 cells. *Proteomics*, **6**, 4732–4738.

10 Li, H.W., Yao, K., Jin, H.Y., Sun, L.X., Lu, D.Q., and Yu, Y.B. (2007) Proteomic analysis of human lens epithelial cells exposed to microwaves. *Jpn. J. Ophthalmol.*, **51**, 412–416.

11 Karinen, A., Heinävaara, S., Nylund, R., and Leszczynski, D. (2008) Mobile phone radiation might alter protein expression in human skin. *BMC Genomics*, **9**, 77–81.

12 Nylund, R., Tammio, H., Kuster, N., and Leszczynski, D. (2009) Proteomic analysis of the response of human endothelial cell line EA.hy926 to 1800 GSM mobile phone radiation. *J. Proteomics Bioinform.*, **2**, 455–462.

13 Gerner, C., Haudek, V., Schnadl, U., Bayer, E., Gundacker, N., Hutter, H.P., and Mosgoeller, W. (2010) Increased protein synthesis by cells exposed to a 1800MHz radio-frequency mobile phone electromagnetic field detected by proteome profiling. *Int. Arch. Occup. Environ. Health*. doi: 10.1007/s00420-010-0513-7.

14 Kim, K.B., Byun, H.O., Han, N.K., Ko, Y.G., Choi, H.D., Kim, N., Pack, J.K., and Lee, J.S. (2010) Two-dimensional electrophoretic analysis of radio frequency radiation-exposed MCF7 breast cancer cells. *J. Radiat. Res.*, **51**, 205–213.

15 Leszczynski, D. and Meltz, ML. (2006) Report: questions and answers concerning applicability of proteomics and transcriptomics in EMF research. *Proteomics*, **6**, 4674–4677.

Part Seven
Challenges for Risk Management

Cancer Risk Evaluation: Methods and Trends,
Edited by Günter Obe, Burkhard Jandrig, Gary E. Marchant, Holger Schütz, and Peter M. Wiedemann.

18
Evaluating the Reliability of Controversial Scientific Results

Alexander Lerchl

18.1
Introduction

Scientific progress depends on reliable data. Before accepting experimental results of any publication that contains relevant data, the experiments are usually independently repeated and only if the original results are confirmed, the effect is called "reproducible." There are many reasons why findings of one experiment are not reproducible, namely, methodological ones or those having to do with statistics. A "significant" difference, for example, may be due to chance since the value of "p" in statistical comparisons means the probability of the difference to be fortuitous. Only if a second experiment, ideally under identical conditions, finds the same significant effect, the chances of having a fortuitous outcome is extremely low.

Scientific progress also depends on honesty. If manipulated data are published, the consequences are manifold, especially in areas that are of great importance for the policy makers, the physicians, or the general public. Manipulation starts with "data polishing," that is, the elimination of "outliers" without a proper justification, in order to get a significant result when the original data do not show a significant difference. The worst kind of manipulation is data fabrication, that is, the production of data that are not based on real experiments. In the past, many examples across scientific disciplines have demonstrated the criminal energy of those who were convicted for fabricating data [1–4].

It goes without saying that the responsibility of scientists is largest when the subject of experiments deals with environmental health effects that are relevant for many people [5]. This responsibility, however, extends also to reviewers, editors, and publishers who must ensure that these publications meet the highest scientific standards. If it turns out after publication that erroneous or manipulated data were published, it is the responsibility of the editors either to publish a note that makes the readers of the respective journal aware of the situation or to retract the publication. What seems straightforward in theory is complicated in reality, for various reasons: editors in general are not in favor of retracting publications because retractions allegedly damage the journal's reputation; authors are likewise not happy when their

Cancer Risk Evaluation: Methods and Trends,
Edited by Günter Obe, Burkhard Jandrig, Gary E. Marchant, Holger Schütz, and Peter M. Wiedemann.

papers are retracted, especially when they were not personally involved in the manipulation but contributed to the publication in another way. Retractions are, however, sometimes the only clean and clear approach toward manipulations since otherwise the manipulated publications continue to exist, are still citable, and thus continue to distort the scientific evidence. The following sections summarize how data manipulations can be detected, before or after publication, what can be done to prevent publication of fraudulent data, and which consequences scientific misconduct should have.

18.2 Detection of Scientific Misconduct

18.2.1 Before Publication

Nowadays, manuscripts usually are submitted to journals electronically, in most cases with the help of specialized web sites that standardize the submission process and thus help authors, referees, and editors to work with a common system. Increasingly, often authors are asked to provide not only the manuscript with figures, tables, and so on but also raw data that are stored for later inspections if questions about the validity of the work arise. This is a very important precondition for detecting scientific misconduct since figures and tables alone, representing in most cases averaged numbers, are less suited for these purposes.

Before such detailed analyses are performed on raw data or derived figures and tables, a plausibility check should be done. This process is hard to define and depends very much on the experience of the editors and referees, but some rules of thumb are helpful. One should be already alerted if one, and especially if more than one, of the following is observed:

- Standard deviations (SD) are very small.
- Sample sizes are very high.
- Significances are extreme.
- Correlation coefficients are close to 1.
- Results are in sharp contrast to previously published data or lack a proper explanation.

On the basis of this first impression, reviewers may come to the preliminary conclusion that something is wrong with the manuscript. The next steps are more or less easy and can be done by either the reviewer or the editors. In case those suspicions exist, the first step is always to ask the authors for raw data, if not submitted already along with the manuscript. If the authors are reluctant to provide those data (i.e., because they argue that those data were confidential and should not be shared with the referees), editors should reject the manuscript without a full review. If raw data are supplied, or if the data contained in the manuscript are sufficient, the following methods are applicable.

Table 18.1 Nonequal distribution of last digits in a manuscript.

Digit	Found (*N*)	Expected (*N*)	Residual
0	4	9.2	−5.2
1	16	9.2	6.8
2	15	9.2	5.8
3	11	9.2	1.8
4	8	9.2	−1.2
5	11	9.2	1.8
6	11	9.2	1.8
7	6	9.2	−3.2
8	8	9.2	−1.2
9	2	9.2	−7.2
Total	92		

Check for Nonequal Distribution of Last Digits

This method is established as a sensitive indicator of data fabrication. It is based on the observation that people, when generating (fabricating) data, unintentionally prefer some digits over others, for example, some have the "favorite" digit 2, others 5 [6, 7]. In real data, in contrast, all last digits should be more or less, within the statistical boundaries, equally distributed. If at least 50 numbers are found in tables or supplementary data files, one can count the last digits and perform a nonparametric test for equal distribution. Table 18.1 illustrates the outcome of such a test that was based on original data submitted to a journal.

The chi-square test reveals that this distribution is significantly nonequal ($p = 0.02$). Although this test is no proof, it is the first and strong hint of data fabrication.

Check for Appropriate Statistics

There are well-established rules that have to be obeyed when statistical tests are used. For parametric tests, for example, the data have to be checked for normal distribution and equal variances before a parametric test can be performed. If the data are not normally distributed, either nonparametric tests have to be performed or the data have to be transformed (e.g., by log or arcsin transformation) before testing for differences. One common mistake is ignoring effects of multiple comparisons. For example, if 20 parameters are compared in two groups, the chance to identify one "significant" – but in fact random – difference is extremely high. Thus, corrections (e.g., the Bonferroni correction of p-values) are mandatory. For reviewers, however, it is difficult to judge whether the reported data sets are actually complete or if only those showing significant differences are provided. This difficulty again highlights the need for all data to be included in any submission.

But not only significant data may be problematic but also nonsignificant differences can be wrong if too few samples are within one group. For example, a 20% difference can be significant or not, depending on the numbers of samples, despite identical means and standard deviations:

Sample 1:	100 ± 10	$N=3$	Sample 2:	80 ± 10	$N=3$	$p=0.07$
Sample 1:	100 ± 10	$N=10$	Sample 2:	80 ± 10	$N=10$	$p=0.0003$

Thus, the first test does not result in a significant difference, while the second does. The proper way to check the robustness of the test is to preferably use a sample size calculation before an experiment is performed or conduct a power analysis after the experiment has been completed, to prevent or detect, respectively, inconclusive results.

Check for Low Standard Deviations

Sometimes, standard deviations (SD) or standard errors of the mean (SEM) are so small that they create a "too-good-to-be-true" feeling. To prove that these deviations are in fact too small to be explained by real biological experiments is, however, often difficult. If data are given in tables, the first check is to calculate the coefficients of variation (SD divided by the mean). If they are consistently low (e.g., below 5%), this may be a sign for manipulation if the respective parameters are known to show higher variability, for example, by comparison with previously published experiments. If only SEM values are given, the SD values can be easily calculated provided the numbers of data (n) per mean are given as well:

$$SD = \sqrt{(SEM^2 \cdot n)}.$$

If data are presented only as graphs, the values for the means, SD, and SEM have to be measured. While the manual approach with a ruler usually requires an enlarged copy of the graph in question, electronic measurements are easier to do and much more precise, too. In particular, the Adobe Acrobat © offers tools that are easy to use and that include the results of the measurements directly in the document. An example is given in Figure 18.1.

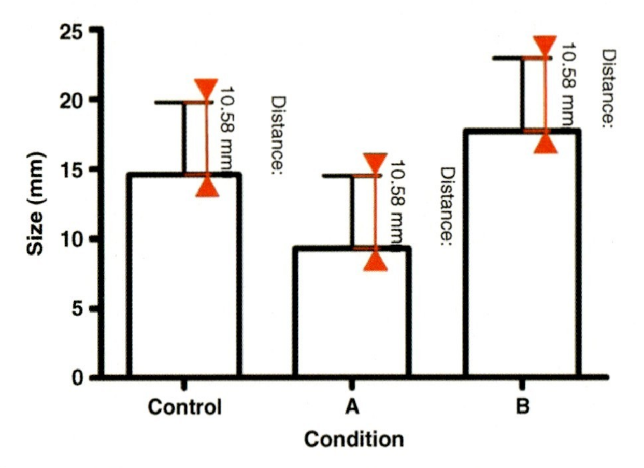

Figure 18.1 Example of data that can be estimated directly in a pdf document with the help of measurement tools.

Here, one can also see that the standard deviations are exactly the same in all three graphs, which is strongly indicative for data manipulation. In extreme cases, the standard deviations are so small that they cannot be readily seen in graphs [8]. This is always suspicious and should lead to a detailed analysis of the original data.

Check for Manipulated Pictures

Often, pictures of blots or gels are presented to support the data visually. An example of a clearly manipulated picture, and how the manipulation was detected, is given here. The manuscript was sent by a journal to the author of this chapter for review. A picture of a gel was included that looked suspicious at first sight. The resolution and size of the picture was, however, not sufficient for a detailed analysis. Consequently, the editors were asked to contact the authors in order to obtain a better picture that is shown in Figure 18.2.

While the original picture led to the impression that lanes 4 and 8 and lanes 2 and 6 were identical, only an elaborate approach using Adobe Photoshop® (contrast changes, pseudocolor) led to the conclusion that the entire picture was grossly manipulated (Figure 18.3).

As one can easily see, lanes 2 and 6 and lanes 4 and 8 are indeed absolutely identical. Furthermore, the background of lanes 2, 4, 6, and 8 is much darker than the remaining background. Both findings are a clear evidence for manipulation, for example, by copying and pasting parts of gel images into another.

18.2.2
After Publication

If papers containing manipulated data have already been published, the situation is more difficult for several reasons. First, the publication is "out," meaning that it is accessible through the journal's web site and databases (such as PubMed, Web of

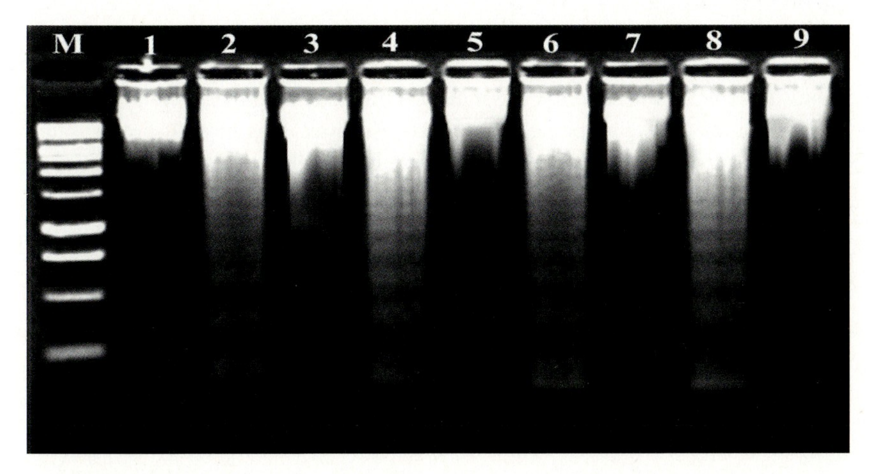

Figure 18.2 Original picture of a submitted manuscript showing DNA damage under various experimental conditions.

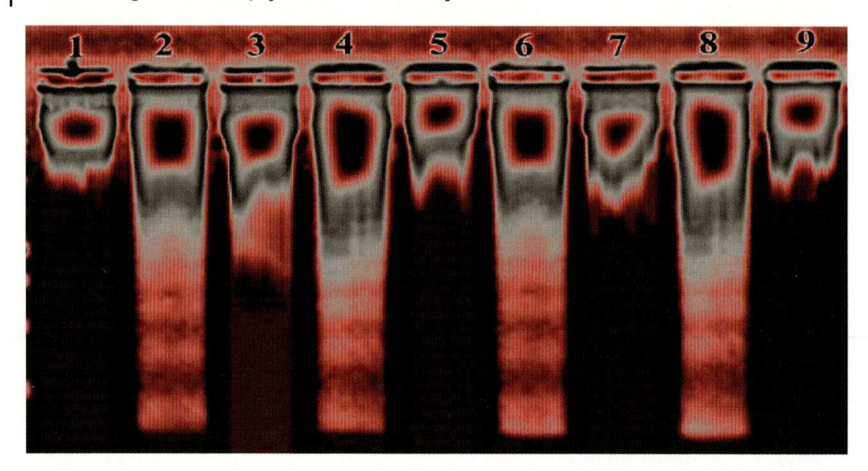

Figure 18.3 The same picture as in Figure 18.2 after contrast enhancement and conversion of gray values into pseudocolor.

Science, etc.) and is citable. Second, the correction or retraction of published papers is administratively complicated and time consuming, to various degrees. The least complicated case is when authors themselves recognize errors in their publications or are informed by their colleagues and send a correction to the journal that has published it. This correction is then published and linked to the original publication in databases so that readers are made aware of these circumstances. Often, however, readers recognize errors and submit a comment to the journal that is sent to the authors of the original publication in order to give them the opportunity to respond. Both the comment and the reply are published in the same issue of the journal and again linked to the original publication. Sometimes, the comments are so critical that an immediate action by the journal's editors – even without contacting the authors of the original publication – is mandatory.

In 2009, a paper was published [9] that allegedly demonstrated harmful effects of microwave radiation on rat leukocytes. The statistical analysis, however, was grossly wrong as stated by the editors: "The statistical methods used by the authors in their publication are not appropriate. The effective sample size in each group is the number of animals (i.e., 8) rather than the number of cells counted (i.e., 800). Thus, the t-tests used by the authors are greatly overstating the statistical significances that were found. We apologize to readers of the journal that this was not detected during the review process." This paper was retracted and the electronic pdf file of the publication was watermarked across the text with "Retracted."

Other examples show that the process of retraction, despite clear evidence of data fabrication, is a lengthy process [10]. Some more recent examples are given here. Within the Fifth European Framework, a project was supported that aimed to investigate the biological effects of magnetic (MF) and high-frequency electromagnetic fields (EMF) (REFLEX: Risk Evaluation of Potential Environmental Hazards from Low Energy Electromagnetic Field Exposure). The study was conducted between 2000 and 2004 by 12 research groups from 7 European countries. The funds by the EU amounted to €2 million, while the total costs were around €3 million. Among the participants, the Vienna group from the Division of Occupational Medicine at the Medical University Vienna (MUV), headed by Hugo Rüdiger, investigated the genotoxic effects of MF and EMF on human and rodent cells. They used the Comet assay that was previously described by Diem and Rüdiger [11]. The first author, Elisabeth Diem (married name Kratochvil), a technician, also contributed to the experiments within the REFLEX project. In total, seven publications resulted from their experiments [12–18]. The last publication [17] resulted from a grant by the Austrian Workers Compensation Board (AUVA) and was not part of the EU project. One key feature of these experiments was the blinding of the exposure conditions. Only after completion of each experiment, according to the authors, the Swiss collaborator broke the exposure code, in exchange with the experimental data.

The results of the experiments were truly remarkable. Both MF and EMF exposure, well below the internationally accepted exposure limits [19], caused highly significant damages to DNA and chromosomes. Since such effects are known to be a consequence of exposure to ionizing radiation and the first strong hint of mutagenic effects, the consequences of the findings would be dramatic. Everyone is exposed to magnetic fields, for example, by household appliances or high-voltage power transmission lines, and also high-frequency electromagnetic fields are increasingly present in our environment, especially since the introduction of mobile phones in the 1990s. For example, the exposure limit for local exposure (i.e., while using a mobile phone) is 2 W/kg SAR (specific absorption rate) for the general public. In one of their experiments [17], DNA damage occurred at 0.05 W/kg, thus at SAR levels 40 times less than the exposure limit.

From a biophysical point of view, these results are difficult to comprehend since the quantum energy of such high-frequency electromagnetic fields is approximately one million times lower than the energy needed for ionization. The only accepted mechanism of such microwave fields to interact with biological systems is heating (thermal effects) that was, however, excluded in those experiments. It was therefore no surprise that the results of the experiments by the Vienna group were not replicable, both for exposure to MF [20] and for exposure to EMF [21], despite the fact that the same cells and the same exposure units were used.

The first doubts about the results were expressed in 2006 by Vijayalaxmi *et al.* [22]. Among the arguments listed were also statistical ones, namely, the "negligible standard deviations" of the data. Indeed, when looking at the raw data that were published by the authors of the original study in their reply [23] to Vijayalaxmi *et al.*, the standard deviations were far too small to be explained by real biological experiments (for details, see Ref. [24]). The lowest standard deviations were reported by Pilger *et al.* [25] in five independent replications with coefficients of variations of approximately 1% in sham-exposed and 0.5% in exposed cells.

The first critical analysis of the last study on EMF effects [17] was published online in 2008 [26]. Here, the low standard deviations, statistical errors, and miscalculations were described in detail. In the same year, the Medical University Vienna started investigating the studies concerning allegations of scientific misconduct. Among other findings, it turned out that the sophisticated, computer-controlled blinding of the exposure conditions was worthless. A simple turn of a knob of one of the controlling units enabled the users to identify which samples were exposed and which were sham exposed. This procedure was described even in the manuals for both the MF and the EMF exposure units [27].

Also in 2008, the MUV published three press releases [28–30], clearly stating that the data of the two publications in question [12, 17] were fabricated: "The data were not measured experimentally but fabricated" [29]. In the second press release, the following was stated: "... said employee was commissioned to conduct test trials within the framework of an in-house quality control in April this year that were also used for the two publications and then supplied data without having made any related microscopic studies and evaluations beforehand. Following her conviction, she immediately confessed to her behaviour and terminated her employment relationship with the MUV with immediate effect. Already before that, a statistical expertise commissioned by the Council for Scientific Ethics had expressed doubts regarding the correct collection of data published in the two works." [28]. The third press release concludes: "The next task of the Council of Scientific Ethics is to determine all further publications in which this author was involved while using the same experimental design and then advise the publishers in charge to retract these publications as well." [30].

In 2010, a detailed statistical analysis of the data was published by Lerchl and Wilhelm [24], including Monte Carlo simulations and multinomial distribution analyses, showing beyond reasonable doubt that the published data were indeed fabricated. Also, highly significant nonequal distributions of last digits, a known hint of data fabrication [6, 7], supported this firm conclusion.

At present, three "Expressions" or "Letters of Concern" have been published [31–33] concerning three different publications [12, 17, 25] from the REFLEX research group in Vienna. Not a single publication of those in question, however, has been retracted so far. It is quite interesting to read the "Letter of Concern" by Drexler and Schaller, the editors of the *International Archives of Occupational and Environmental Health*: "We conclude that an essential part of the Methods section (an externally imposed blind) of the Schwarz *et al.* paper is unreliable because of the undisclosed opportunity for fraud. Therefore, all subsequent parts of the paper (results, discussion) cannot safely be relied on. The editors of IAOEH wish to express their doubts about the results reported in the paper by Schwarz *et al.* (2008) in this *Expression of Concern* and to apologize to the readers of IAOEH for publishing this paper." [32]. The editors *apologize* for publishing this paper, but they do not retract it, leaving a puzzling and disturbing impression about the editors' responsibility.

Remarkably different is the "Editorial Expression of Concern" by Friedl and Rühm, the editors of *Radiation and Environmental Biophysics*. They provide the readers with a detailed description of the background and a rationale why the paper in question has not yet been retracted: "We acknowledge that the paper in question is currently being investigated. In line with guidance from COPE, we decided to issue this expression of concern since the results of the investigation by the Austrian Agency for Scientific Integrity may not become available for a considerable time. This Editorial Expression of Concern is intended to make all readers of the paper by Pilger *et al.* (2004) aware of the fact that an investigation on the validity of the DNA strand break data published therein is currently being conducted. Finally, we will inform our readers about the outcome of the investigation in due time." [33].

As mentioned in this "Expression of Concern," all REFLEX publications from the group of Rüdiger are being investigated by the Austrian Agency for Scientific Integrity, whose final report is expected for fall 2010.

18.3
Committee on Publication Ethics

The Committee on Publication Ethics (COPE) was founded in 1997 and has now (as of August 2010) more than 5200 members worldwide (http://publicationethis.org), mostly editors of journals and publishers (e.g., Elsevier, Springer, and Wiley-Blackwell) who have signed up all their journals as COPE members. COPE's main tasks are given on their web site: "COPE provides a forum for editors of academic journals to discuss issues relating to the integrity of the work submitted to, or published in, their journals. Examples include conflicts of interest, falsification and fabrication of data, plagiarism, unethical experimentation, redundant publication and authorship disputes. COPE encourages its members to seek investigation into possible misconduct by universities, hospitals or other funders." The forum of COPE regularly meets, discusses cases that have been submitted, and gives advice on how to deal with such cases. Each year, around 30–40 cases are submitted. The most important part of the work of COPE is summarized in flowcharts that serve as orientation for readers, referees, and editors. For example, if readers of publications identify data that seem to indicate data fabrication, they should write a note to the editors who then have to get a second opinion from another reviewer. After this, evidence for fabrication has to be assembled, ideally by using original data if available. The next step is to contact the corresponding author, explain the concerns, and ask for raw data or lab notebooks as appropriate. If the response of the author is satisfactory (i.e., if it allays the concerns), the case is closed. If not, the other authors of the publication, if any, have to be informed, and the authors' host institution is contacted requesting an investigation. If the authors are found guilty of data fabrication, the publication must be retracted.

It is important to note that these guidelines are binding on COPE members. If the members do not follow these guidelines, they can be sanctioned. It is also important for editors, being members of COPE, that they *have to* pursue misconduct, as outlined in the Code of Conduct (http://publicationethics.org/files/u2/New_Code.pdf):

- Editors have a duty to act if they suspect misconduct. This duty extends to both published and unpublished papers.
- Editors should not simply reject papers that raise concerns about possible misconduct.
- They are ethically obliged to pursue alleged cases.
- Editors should first seek a response from those accused. If they are not satisfied with the response, they should ask the relevant employers or some appropriate body (perhaps a regulatory body) to investigate.

- Editors should follow the COPE flowcharts where applicable.
- Editors should make all reasonable efforts to ensure that a proper investigation is conducted; if this does not happen, editors should make all reasonable attempts to persist in obtaining a resolution to the problem. This is an onerous but important duty.

18.4 Conclusions

Scientific misconduct is not a trivial offense, but a severe breach of the mutual trust within the scientific community. It is the task of referees, editors, and publishers to handle such cases appropriately. It goes without saying that these efforts are sometimes time consuming, but important in order to keep science clean.

A precondition for such actions is knowledge about how to detect scientific misconduct in papers that are submitted and sent for review or that have been published already. Apart from experience in the respective field of research that is useful to do a first plausibility check, more knowledge is needed when it comes to statistical calculations and suspected data manipulations. As described in the chapter, there are some more or less simple tests that can serve as a starting point for investigations that should be initiated at the request of the editors, that is, after hints pointing to scientific misconduct have been identified. As stated by COPE, it is not the responsibility of the editors to perform such investigations, but to formally request such an investigation by the institution where the data have been generated. Simply rejecting a paper that is suspected to contain manipulated data is clearly insufficient since the same data may be published later elsewhere.

At any rate, it should become standard to submit raw data together with manuscripts in order to enable referees and editors to do (re-)calculations whenever deemed appropriate. These raw data should also be published online – as supplementary data – after acceptance of the manuscript so that other scientists can evaluate them.

Besides scientific misconduct, studies often contain data or interpretations of data, which readers may consider worth to be discussed. The appropriate platform for such discussions is, of course, the journal in which the study has been published, and letters to editors and the replies by the authors of the original study are the mode of communication. It is trivial to mention that such correspondence should be processed without delay, especially when issues discussed are relevant to many people, for example, environmental hazards.

In practice, unfortunately, these communications are sometimes impeded by the reluctance of editors and publishers to publish critical comments on articles that have appeared in their journals. As an example, a recent critical letter [34] was accepted more than 8 months after submission since the corresponding author of the criticized publication was unable to supply a reply earlier. It goes without saying that such long processing times do not foster a timely discussion.

In summary, this chapter has described some examples of scientific misconduct, how to detect it, and how to handle such cases appropriately. It is to be hoped that all parties involved are aware of the severity of scientific misconduct and to be critical when evaluating scientific publications. To this end, this chapter may have contributed to some extent.

References

1 Brumfiel, G. (2002) Investigation into nanotechnology papers expands. *Nature*, **417**, 473.

2 Hagmann, M. (2000) Scientific misconduct. Panel finds scores of suspect papers in German fraud probe. *Science*, **288**, 2106–2107.

3 Reynolds, S.M. (2004) ORI findings of scientific misconduct in clinical trials and publicly funded research, 1992–2002. *Clin. Trials*, **1**, 509–516.

4 Marcus, A. (2009) Fraud case rocks anesthesiology community. *Anesthesiol. News*, **35**. http://www.anesthesiologynews.com/index.asp?section_id=3&show=dept&article_id=12634.

5 Merlo, D.F., Vahakangas, K., and Knudsen, L.E. (2008) Scientific integrity: critical issues in environmental health research. *Environ. Health*, **7** (Suppl. 1), S9.

6 Al-Marzouki, S., Evans, S., Marshall, T., and Roberts, I. (2005) Are these data real? Statistical methods for the detection of data fabrication in clinical trials. *BMJ*, **331**, 267–270.

7 Mosimann, J.E., Wiseman, C.V., and Edelman, R.E. (1995) Data fabrication: can people generate random digits? *Accountability Res.*, **4**, 31–55.

8 Lerchl, A. (2010) Are these data real? Comments on "No effects of intermittent 50Hz EMF on cytoplasmic free calcium and on the mitochondrial membrane potential in human diploid fibroblasts." by Pilger *et al.* (*Radiat Environ Biophys* 43:203–7 (2004)). *Radiat. Environ. Biophys.*, **49** (3), 491–493.

9 Garaj-Vrhovac, V., Gajski, G., Trosic, I., and Pavicic, I. (2009) Evaluation of basal DNA damage and oxidative stress in Wistar rat leukocytes after exposure to microwave radiation. *Toxicology*, **259**, 107–112.

10 Cooper-Mahkorn, D. (1998) Many journals have not retracted "fraudulent" research. *BMJ*, **316**, 1850.

11 Diem, E. and Rüdiger, H. (1999) Mikrokerntest und Comet assay: ein ergebnisvergleich bei normalprobanden. *Arbeitsmed. Sozialmed. Umweltmed.*, **34**, 437–441.

12 Diem, E., Schwarz, C., Adlkofer, F., Jahn, O., and Rüdiger, H. (2005) Non-thermal DNA breakage by mobile-phone radiation (1800MHz) in human fibroblasts and in transformed GFSH-R17 rat granulosa cells *in vitro*. *Mutat. Res.*, **583**, 178–183.

13 Ivancsits, S., Diem, E., Jahn, O., and Rüdiger, H.W. (2003) Age-related effects on induction of DNA strand breaks by intermittent exposure to electromagnetic fields. *Mech. Ageing Dev.*, **124**, 847–850.

14 Ivancsits, S., Diem, E., Jahn, O., and Rüdiger, H.W. (2003) Intermittent extremely low frequency electromagnetic fields cause DNA damage in a dose-dependent way. *Int. Arch. Occup. Environ. Health*, **76**, 431–436.

15 Ivancsits, S., Diem, E., Pilger, A., Rüdiger, H.W., and Jahn, O. (2002) Induction of DNA strand breaks by intermittent exposure to extremely-low-frequency electromagnetic fields in human diploid fibroblasts. *Mutat. Res.*, **519**, 1–13.

16 Ivancsits, S., Pilger, A., Diem, E., Jahn, O., and Rüdiger, H.W. (2005) Cell type-specific genotoxic effects of intermittent extremely low-frequency electromagnetic fields. *Mutat. Res.*, **583**, 184–188.

17 Schwarz, C., Kratochvil, E., Pilger, A., Kuster, N., Adlkofer, F., and Rüdiger, H.W. (2008) Radiofrequency electromagnetic fields (UMTS, 1,950MHz) induce genotoxic effects *in vitro* in human fibroblasts but not in lymphocytes. *Int. Arch. Occup. Environ. Health*, **81**, 755–767.

18 Winker, R., Ivancsits, S., Pilger, A., Adlkofer, F., and Rudiger, H.W. (2005) Chromosomal damage in human diploid fibroblasts by intermittent exposure to extremely low-frequency electromagnetic fields. *Mutat. Res.*, **585**, 43–49.
19 ICNIRP (1998) Guidelines for limiting exposure to time-varying electric, magnetic, and electromagnetic fields (up to 300GHz). *Health Phys.*, **74**, 494–522.
20 Scarfi, M.R., Sannino, A., Perrotta, A., Sarti, M., Mesirca, P., and Bersani, F. (2005) Evaluation of genotoxic effects in human fibroblasts after intermittent exposure to 50Hz electromagnetic fields: a confirmatory study. *Radiat. Res.*, **164**, 270–276.
21 Speit, G., Schutz, P., and Hoffmann, H. (2007) Genotoxic effects of exposure to radiofrequency electromagnetic fields (RF-EMF) in cultured mammalian cells are not independently reproducible. *Mutat. Res.*, **626**, 42–47.
22 Vijayalaxmi, McNamee, J.P., and Scarfi, M.R. (2006) Comments on: "DNA strand breaks" by Diem *et al.* [*Mutat. Res.* 583 (2005) 178–183] and Ivancsits *et al.* [*Mutat. Res.* 583 (2005) 184–188]. *Mutat. Res.*, **603**, 104–106, author reply 107–9.
23 Rüdiger, H.W., Kratochvil, E., and Pilger, A. (2006) Reply to the letter of Vijayalaxmi *et al. Mutat. Res.*, **603**, 107–109.
24 Lerchl, A. and Wilhelm, A. (2010) Critical comments on DNA breakage by mobile phone electromagnetic fields (Diem *et al.*, *Mutation Research* 2005, 583, 178–183). *Mutat. Res.*, **697**, 60–65.
25 Pilger, A., Ivancsits, S., Diem, E., Steffens, M., Kolb, H.A., and Rüdiger, H.W. (2004) No effects of intermittent 50Hz EMF on cytoplasmic free calcium and on the mitochondrial membrane potential in human diploid fibroblasts. *Radiat. Environ. Biophys.*, **43**, 203–207.
26 Lerchl, A. (2009) Comments on "Radiofrequency electromagnetic fields (UMTS, 1,950MHz) induce genotoxic effects *in vitro* in human fibroblasts but not in lymphocytes" by Schwarz *et al.* (*Int Arch Occup Environ Health* 2008: doi: 10.1007/s00420-008-0305-5). *Int. Arch. Occup. Environ. Health*, **82**, 275–278.
27 Wolf, C. (2008) Security considerations in blinded exposure experiments using electromagnetic waves. *Bioelectromagnetics*, **29**, 658–659.
28 MUV (2008) Press release, July 29, http://www.meduniwien.ac.at/homepage/news-and-topstories/en/?Fsize=0&tx_ttnews%5Btt_news%5D=231&cHash=8e7b70be11.
29 MUV (2008) Press release, May 23, http://www.meduniwien.ac.at/homepage/news-and-topstories/en/?Fsize=0&tx_ttnews%5Btt_news%5D=204&cHash=825ebb3101.
30 MUV (2008) Press release, September 1, http://www.meduniwien.ac.at/homepage/news-and-topstories/en/?Fsize=0&tx_ttnews%5Btt_news%5D=243&cHash=d289b91bc7.
31 Baan, R.A. (2010) Letter of concern. *Mutat. Res.*, **695**, 1.
32 Drexler, H. and Schaller, K.H. (2009) Expression of concern. *Int. Arch. Occup. Environ. Health*, **82**, 143–144.
33 Friedl, A.A. and Rühm, W. (2010) Editorial expression of concern regarding: Pilger A *et al.* (2004) No effects of intermittent 50Hz EMF on cytoplasmic free calcium and on the mitochondrial membrane potential in human diploid fibroblasts. *Radiat. Environ. Biophys.*, **43**, 203–207.
34 Lerchl, A. (2010) Inappropriate statistical analyses lead to erroneous results [Focke et al., Mutatation Research 683 (2010) 74-83]. *Mutat. Res.*, **692**, 61–62.

19
Comparative Risk Assessment with Ionizing and Nonionizing Radiations

Jürgen Kiefer

19.1
Introduction

New technological developments have always been met with anxiety in the general public and connected with possible risks to human health. Mobile communication is no exception in this respect, but it represents a special case as electromagnetic fields are involved. There is already quite solid knowledge about the hazards of X- or γ-rays and it is tempting to generalize the results obtained with them to all kinds of electromagnetic radiations. Scientists, especially physicists and radiobiologists, are of course not prone to this fallacy, but the cancer deaths of Hiroshima and Nagasaki are subconscientiously still part of the public concern. Furthermore, there is a wealth of scientific information about the action of ionizing radiation (as detailed to some extent below) that lead to a sophisticated and well-founded system of radiation protection. The present status is described in publication 107 of the ICRP [1]. Such a comprehensive approach is clearly lacking for nonionizing radiations, although differences between wavelength regions have to be acknowledged.

Electromagnetic radiations span a wide range of frequencies, wavelengths, and quantum energies as schematically summarized in Table 19.1.

A word of clarification is indicated when dealing with risk assessment. A certain agent presents a risk only if it constitutes a hazard (which "means the observed toxic manifestation(s) induced by a known quantity of a substance under known exposure conditions" [2]) and if there is an exposure. This statement can be summarized in a very simple formula:

$$\text{Risk} = \text{Hazard (as a function of exposure)} \times \text{Exposure}$$

The first step of risk assessment is thus to establish whether the agent under consideration constitutes a hazard. Quantitative evaluation requires the knowledge of hazard–exposure relationship and also the determination of the actual exposure (commonly called dosimetry). Most of the present discussion is centered on the question if a hazard exists at all for a given – very often not very well-defined – exposure scenario.

Cancer Risk Evaluation: Methods and Trends,
Edited by Günter Obe, Burkhard Jandrig, Gary E. Marchant, Holger Schütz, and Peter M. Wiedemann.

Table 19.1 Schematic overview of the range of electromagnetic radiations and their applications.

Frequency (Hz)	Wavelength (m)		Quantum energy (eV)
$> 3 \times 10^{15}$	$<10^{-7}$	Ionizing radiation	>12.4
3×10^{15}	10^{-7}	Ultraviolet	12.4
7.5×10^{14}	4×10^{-7}		3.1
7.5×10^{14}	4×10^{-7}	Visible light	3.1
3.75×10^{14}	8×10^{-7}		1.6
3.75×10^{14}	8×10^{-7}	Infrared	1.6
3×10^{12}	10^{-4}		1.2×10^{-2}
3×10^{12}	10^{-4}	Terahertz	1.2×10^{-2}
3×10^{10}	10^{-2}		1.2×10^{-4}
3×10^{10}	10^{-2}	Mobile communication,	1.2×10^{-4}
3×10^{9}	10^{-1}	microwaves	1.2×10^{-5}
3×10^{9}	10^{-1}	Radio frequency	1.2×10^{-5}
3×10^{4}	10^{4}		1.2×10^{-10}
$<3 \times 10^{2}$	$> 10^{2}$	Extremely low frequency (ELF)	$< 10^{-12}$

Hazard identification and risk assessment follows essentially three approaches (Figure 19.1): epidemiology (with humans), animal experiments, and *in vitro* investigations.

Epidemiology plays an important role, but studies have to be performed with scrutiny in order to avoid premature conclusions. A critical discussion on epidemiological methods was published in the 2006 UNSCEAR report [3], and the International Association of Epidemiology (IAE) has provided "guidelines for good epidemiological practice" that are regularly updated and accessible at the IEA web site [4].

Epidemiological studies are usually of two types: cohort and case–control studies. Both are liable to "confounders" and "bias," which can never be excluded and require careful consideration and discussion. A third type that is unfortunately quite popular

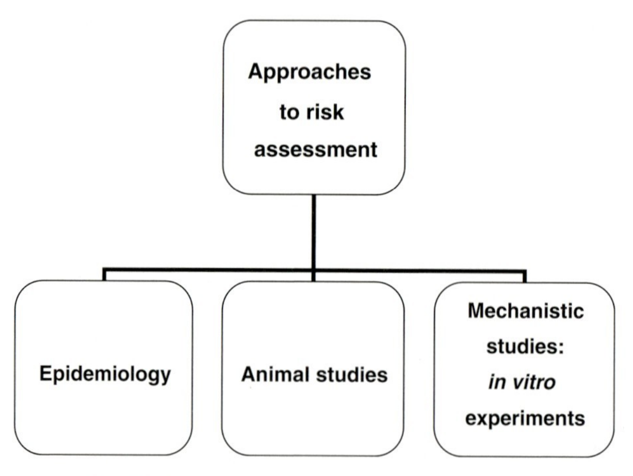

Figure 19.1 Methods of risk assessment.

among nonspecialists is the so-called ecological study. While in both cohort and case–control studies the individual exposure determination (or at least reasonable estimates) is used for the comparison of persons with and without the disease in question, in ecological studies commonly disease incidences of populations living in areas with different average exposures are compared. This approach is dangerous as cofactors (e.g., living conditions, social standard, life style, etc.) cannot be properly weighted and erroneous conclusions are not rare.

Hill [5] formulated "viewpoints" (known as Bradford Hill criteria) that can serve as a guideline when comparing risks of different agents: *strength, consistency, specificity, temporality, biological gradient, plausibility, coherence, experiment, and analogy.*

Following are the most important items in the present context:

- **Consistency**: Consistent findings observed by different persons in different places with different samples strengthen the likelihood of an effect.
- **Biological gradient**: Greater exposure should generally lead to greater incidence of the effect. However, in some cases, the mere presence of the factor can trigger the effect (*dose–effect relationship*).
- **Plausibility**: A plausible mechanism between cause and effect is helpful (but Hill noted that knowledge of the mechanism is limited by the current knowledge).
- **Coherence**: Coherence between epidemiological and laboratory findings increases the likelihood of an effect. *But different agents claimed to have the same effect should also display an analogous pattern of reactions, both in vivo and in vitro.*

Although these criteria are certainly not comprehensive, they constitute some kind of "yardstick" to review the available data with different radiation types.

19.2 Review of Different Radiation Types

19.2.1 Ionizing Radiation

Nonstochastic effects are less of a problem for radiation protection (with the exception of eye cataracts) [6] because there is no risk if the exposure remains clearly below the threshold doses. The remaining part of the chapter will, therefore, mainly concentrate on stochastic effects with emphasis on cancer.

There is ample epidemiological evidence showing the relationship between radiation dose and cancer. The most recent UNSCEAR report [3] lists nearly 100 cohort and/or case–controls studies dealing with this subject, although the quality of the publications is admittedly variable. The most important source of information for more than 60 years has been the investigation of bomb survivors in Hiroshima and Nagasaki. Results were summarized for solid cancer mortality [7] and incidence [8] as well as for leukemia mortality [9]. Little [10] gives a useful and critical review of the results so far obtained, including noncancer diseases. Cancer incidence in nuclear

Radiation effects in humans

Acute (nonstochastic) Effects

Impairment of organ function:

Blood forming system, alimentary tract, fertility, and skin

Eye cataracts

Acute radiation syndrome and acute radiation death

Teratogenic Effects

Characteristics:

Extent is dose dependent

Effects are ***deterministic***

There are threshold doses

Late Effects

Genetic risk by germ line mutations, cancer

Characteristics:

Probability is dose dependent

Effects are ***stochastic***

Threshold doses are not known

Figure 19.2 A summary of the effects of ionizing radiations.

workers has also been analyzed yielding results largely compatible with those of the bomb survivors [11]. A clear dose–effect relationship has been observed in all studies, thus fulfilling the criterion of a "biological gradient." There is also "coherence" as similar results were independently obtained by different authors.

A necessary prerequisite for all risk estimates based on epidemiological investigations is reliable dosimetry of the included individuals. In the case of bomb survivors, no measurements are available, but doses have been estimated on the basis of bomb data and distance from the epicenter. The first data were published in 1965 (for review, see Ref. [18]), but they underwent a number of revisions [17]. In spite of some uncertainties remaining, there appears to be now a solid ground for risk estimation. The situation is clearly better with nuclear workers as they are continuously monitored and the records are available. This makes the nuclear workers study [11] particularly valuable, although the statistical errors are fairly large because of the comparatively small doses (and hence low effect levels) involved (see Figure 19.2 and Chapter 6).

Radiation carcinogenesis in experimental animals has already been consistently demonstrated many years ago [12–14]. The results were summarized by Fry and Storer [15] and Grahn *et al.* [16]. Solid tumors and leukemia were found after exposure to doses in the range around 1 Gy, irrespective of whether they were given acutely or over long times.

The number of papers dealing with the mechanism of cancer induction by ionizing radiations is virtually uncountable. A general scheme, depicted in Figure 19.3 in a very simplified manner, has evolved over the years supported by many experiments.

The primary ionizations practically involve all cellular molecules, but alterations of the genetic material, the DNA, are certainly the most important. There are many radiation products in DNA, but it is now widely accepted that double-strand breaks (DSB) play the most important role. They are formed in a linear dose response over five orders of magnitude [20], covering even exposures to diagnostic X-rays [21] and initiating a multitude of cellular reactions. Many genes are activated and repair processes start. Repair can act in an error-free or in an error-prone manner. Misrepair leads to chromosomal aberrations, loss of colony-forming ability, and neoplastic transformation, which are the first steps in carcinogenesis. Cell cycle progression is delayed. Depending on dose and cell type, the irradiated cells may undergo apoptosis that has to be distinguished from the loss of colony-forming ability. Also, the methods of detection differ: the standard vitality tests, for example, dye exclusion tests, do not give a measure of colony-forming ability but may indicate apoptosis, although more specific methods are now standard. This is not always appreciated in many investigations with nonionizing radiations.

The cellular reactions to ionizing radiation are very complex and definitely not yet fully explored. Modern techniques like genomics, proteomics, and others will yield new results in the future. A useful summary of the recent status is provided by Harper and Elledge [19].

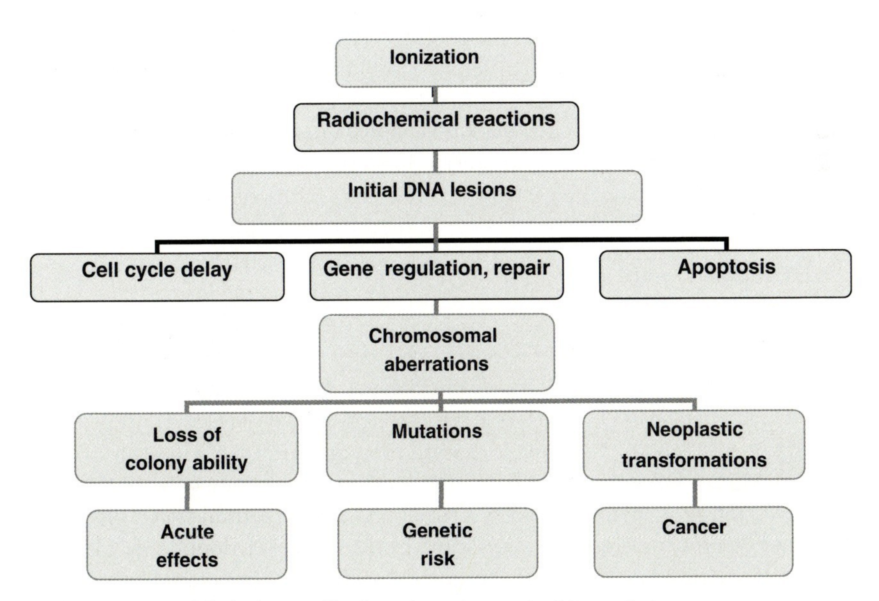

Figure 19.3 Simplified scheme of biological reactions to ionizing radiations.

Let us now judge the situation of risk assessment with ionizing radiations by applying the Bradford Hill criteria already mentioned:

- **Consistency**: The observations summarized above have been reproduced by numerous investigators.
- **Biological gradient**: There is a clear dose–effect relationship sometimes over many orders of magnitude.
- **Plausibility**: Although the mechanism of cancer formation following ionizing radiation exposure is not yet fully explored, the main pathway is clear and biologically plausible.
- **Coherence**: Epidemiological and laboratory investigations, both with animals and *in vitro*, yield quite compatible results.

In summary, despite a number of still open questions, the investigations with ionizing radiations fulfill definitely the Bradford Hill criteria and may thus serve as a model for risk assessment.

19.2.2
Ultraviolet Radiation

There is no doubt that skin cancer is related to ultraviolet radiation (UVR) [22, 23]. Evidence for this statement comes from epidemiology, animal experiments, and numerous laboratory investigations. Other effects include erythema (sunburn), skin aging, and immune suppression and actions on the eye; however, they are not further discussed here. The formation of vitamin D also depends on UVR that is vital for human health as its deficiency causes rickets, a deformation of bones. It would hence be desirable to be able to analyze the risk–benefit relation in quantitative terms. Unfortunately, this is not yet possible because the data available are not sufficient.

There are three types of skin cancer: basal cell carcinoma, squamous cell carcinoma, and malignant melanoma. The first two have a low mortality, while the third, that is, melanomas, are fatal when not treated at a very early stage.

UVR in the terrestrial environment extends from 280 to 380 nm wavelength. It is commonly subdivided as under:

UV A: 315–400 nm
UV B: 280–315 nm
UV C: 200–280 nm

UV C is the part of the solar spectrum that does not reach the earth's surface because it is absorbed by ozone in the atmosphere. The biological action of UVR is mediated through the specific absorption of essential biomolecules and thus depends strongly on wavelength. The most important component is DNA whose absorption peaks at 265 nm but extends to about 320 nm. There are, however, other chromophores with absorptions also in the UV A region that could indirectly lead to biological effects via energy transfer processes or the formation of reactive species like free radicals. The dependence of bioeffects on wavelength is called an "action spectrum" that may indicate the molecular starting point for the particular action.

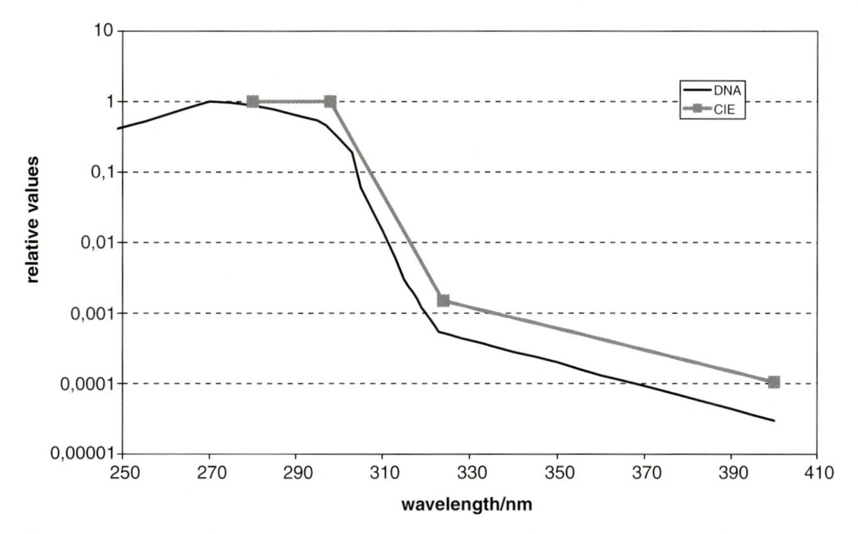

Figure 19.4 The normalized CIE erythema action spectrum [25] compared to DNA absorption [26].

An action spectrum for cancer induction in humans has not yet been established. It is assumed, also on the basis of animal experiments [24], that it closely resembles that for erythema formation. The Commission internationale de l'eclairage (CIE) suggested a normalized action spectrum that could also be used as weighting function for multichromatic UV sources (Figure 19.4) [25].

It is thus clear that UV B plays the most important role. It is part of the solar terrestrial spectrum, but its relative contribution depends on geographical latitude. Many years ago it was noted that skin cancer incidence also varies with this parameter [27]. Although this finding was supported by a number of investigators [22, 23], there has not yet been established a dose–effect relationship for any type of skin cancer. In most cases, UV exposure levels were inferred from solar spectra at different geographical positions. In one case they were actually measured [28] and related to the incidence of different kinds of skin cancer. But even this work does not fulfill the strong conditions of epidemiology. As long as it is not possible to determine retrospectively individual doses in cancer cases, a quantitative risk assessment will not be possible. This is not only desirable for solar exposure – be it at work or with leisure activities – but also for artificial sources, for example, sunbeds [29]. Contrary to ionizing radiation with UV, it is possible to relate cancer induction to exposure at the molecular level as mutations typical for UV could be detected in human squamous skin carcinoma [30].

There is ample evidence that UV is able to induce skin tumors also in animals, although there is no good animal model for human skin. Most studies were performed with nude mice (for review, see Ref. [24]); melanoma was also investigated in fish [31], but here the action spectrum showed an unusually high effectiveness for UV A that was not seen in the mouse model.

The cellular reactions of UV are known in considerable detail and impossible to review here. Recent summaries with special emphasis on skin cancer were, for example, provided by Greinert *et al.* [32] and de Gruijl and Rebel [33]. The absorption of UV in DNA causes the formation of specific photoproducts [34] among which cyclobutyl pyrimidine dimers (CPD) are the most important. The action spectrum *in vitro* closely resembles the absorption spectrum of DNA [35]. CPD have also been found in UV-exposed human skin [36]. All photoproducts are subject to repair [37, 38, 41]; if repair fails, mutations may result. CPDs lead to a specific type of mutations, namely, cytosine to thymine transitions, which are rarely found with other mutagens, and thus constitute a "UV signature." As already said, they are found in biopsies of human skin cancers [30], thus proving the relationship between UV exposure and skin cancer induction. This is a unique situation that is not paralleled with any other kind of radiation.

Another line of evidence comes from studies dealing with repair deficient organisms [40]. As early as in 1968, Cleaver [42] noted that people suffering from the disease xeroderma pigmentosum that lack certain parts of the repair system are particularly sensitive to environmental UV and prone to develop skin cancer at a very early age. More recent developments were summarized again by Cleaver [43] and Daya-Grosjean [39].

The UV A part of the spectrum deserves a special discussion. As DNA only very weakly absorbs in this region, it has traditionally been considered as unimportant in terms of skin damage. This view has rather recently been drastically changed. It is now clear that UV A contributes significantly to the skin cancer risk, and the molecular and cellular mechanisms are fairly well known [44].

In summary, it can be stated that ultraviolet carcinogenesis fulfils all the criteria mentioned in Section 19.1. The findings are consistent, plausible, and coherent, and a biological gradient has also been demonstrated both *in vivo* and *in vitro*. There is so far, however, no established dose–effect relationship for the induction of any type of skin cancer, particularly for malignant melanoma. The influence of the time pattern of exposure has also not yet been clarified [45, 46], and studies of sunbed users yielded inconsistent results [47]. One has to conclude finally that there is a very clear picture concerning the relationship between skin cancer and UV exposure, but a quantitative risk assessment for humans is not yet possible because of the unclear dose dependence.

UV also possesses a beneficial effect as it is necessary to produce vitamin D. Lack of sunlight leads to vitamin D deficiency with a number of health symptoms [48]. A risk–benefit analysis would be highly desirable, but it is not yet feasible because of the lacking dose–effect relationship.

19.2.3
Visible Light

Visible light (400–800 nm) is considered to be noncarcinogenic [49] and there are no animal data to suggest the contrary. *In vitro* studies, however, demonstrate that possible toxic photoproducts can be formed upon exposure to short-wavelength visible light [50] and micronuclei [51] that are commonly taken as indicators of genotoxicity. There thus appears to be a – presumably small – hazard potential, but quantification is not possible.

19.2.4 Infrared

Infrared (IR) spans the wavelength region from about 1 μm to 1 mm. It is subdivided into

IR A: 800–1400 nm
IR B: 1.4–3 μm
IR C: 3 μm–1 mm.

IR C overlaps to a large extent with "terahertz radiation" (see below).

IR is part of solar radiation, can be found at working places where high temperatures are involved, and its application has a long-standing tradition in medicine. More recently, warming cabins and similar devices have become popular as substitutes for saunas. IR is generally considered to be noncarcinogenic, but a cocarcinogenic or promoting action of heat cannot be fully excluded [52]. It has also been suggested that higher temperatures may enhance the effect of UV exposure [53]. In hairless mice, more skin tumors were induced by ultraviolet radiation at higher ambient temperatures [54]. On the basis of these findings and human epidemiological data, van der Leun *et al.* [53] concluded that the incidence of nonmelanoma cancers may rise by a few percent per degree Celsius.

Although it seems clear that IR does not directly induce DNA damage in normal cells [55], there are other photochemical reactions that may be of concern. IR A is selectively absorbed by the enzyme cytochrome *c* oxidase, which is an essential part of mitochondria [56]. Its alteration may change the cell's redox state and influence important regulatory metabolic pathways [57]; reactive oxygen species (ROS) may also be generated. Whether these nonthermal processes lead to serious health consequences (other than skin aging) is not yet clear. It is not very likely that radicals formed in mitochondria cause DNA damage because there are many substances in the cell on their way to the nucleus to react with.

19.2.5 Terahertz Radiation

"Terahertz radiation" ranges from about 0.1 to 10 THz. Its application was limited in the past because there were no powerful sources available. This situation has changed recently. Apart from its use as analytical tool, terahertz radiation has now gained actual interest because it has been suggested to use it for "body scanners" in security checks [58], for example, in airports. Also, the use of terahertz radiation for medical imaging is discussed [59, 60]. Risk assessment has not yet been an issue, but as there is now some public concern, a critical evaluation is indicated.

The penetration depth is very small (less than 1 mm in tissue) [52], so only surface effects in the skin have to be considered. Another limitation is that terahertz radiation is strongly absorbed by water. The absorption coefficient lies between 100 and 1000 cm^{-1} [61]. There are neither epidemiological nor animal data. *In vitro* experiments generally did not show any cytotoxic or genotoxic effects [62, 63]; the only exception is a report on induced genetic instability [64]. A collaborative EU project has addressed the issue, the final report is available at www.frascati.enea.it/THz-BRIDGE/.

In summary, it can be stated that there is no evidence suggesting a biological or health hazard related to the application of terahertz radiation.

19.2.6 Mobile Communication

Mobile communication operates in the microwave region from 0.9 to about 3 GHz (microwave ovens use typically 2.45 GHz). There has been a vigorous public discussion about possible health effects not only in the scientific literature but also in the media, which is by no means closed. National governments and international institutions and agencies have launched numerous programs and research projects but a final clarification does not yet seem in sight. It is clear that the quantum energy of microwaves is far too small to lead to ionizations or electronic excitations. Molecular vibrations and rotations possess energy levels that lie at considerably higher frequencies. The physical interaction is described by the Maxwell equations and dominated by the dielectric properties of the exposed tissue. Conductivity and permittivity of a number of human organs were determined by Gabriel and coworkers [65–70]. The main effect is undoubtedly tissue heating, but it is often postulated that so-called athermal interactions exist; plausible physical mechanisms, however, have not been suggested.

There is quite a large number of epidemiological investigations, mostly case–control studies, that lack individual exposure assessments. Time and frequency of cell phone use, explored by personal interviews, are used as "proxies." As the head is the most exposed organ, brain and neck tumors are in the center of interest (glioma, meningioma, acoustic neurinoma, and tumors of the salivary glands). The results of the studies are varied, in most cases no significant association between tumor induction and cell phone use was found but this view has been challenged [71–74]. A comprehensive review was given by Ahlbom *et al.* [75]. They analyzed all available studies and concluded that "Despite ... methodologic shortcomings and the still limited data on long latency and long-term use, the available data do not suggest a causal association between mobile phone use and fast-growing tumors such as malignant glioma in adults, at least those tumors with short induction periods. For slow-growing tumors such as meningioma and acoustic neuroma, as well as for glioma among long-term users, the absence of associations reported thus far is less conclusive because the current observation period is still too short." Examples from their analysis are in Table 19.2.

Table 19.2 Summary of epidemiological studies (mostly case–control studies) to find an association between head tumors after "long-term" (more than 6 years) use of mobile phones.

	Number of individual studies	Number of studies with significant positive association	Total number of cases	Odds ratio of all pooled studies (95% CI)	Odds ratio of pooled INTERPHONE studies (95% CI)
Glioma	12	2	360	1.1 (0.8–1.4)	1.0 (0.7–1.2)
Meningioma	6	1	87	1.2 (0.7–2.2)	0.9 (0.7–1.3)
Neurinoma	8	0	46	1.4 (0.7–2.5)	1.0 (0.7–1.5)
Salivary gland	5	0	38	0.9 (0.5–1.4)	—

Pooled analyses are not included in the total number of studies and cases to avoid double counts. Data from Ref. [75].

A large multinational study, the "INTERPHONE study," coordinated by the International Agency for Research on Cancer was started some time ago. It comprises 16 case–control studies performed in 13 countries. Some of the contributions by different groups are already published, and they are included in the review by Ahlbom *et al.* [75].

The complete report on brain cancers has just appeared in an electronic version [76]; it confirms essentially the previous findings, although a significant increase of gliomas is reported for very excessive use that is, however, based on rather improbable figures of speech durations. A final assessment at present seems to be premature.

Table 19.2 demonstrates that in the majority of studies, no significant association was found and the odds ratios of the pooled studies are close to 1 that lies always within the confidence interval. The few investigations showing positive associations come from the same group [74].

In summary, it can be said that overall an association between tumor formation and cell phone use could not consistently been found, but one has to consider that some of the tumor entities have very long latency times (meningioma, neurinoma) and as such the observation period might be too short. But even if one accepts the few positive findings, they cannot be used for a quantitative risk assessment because no real exposure data are available.

Animal experiments could – in theory – complement the human studies. They can be performed under well-controlled conditions so that bias and confounders that are a problem in epidemiology can be avoided. The well-defined exposure of large animal populations, however, is by no means an easy task and requires some technical effort. This was not always realized in earlier investigations. Dasenbrock [77] reviewed the experiments related to carcinogenicity in animals available at the time and concluded: "Under the described experimental circumstances and with the shortcomings listed below, the animal cancer studies reviewed and published until now did not show a significant tumor-promoting or co-carcinogenic effect due to mobile phone-relevant

RF radiation. The only exception was Repacholi's study of 1997." The last mentioned study [78] prompted replication attempts that could not confirm the original findings [79, 80]. Similar investigations [81, 82] also yielded negative results. To further clarify the issue, the European Union funded a multicenter study (Perform A). The results of the different subgroups are published [83–86]. The overall result is that the carefully performed animal experiments do not support the assumption that electromagnetic fields as used for mobile communication may cause cancer (see Chapter 7).

Although human and animal data do not show consistently that the use of mobile phones may induce tumors, there is still the not yet finally answered question whether the situation may be different with very long exposure times. As the latency times of many tumors extend to decades, the observation time may still be too short. *In vitro* experiments on cells could be helpful. If it could be shown that these electromagnetic fields are able to induce alterations in the genetic material similar to those induced by ionizing radiations, they are also likely to be carcinogenic. This is the main reason for numerous studies that aim to show that microwaves lead to critical alterations in DNA. The field cannot be summarized here, but one may refer to a number of recent reviews [87–92]. The German Radiation Protection Commission (Strahlenschutzkommission (SSK)) addressed the issue of genotoxicity in an extensive statement [93] and concluded that there is no consistent evidence for a genotoxic action of electromagnetic fields in the microwave range. It was also noted that only few experimental approaches were applied (mainly comet and micronucleus assays). There is so far no investigation on colony-forming ability that is standard with ionizing radiation. Determinations of mutation induction and neoplastic transformations are rare. The few available recent publications [94–96] reported negative results, at least at field intensities around current recommended limits. There are also two multicenter EU projects, REFLEX (final report available at www.verum-foundation.de.admin.excellent-ms.net/www2004/html/pdf/euprojekte01/REFLEX_final%20report.pdf) and PERFORM B [97] that gave controversial results. Although in some subprojects of REFLEX clear genotoxic effects were reported, they were clearly excluded in PERFORM B. It appears that there is by no means sufficient evidence to assume that microwave fields cause deleterious genetic effects *in vitro* at intensities below or around current recommended limits. This is also unlikely on biophysical grounds [98, 99].

19.2.7
Radio Frequency

Here, the emissions of radio and television transmitters in the region below 300 MHz are discussed. As their relevance is presently diminishing because of the new techniques (satellite, cable), public concern is obviously less than with the mobile communication.

There are a few epidemiological studies suggesting an increased risk of leukemia in the vicinity of powerful radio or TV stations, for example, around Radio Vatican in Rome [100], but they suffer from the small number of cases and inadequate exposure assessment. A larger study [101] in Germany could not substantiate the presumed health risk. A recent review [105] also comes to the conclusion that there is no reliable evidence for an association between TV or radio stations and childhood leukemia. There are also no animal or *in vitro* data to support the suspicion that these radio waves might be carcinogenic.

19.2.8
Extremely Low Frequencies

In the extremely low-frequency (ELF) regions, electric and magnetic fields are decoupled and thus act separately on biological systems. As the penetration of electric fields is very small (due to the "skin effect"), only magnetic fields have to be considered [102].

Since the first report [103] that residential exposure to increased magnetic fields is associated with childhood leukemia, a number of epidemiological studies were conducted that essentially confirmed this surprising finding [104]. Figure 19.5 gives an overview of the odds ratios for a range of magnetic field strengths taken from the recent review by Schüz and Ahlbom [105].

These data are based on carefully conducted epidemiological studies and mostly also on measurements of residential exposures. There is a clear dose–effect relationship (Figure 19.5) that could be used for risk assessment.

The findings reported are still an enigma. They are not consistently supported by animal experiments [106]; recent investigations involving large number of mice and careful exposure conditions also gave negative results [107, 108].

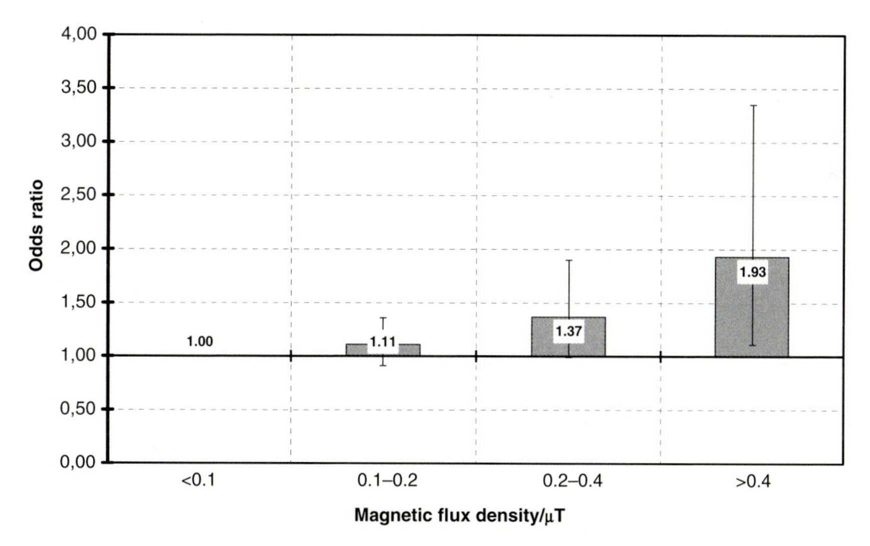

Figure 19.5 Odds ratios with 95% confidence intervals from a pooled analysis of epidemiological studies on the relationship between childhood leukemia and residential exposure to magnetic fields. (Data from Ref. [105].)

In vitro experiments do not suggest that ELF are genotoxic or carcinogenic [109, 117], although the issue is similarly controversial as with the microwaves [110]. Ivancsits *et al.* [111–113] reported DNA lesions to be induced by ELF, but these studies were heavily criticized [114] and could not be independently replicated [115]. Focke *et al.* [116] reported the induction of DNA lesions using the comet assay, but the results were only just significant. They are also at variance with the findings of Burdak-Rothkamm *et al.* [117] who used a number of different experimental approaches.

There is at present no plausible explanation for the association between childhood leukemia and ELF. A large number of epidemiological studies make the possibility of chance effects unlikely, but exposure misclassification, confounders, and bias cannot be rigorously excluded.

If the association is taken as real, the risk of additional childhood leukemia may be estimated. There are only few residences with magnetic flux densities above 0.4 μT. The World Health Organization (WHO) concludes on the basis of present knowledge that only 0.2–4.95% of the yearly incidence may be attributed to this environmental influence, that is, 100–2400 cases worldwide compared to a total incidence of 49 000 cases per year [118].

19.3 Discussion

Obviously, the possibilities of risk assessment vary considerably with the different types of electromagnetic radiation. In many cases it is not even clear whether they present a hazard, in others there are too few data to decide if there is a hazard and there is no established dose–effect relationship, except for ionizing radiation where, however, the dependence is not really known at low doses.

In Table 19.3 a summary is attempted that is, however, not free of subjective judgment.

Table 19.3 Summary of results for cancer-related hazard identification and dose–effect relationships for different regions of electromagnetic radiations.

Region	Cancer-related hazard identification			Dose dependence
	Epidemiological	Animals	*In vitro*	
Ionizing	+	+	+	+
Ultraviolet	+	+	+	–
Infrared	–	–	–	–
Terahertz	–	–	–	–
Microwaves	–	–	Conflicting	–
ELF	+	–	–	(+)

It is seen that only ionizing radiation fulfils all requirements for hazard identification, although even here there are still numerous open questions. The carcinogenic potency of ultraviolet is proven at all levels, but a quantitative risk assessment for humans is not possible as there is no established dose response: Infrared definitely constitutes a thermal hazard but as far as cancer is concerned, there are no reliable indications, Some *in vitro* investigations may be interpreted to mean that infrared might be dangerous in this respect but they are not supported either by human or by animal data. An evaluation of terahertz radiation is not yet possible because this technology is quite new, and there are only very few investigations that do not point to a real hazard.

The case of microwaves and mobile communication will not be closed for quite some time, but a survey of the literature does not provide a proof that they really constitute a hazard. There are, however, still some studies to be completed before a final statement can be made. But even if a hazard exists, a proper assessment will be impossible because even the carefully performed international studies do not provide real exposure determinations and are essentially based on only proxies, that is, length of use, which are also subject to considerable bias. This uncertainty has obviously some bearing on the ongoing discussion of limit values. The situation with extremely low frequencies is puzzling. Several epidemiological studies indicate clearly that leukemia in children is associated with low-frequency magnetic fields and there is even a dose–effect relationship, although rather crude. But, on the other hand, neither *in vitro* nor animal data reflect the situation in humans. Based on epidemiology alone, a risk estimate is possible that has been performed, showing fortunately that on a population basis the risk is comparatively low. It may be speculated that the discussion will gain momentum if new large electricity networks will be built.

At the end, one is thus left with the sobering conclusion that a real risk assessment is possible only with ionizing radiations. The situation could change with ultraviolet if individual exposure determinations could be introduced in epidemiological studies. For all other radiation types, it is by no means clear whether "more research" is able to change the situation unless it is based on scientifically sound hypotheses and very careful experimental investigations involving the whole scale of present-day possibilities. Studies following the "shot-gun approach" will no longer do.

References

1 International Commission on Radiological Protection (2007) The 2007 Recommendations of the International Commission on Radiological Protection, publication 103. *Ann. ICRP*, **37**, 1–332.

2 OECD (2002) Guidance notes for analysis and evaluation of chronic toxicity and carcinogenicity studies. OECD Environment, Health and Safety Publications, Series on Testing and Assessment No. 35 and Series on Pesticides No. 14, Organisation for Economic Cooperation and Development, Paris.

3 UNSCEAR (2008) UNSCEAR 2006 Report: "Effects of Ionizing Radiation," United Nations, New York.

4 IEA (2010) Good epidemiological practice: IEA guidelines for proper conduct of epidemiological research,

http://www.ieatemp.com/goodEpiPractice.aspx (accessed May 10, 2010).

5 Hill, A.B. (1965) The environment and disease: association or causation? *Proc. R. Soc. Med.*, **58**, 295–300.

6 Ainsbury, E.A., Bouffler, S.D., Dörr, W., Graw, J., Muirhead, C.R., Edwards, A.A., and Cooper, J. (2009) Radiation cataractogenesis: a review of recent studies. *Radiat. Res.*, **172**, 1–9.

7 Preston, D.L., Shimizu, Y., Pierce, D.A., Suyama, A., and Mabuchi, K. (2003) Studies of mortality of atomic bomb survivors. Report 13: solid cancer and noncancer disease mortality: 1950–1997. *Radiat. Res.*, **160**, 381–407.

8 Preston, D.L., Ron, E., Tokuoka, S., Funamoto, S., Nishi, N., Soda, M., Mabuchi, K., and Kodama, K. (2007) Solid cancer incidence in atomic bomb survivors: 1958–1998. *Radiat. Res.*, **168**, 1–64.

9 Richardson, D., Sugiyama, H., Nishi, N., Sakata, R., Shimizu, Y., Grant, E.J., Soda, M., Hsu, W.L., Suyama, A., Kodama, K., and Kasagi, F. (2009) Ionizing radiation and leukemia mortality among Japanese atomic bomb survivors, 1950–2000. *Radiat. Res.*, **172**, 368–382.

10 Little, M.P. (2009) Cancer and non-cancer effects in Japanese bomb survivors. *Radiol. Prot.*, **29**, A43–A59.

11 Cardis, E., Vrijheid, M., Blettner, M., Gilbert, E., Hakama, M., Hill, C., Howe, G., Kaldor, J., Muirhead, C.R., Schubauer-Berigan, M., Yoshimura, T., Bermann, F., Cowper, G., Fix, J., Hacker, C., Heinmiller, B., Marshall, M., Thierry-Chef, I., Utterback, D., Ahn, Y.O., Amoros, E., Ashmore, P., Auvinen, A., Bae, J.M., Bernar, J., Biau, A., Combalot, E., Deboodt, P., Diez Sacristan, A., Eklöf, M., Engels, H., Engholm, G., Gulis, G., Habib, R.R., Holan, K., Hyvonen, H., Kerekes, A., Kurtinaitis, J., Malker, H., Martuzzi, M., Mastauskas, A., Monnet, A., Moser, M., Pearce, M.S., Richardson, D.B., Rodriguez-Artalejo, F., Rogel, A., Tardy, H., Telle-Lamberton, M., Turai, I., Usel, M., and Veress, K. (2007) The 15-country collaborative study of cancer risk among radiation workers in the nuclear industry: estimates of radiation-related cancer risks. *Radiat. Res.*, **167**, 396–416.

12 Ullrich, R.L. and Storer, J.B. (1979) Influence of gamma irradiation on the development of neoplastic disease in mice. III. Dose-rate effects. *Radiat. Res.*, **80**, 325–342.

13 Ullrich, R.L. and Storer, J.B. (1979) Influence of gamma irradiation on the development of neoplastic disease in mice. II. Solid tumors. *Radiat. Res.*, **80**, 317–324.

14 Ullrich, R.L. and Storer, J.B. (1979) Influence of gamma irradiation on the development of neoplastic disease in mice. I. Reticular tissue tumors. *Radiat. Res.*, **80**, 303–316.

15 Fry, R.J.M. and Storer, J.B. (1987) External radiation carcinogenesis, in *Advances in Radiation Biology*, vol. **13** (ed. J.T. Lett), Academic Press, New York, pp. 31–90.

16 Grahn, D., Lombard, L.S., and Carnes, B.A. (1992) The comparative tumorigenic effects of fission neutrons and cobalt-60 gamma rays in the B6CF1 mouse. *Radiat. Res.*, **129**, 19–36.

17 Cullings, H.M., Fujita, S., Funamoto, S., Grant, E.J., Kerr, G.D., and Preston, D.L. (2006) Dose estimation for atomic bomb survivor studies: its evolution and present status. *Radiat. Res.*, **166**, 219–254.

18 Auxier, J.A. (1977) *ICHIBAN: Radiation Dosimetry for Survivors of the Bombing of Hiroshima and Nagasaki*, U.S. Department of Energy, Washington, DC, ID-27080.

19 Harper, J.W. and Elledge, S.J. (2007) The DNA damage response: ten years after. *Mol. Cell*, **28**, 739–745.

20 Rothkamm, K. and Löbrich, M. (2003) Evidence for a lack of DNA double-strand break repair in human cells exposed to very low X-ray doses. *Proc. Natl. Acad. Sci. USA*, **100**, 5057–5062.

21 Löbrich, M. and Kiefer, J. (2006) Assessing the likelihood of severe side effects in radiotherapy. *Int. J. Cancer*, **118**, 2652–2656.

22 AGNIR (2002) Health effects from ultraviolet radiation. Report of an advisory group on non-ionising radiation. *Doc. NRPB*, **13** (1), 1–276.

23 Lucas, R., McMichael, T., Smith, W., and Armstrong, B. (2006) *Solar Ultraviolet Radiation: Global Burden of Disease from Solar Ultraviolet Radiation*, Environmental Burden of Disease Series No. 13, World Health Organization.

24 de Gruijl, F.R. (2000) Biological action spectra. *Radiat. Prot. Dosimetry*, **91**, 57–63.

25 McKinlay., A.F. and Diffey, B.L. (1987) A reference action spectrum for ultraviolet induced erythema in human skin. *CIE J.*, **6**, 17–22.

26 Sutherland, J.C. and Griffin, K.P. (1981) Absorption spectrum of DNA for wavelengths greater than 300nm. *Radiat. Res.*, **86**, 399–410.

27 Gordon, D. and Silverstein, H. (1976) Actinic effects of light and biological implications. *Photochem. Photobiol*, **32**, 117–129.

28 Armstrong, B.K. and Kricker, A. (2001) The epidemiology of UV induced skin cancer. *J. Photochem. Photobiol. B*, **63**, 8–18.

29 International Commission on Non-Ionizing Radiation Protection (2004) Guidelines on limits of exposure to ultraviolet radiation of wavelengths between 180 nm and 400 nm (incoherent optical radiation). *Health Phys.*, **87**, 171–186.

30 Brash, D.E., Rudolph, J.A., Simon, J.A., Lin, A., McKenna, G.J., Baden, H.P., Halperin, A.J., and Ponten, J. (1991) A role for sunlight in skin cancer: UV-induced p53 mutations in squamous cell carcinoma. *Proc. Natl. Acad. Sci. USA*, **88**, 10124–10128.

31 Setlow, R.B., Grist, E., Thompson, K., and Woodhead, A.D. (1993) Wavelengths effective in induction of malignant melanoma. *Proc. Natl. Acad. Sci. USA*, **90**, 6666–6670.

32 Greinert, R., Breitbart, E.W., and Volkmer, B. (2004) UV-radiation biology as part of cancer research, in *Life Sciences and Radiation* (ed. J. Kiefer), Springer, Berlin, pp. 139–156.

33 de Gruijl F.R. and Rebel, H. (2008) Early events in UV carcinogenesis: DNA damage, target cells and mutant p53 foci. *Photochem. Photobiol.*, **84**, 382–387.

34 Kiefer, J. (2007) Effects of ultraviolet on DNA, in *Chromosomal Aberrations* (eds G. Obe and Vijayalaxmi), Springer, Berlin, pp. 39–54.

35 Enninga, I.C., Groenendijk, R.T.L., Filon, A.R., Van Zeeland, A.A., and Simons, J.W.I.M. (1986) The wavelength dependence of U.V.-induced pyrimidine dimer formation, cell killing and mutation induction in human diploid skin fibroblasts. *Carcinogenesis*, **7**, 1829–1833.

36 Mitchell, D.L., Volkmer, B., Breitbart, E.W., Byrom, M., Lowery, M.G., and Greinert, R. (2001) Identification of a non-dividing subpopulation of mouse and human epidermal cells exhibiting high levels of persistent ultraviolet photodamage. *J. Invest. Dermatol.*, **117**, 590–595.

37 Wood, R.D., Mitchell, M., Sgouros, J., and Lindahl, T. (2001) Human DNA repair genes. *Science*, **291**, 1284–1289.

38 Wood, R.D., Mitchell, M., and Lindahl, T. (2005) Human DNA repair genes. *Mutat. Res.*, **577**, 275–283.

39 Daya-Grosjean, L. (2008) Xeroderma pigmentosum and skin cancer. *Adv. Exp. Med. Biol.*, **637**, 19–27.

40 Li, C., Wang, LE., and Wei, Q. (2009) DNA repair phenotype and cancer susceptibility: a minireview. *Int. J. Cancer*, **124**, 999–1007.

41 Cleaver, J.E. and Crowley, E. (2002) UV damage, DNA repair and skin carcinogenesis. *Front. Biosci.*, **7**, d1024–d1043.

42 Cleaver, J.E. (1968) Defective repair replication in xeroderma pigmentosum. *Nature*, **218**, 652–656.

43 Cleaver, J. (2000) Common pathways for ultraviolet skin carcinogenesis in the repair and replication defective groups of xeroderma pigmentosum. *J. Dermatol. Sci.*, **23**, 1–11.

44 Ridley, A.J., Whiteside, J.R., McMillan, T.J., and Allinson, S.L. (2009) Cellular and sub-cellular responses to UVA in relation to carcinogenesis. *Int. J. Radiat. Biol.*, **85**, 177–195.

45 Leiter, U. and Garbe, C. (2008) Epidemiology of melanoma and nonmelanoma skin cancer: the role of sunlight. *Adv. Exp. Med. Biol.*, **624**, 89–103.

46 MacKie, R.M. (2006) Long-term health risk to the skin of ultraviolet radiation. *Prog. Biophys. Mol. Biol.*, **92**, 92–96.

47 International Agency for Research on Cancer Working Group on Artificial Ultraviolet (UV) Light and Skin Cancer (2007) The association of use of sunbeds with cutaneous malignant melanoma and other skin cancers: a systematic review. *Int. J. Cancer*, **120**, 1116–1122.

48 Reichrath, J. (2006) The challenge resulting from positive and negative effects of sunlight: how much solar UV exposure is appropriate to balance between risks of vitamin D deficiency and skin cancer? *Prog. Biophys. Mol. Biol.*, **92**, 9–16.

49 IARC (1992) Solar and ultraviolet radiation, in *IARC Monographs on the Evaluation of Carcinogenic Risks to Humans*, vol. **55**, International Agency for Research on Cancer, Lyon.

50 Kielbassa, C. and Epe, B. (2000) DNA damage induced by ultraviolet and visible light and its wavelength dependence. *Methods Enzymol.*, **319**, 436–445.

51 Hoffmann-Dörr, S., Greinert, R., Volkmer, B., and Epe, B. (2005) Visible light (>395 nm) causes micronuclei formation in mammalian cells without generation of cyclobutane pyrimidine dimers. *Mutat. Res.*, **572**, 142–149.

52 International Commission on Non-Ionizing Radiation Protection (2006) ICNIRP statement on far infrared radiation exposure. *Health Phys.*, **91**, 630–645.

53 van der Leun, J.C., Piacentini, R.D., and de Gruijl, F.R. (2008) Climate change and human skin cancer. *Photochem. Photobiol. Sci.*, **7**, 730–733.

54 Freeman, R.G. and Knox, J.M. (1964) Influence of temperature on ultraviolet injury. *Arch. Dermatol.*, **89**, 858–864.

55 Dewhirst, M.W., Lora-Michiels, M., Viglianti, B.L., Dewey, W.C., and Repacholi, M. (2003) Carcinogenic effects of hyperthermia. *Int. J. Hyperthermia*, **19**, 236–251.

56 Karu, T.I. (2008) Mitochondrial signaling in mammalian cells activated by red and near-IR radiation. *Photochem. Photobiol.*, **84**, 1091–1099.

57 Schieke, S.M., Schroeder, P., and Krutmann, J. (2003) Cutaneous effects of infrared radiation: from clinical observations to molecular response mechanisms. *Photodermatol. Photoimmunol. Photomed.*, **19**, 228–234.

58 Choi, M.K., Bettermann, A., and van der Weide, D.W. (2004) Potential for detection of explosive and biological hazards with electronic terahertz systems. *Philos. Trans. A Math. Phys. Eng. Sci.*, **362**, 337–347.

59 Humphreys, K., Loughran, J.P., Gradziel, M., Lanigan, W., Ward, T., Murphy, J.A., and O'Sullivan, C. (2004) Medical applications of terahertz imaging: a review of current technology and potential applications in biomedical engineering. *Conf. Proc. IEEE Eng. Med. Biol. Soc.*, **2**, 1302–1305.

60 Wolbarst, A.B. and Hendee, W.R. (2006) Evolving and experimental technologies in medical imaging. *Radiology*, **238**, 16–39.

61 Palik, E.G. (1985) *Handbook of Optical Constants of Solids*, Academic Press, New York.

62 Smye, S.W., Chamberlain, J.M., Fitzgerald, A.J., and Berry, E. (2001) The interaction between terahertz radiation and biological tissue. *Phys. Med. Biol.*, **46**, R101–R112.

63 Zeni, O., Gallerano, G.P., Perrotta, A., Romanò, M., Sannino, A., Sarti, M., D'Arienzo, M., Doria, A., Giovenale, E., Lai, A., Messina, G., and Scarfi, M.R. (2007) Cytogenetic observations in human peripheral blood leukocytes following *in vitro* exposure to THz radiation: a pilot study. *Health Phys.*, **92**, 349–357.

64 Korenstein-Ilan, A., Barbul, A., Hasin, P., Eliran, A., Gover, A., and Korenstein, R. (2008) Terahertz radiation increases genomic instability in human lymphocytes. *Radiat. Res.*, **170**, 224–234.

65 Gabriel, C., Gabriel, S., and Corthout, E. (1996) The dielectric properties of biological tissues: I. Literature survey. *Phys. Med. Biol.*, **41**, 2231–2249.

66 Gabriel, S., Lau, R.W., and Gabriel, C. (1996) The dielectric properties of biological tissues: II. Measurements in the frequency range 10 Hz to 20GHz. *Phys. Med. Biol.*, **41**, 2251–2269.

67 Gabriel, S., Lau, R.W., and Gabriel, C. (1996) The dielectric properties of biological tissues: III. Parametric models for the dielectric spectrum of tissues. *Phys. Med. Biol.*, **41**, 2271–2293.

68 Gabriel, C. (2005) Dielectric properties of biological tissue: variation with age. *Bioelectromagnetics*, **26** (Suppl. 7), S12–S18.

69 Gabriel, C. and Peyman, A. (2006) Dielectric measurement: error analysis and assessment of uncertainty. *Phys. Med. Biol.*, **51**, 6033–6046.

70 Peyman, A., Gabriel, C., Grant, E.H., Vermeeren, G., and Martens, L. (2009) Variation of the dielectric properties of tissues with age: the effect on the values of SAR in children when exposed to walkie-talkie devices. *Phys. Med. Biol.*, **54**, 227–241.

71 Khurana, V.G., Teo, C., Kundi, M., Hardell, L., and Carlberg, M. (2009) Cell phones and brain tumors: a review including the long-term epidemiologic data. *Surg. Neurol.*, **72**, 205–214.

72 Kundi, M. (2009) The controversy about a possible relationship between mobile phone use and cancer. *Environ. Health Perspect.*, **117**, 316–324.

73 Hardell, L. and Carlberg, M. (2009) Mobile phones, cordless phones and the risk for brain tumours. *Int. J. Oncol.*, **35**, 5–17.

74 Hardell, L., Carlberg, M., and Hansson Mild, K. (2009) Epidemiological evidence for an association between use of wireless phones and tumor diseases. *Pathophysiology*, **16**, 113–122.

75 Ahlbom, A., Feychting, M., Green, A., Kheifets, L., Savitz, D.A., and Swerdlow, A.J. (2009) Epidemiologic evidence on mobile phones and tumor risk: a review. *Epidemiology*, **20**, 639–652.

76 INTERPHONE Study Group (2010) Brain tumour risk in relation to mobile telephone use: results of the INTERPHONE international case–control study. *Int. J. Epidemiol.*, **39** (3), 675–694.

77 Dasenbrock, C. (2005) Animal carcinogenicity studies on radiofrequency fields related to mobile phones and base stations. *Toxicol. Appl. Pharmacol.*, **207** (2 Suppl), 342–346.

78 Repacholi, M.H., Basten, A., Gebski, V., Noonan, D., Finnie, J., and Harris, A.W. (1997) Lymphomas in EA-Pim1 transgenic mice exposed to pulsed 900MHz electromagnetic fields. *Radiat. Res.*, **147**, 631–640.

79 Utteridge, T.D., Gebski, V., Finnie, J.W., Vernon-Roberts, B., and Kuchel, T.R. (2002) Long-term exposure of E-mu-Pim1 transgenic mice to 898.4MHz microwaves does not increase lymphoma incidence. *Radiat. Res.*, **158**, 357–364.

80 Utteridge, T.D., Gebski, V., Finnie, J.W., Vernon-Roberts, B., and Kuchel, T.R. (2003) Response to the letters to the Editor sent by (1) Kundi, (2) Goldstein/Kheifets/van Deventer/ Repacholi, and (3) Lerchl. *Radiat. Res.*, **159**, 276–278.

81 Sommer, A.M., Streckert, J., Bitz, A.K., Hansen, V.W., and Lerchl, A. (2004) No effects of GSM-modulated 900MHz electromagnetic fields on survival rate and spontaneous development of lymphoma in female AKR/J mice. *BMC Cancer*, **4**, 77.

82 Sommer, A.M., Bitz, A.K., Streckert, J., Hansen, V.W., and Lerchl, A. (2007) Lymphoma development in mice chronically exposed to UMTS-modulated radiofrequency electromagnetic fields. *Radiat. Res.*, **168**, 72–80.

83 Tillmann, T., Ernst, H., Ebert, S., Kuster, N., Behnke, W., Rittinghausen, S., and Dasenbrock, C. (2007) Carcinogenicity study of GSM and DCS wireless communication signals in

B6C3F1 mice. *Bioelectromagnetics*, **28**, 173–187.

84 Smith, P., Kuster, N., Ebert, S., and Chevalier, H.J. (2007) GSM and DCS wireless communication signals: combined chronic toxicity/ carcinogenicity study in the Wistar rat. *Radiat. Res.*, **168**, 480–492.

85 Oberto, G., Rolfo, K., Yu, P., Carbonatto, M., Peano, S., Kuster, N., Ebert, S., and Tofani, S. (2007) Carcinogenicity study of 217 Hz pulsed 900 MHz electromagnetic fields in Pim1 transgenic mice. *Radiat. Res.*, **168**, 316–326.

86 Hruby, R., Neubauer, G., Kuster, N., and Frauscher, M. (2008) Study on potential effects of "902-MHz GSM-type Wireless Communication Signals" on DMBA-induced mammary tumours in Sprague-Dawley rats. *Mutat. Res.*, **649**, 34–44.

87 Heynick, L.N., Johnston, S.A., and Mason, P.A. (2003) Radio frequency electromagnetic fields: cancer, mutagenesis, and genotoxicity. *Bioelectromagnetics*, **24** (Suppl. 6), S74–S100.

88 Meltz, M.L. (2003) Radiofrequency exposure and mammalian cell toxicity, genotoxicity, and transformation. *Bioelectromagnetics*, **24** (Suppl. 6), S196–S213.

89 Vijayalaxmi and Obe, G. (2005) Controversial cytogenetic observations in mammalian somatic cells exposed to extremely low frequency electromagnetic radiation: a review and future research recommendations. *Bioelectromagnetics*, **26**, 412–430.

90 Vijayalaxmi and Obe, G. (2004) Controversial cytogenetic observations in mammalian somatic cells exposed to radiofrequency radiation. *Radiat. Res.*, **162**, 481–496.

91 Vijayalaxmi and Prihoda, T.J. (2008) Genetic damage in mammalian somatic cells exposed to radiofrequency radiation: a meta-analysis of data from 63 publications (1990–2005). *Radiat. Res.*, **169**, 561–574.

92 Verschaeve, L. (2009) Genetic damage in subjects exposed to radiofrequency radiation. *Mutat. Res.*, **681**, 259–320.

93 SSK, Strahlenschutzkommission (2009) *Wirkung hochfrequenter Felder auf das Genom: Genotoxizität und Genregulation (Effects of Radiofrequency Fields on the Genome: Genotoxicity and Gene Regulation)* (in German, with English summary) Berichte der Strahlenschutzkommission, Heft 62, Verlag H. Hoffmann GmbH, Berlin.

94 Hirose, H., Suhara, T., Kaji, N., Sakuma, N., Sekijima, M., Nojima, T., and Miyakoshi, J. (2008) Mobile phone base station radiation does not affect neoplastic transformation in BALB/3T3 cells. *Bioelectromagnetics*, **29**, 55–64.

95 Koyama, S., Takashima, Y., Sakurai, T., Suzuki, Y., Taki, M., and Miyakoshi, J. (2007) Effects of 2.45 GHz electromagnetic fields with a wide range of SARs on bacterial and HPRT gene mutations. *J. Radiat. Res. (Tokyo)*, **48**, 69–75.

96 Ono, T., Saito, Y., Komura, J., Ikehata, H., Tarusawa, Y., Nojima, T., Goukon, K., Ohba, Y., Wang, J., Fujiwara, O., and Sato, R. (2004) Absence of mutagenic effects of 2.45 GHz radiofrequency exposure in spleen, liver, brain, and testis of lacZ-transgenic mouse exposed *in utero*. *Tohoku J. Exp. Med.*, **202**, 93–103.

97 Stronati, L., Testa, A., Moquet, J., Edwards, A., Cordelli, E., Villani, P., Marino, C., Fresegna, A.M., Appolloni, M., and Lloyd, D. (2006) 935 MHz cellular phone radiation. An *in vitro* study of genotoxicity in human lymphocytes. *Int. J. Radiat. Biol.*, **82**, 339–346.

98 Adair, R.K. (2003) Biophysical limits on athermal effects of RF and microwave radiation. *Bioelectromagnetics*, **24**, 39–48.

99 Sheppard, A.R., Swicord, M.L., and Balzano, Q. (2008) Quantitative evaluations of mechanisms of radiofrequency interactions with biological molecules and processes. *Health Phys.*, **95**, 365–396.

100 Michelozzi, P., Capon, A., Kirchmayer, U., Forastiere, F., Biggeri, A., Barca, A., and Perucci, C.A. (2002) Adult and childhood leukemia near a high-power radio station in Rome, Italy. *Am. J. Epidemiol.*, **155**, 1096–1103.

101 Merzenich, H., Schmiedel, S., Bennack, S., Brüggemeyer, H., Philipp, J., Blettner, M., and Schüz, J. (2008) Childhood leukemia in relation to radio frequency electromagnetic fields in the vicinity of TV and radio broadcast transmitters. *Am. J. Epidemiol.*, **168**, 1169–1178.

102 International Commission on Non-Ionizing Radiation Protection (1998) Guidelines for limiting exposure to time-varying electric, magnetic, and electromagnetic fields (up to 300 GHz). *Health Phys.*, **74**, 494–522 (erratum in *Health Phys.*, 75, 442).

103 Wertheimer, N. and Leeper, E. (1979) Electrical wiring configurations and childhood cancer. *Am. J. Epidemiol.*, **109**, 273–284.

104 Ahlbom, A., Day, N., Feychting, M., Roman, E., Skinner, J., Dockerty, J., Linet, M., McBride, M., Michaelis, J., Olsen, J.H., Tynes, T., and Verkasalo, P.K. (2000) A pooled analysis of magnetic fields and childhood leukaemia. *Br. J. Cancer*, **83**, 692–698.

105 Schüz, J. and Ahlbom, A. (2008) Exposure to electromagnetic fields and the risk of childhood leukaemia: a review. *Radiat. Prot. Dosimetry*, **132**, 202–211.

106 World Health Organization (2007) *Environmental Health Criteria: Extremely Low Frequency (ELF) Fields*, World Health Organization, Geneva.

107 Sommer, A.M. and Lerchl, A. (2004) The risk of lymphoma in AKR/J mice does not rise with chronic exposure to 50Hz magnetic fields (1 μT and 100 μT). *Radiat. Res.*, **162**, 194–200.

108 Sommer, A.M. and Lerchl, A. (2006) 50 Hz magnetic fields of 1 mT do not promote lymphoma development in AKR/J mice. *Radiat. Res.*, **165**, 343–349.

109 Vijayalaxmi and Prihoda, T.J. (2009) Genetic damage in mammalian somatic cells exposed to extremely low frequency electro-magnetic fields: a meta-analysis of data from 87 publications (1990–2007). *Int. J. Radiat. Biol*, **85**, 196–213.

110 Santini, M.T., Rainaldi, G., and Indovina, P.L. (2009) Cellular effects of extremely low frequency (ELF) electromagnetic fields. *Int. J. Radiat. Biol.*, **85**, 294–313.

111 Ivancsits, S., Diem, E., Jahn, O., and Rüdiger, H.W. (2003) Age-related effects on induction of DNA strand breaks by intermittent exposure to electromagnetic fields. *Mech. Ageing Dev.*, **124**, 847–850.

112 Ivancsits, S., Diem, E., Jahn, O., and Rüdiger, H.W. (2003) Intermittent extremely low frequency electromagnetic fields cause DNA damage in a dose-dependent way. *Int. Arch. Occup. Environ. Health*, **76**, 431–436.

113 Ivancsits, S., Pilger, A., Diem, E., Jahn, O., and Rüdiger, H.W. (2005) Cell type-specific genotoxic effects of intermittent extremely low-frequency electromagnetic fields. *Mutat. Res.*, **583**, 184–188.

114 Vijayalaxmi, McNamee, J.P., and Scarfi, M.R. (2006) Comments on: "DNA strand breaks" by Diem *et al.* [*Mutat. Res.* 583 (2005) 178–183] and Ivancsits *et al.* [*Mutat. Res.* 583 (2005) 184–188]. *Mutat. Res.*, **603** 104–106.

115 Scarfi, M.R., Sannino, A., Perrotta, A., Sarti, M., Mesirca, P., and Bersani, F. (2005) Evaluation of genotoxic effects in human fibroblasts after intermittent exposure to 50 Hz electromagnetic fields: a confirmatory study. *Radiat. Res.*, **164**, 270–276.

116 Focke, F., Schuermann, D., Kuster, N., and Schär, P. (2010) DNA fragmentation in human fibroblasts under extremely low frequency electromagnetic field exposure. *Mutat. Res.*, **683**, 74–83.

117 Burdak-Rothkamm, S., Rothkamm, K., Folkard, M., Patel, G., Hone, P., Lloyd, D., Ainsbury, L., and Prise, KM. (2009) DNA and chromosomal damage in response to intermittent extremely low-frequency magnetic fields. *Mutat. Res.*, **672**, 82–89.

118 World Health Organization (2007) Electromagnetic fields and public health, fact sheet 322, www.who.int/mediacentre/factsheets/fs322/en/index.html (accessed February 25, 2010).

20
Communicating about Uncertainties in Cancer Risk Assessment

Peter Wiedemann and Holger Schütz

20.1
Introduction

Assessments of environmental health risks are often fraught with uncertainties. This is particularly true for cancer risk assessment. Unlike in risk assessment of many noncancer toxic agents, where thresholds for safe dose and appropriate uncertainty factors can be established, this is not possible for carcinogens. In addition, cancer risks are typically long-term risks, and the process of carcinogenesis is still only incompletely understood (see Chapter 2).

How can the results of cancer risk assessments best be communicated to a nonexpert audience? While a lot of research has been devoted to the question of how to best communicate risk estimates to a lay public (see, for example, the reviews by Lipkus [1] or Visschers *et al.* [2]), the question of how to communicate the uncertainty that is associated with such risk estimates has received much less attention. This is somewhat surprising, because the importance of informing the public about uncertainties in risk assessment has been stressed repeatedly. For example, the National Research Council [3] underlines that proper characterization of uncertainty is essential in risk assessment because the omission or underestimation of uncertainty can leave decision makers with a false sense of confidence in estimates of risk. In addition, it is assumed that communicating uncertainty has positive effects. Thus, the WHO handbook for crisis communication states: "... trust-building measures are often counter-intuitive (such as acknowledging uncertainty or avoiding excessive reassurance)" ([4], p. 2). Informing about uncertainties is thought to be a trust-building measure. This view is supported by many others; see, for instance, Holmes *et al.* [5] for their opinion on gaining trust through revealing uncertainty in health crisis situations and Wibeck [6] and Gross [7] on improving public confidence in science by communicating uncertainty of environmental risk assessments. Similarly, the EPA assumes that "...scientific uncertainty is a fact of life (and) ... a balanced discussion of reliable conclusions and related uncertainties enhances, rather than detracts, from the overall credibility of each assessment ..." ([8], p. VII). However, whether the disclosure of uncertainty has these intended

Cancer Risk Evaluation: Methods and Trends,
Edited by Günter Obe, Burkhard Jandrig, Gary E. Marchant, Holger Schütz, and Peter M. Wiedemann.

effects cannot be taken for granted. This, of course, is an empirical question. Risk communication cannot simply rely on good intentions, even if they appear plausible.

In the following, we will review studies that investigated uncertainty communication and its effects on risk perception and other relevant issues (such as trust in the information source). Most of these studies do not specifically address cancer risks but deal with varying human health risks. This should be taken into account when drawing conclusions: It could well be that the effects of communicating uncertainty for a highly dreaded risk such as cancer are different from less dreaded risks such as high blood pressure.

Before we start the review, we will briefly expound some facets of the concept of uncertainty, and will then have a closer look at the reasons that are given for communicating uncertainties in risk, as these provide criteria for judging whether uncertainty communication is successful or not.

20.2 The Concept of Uncertainty

In 1983, the National Research Council (NRC) report Risk Assessment in the Federal Government has underlined an issue that remains a crucial challenge: "The dominant analytic difficulty [in decision-making based on risk assessments] is pervasive uncertainty. . . . there is often great uncertainty in estimates or the types, probability, and magnitude of health effects . . . These problems have no immediate solutions, given the many gaps in our understanding of the causal mechanisms of carcinogenesis and other health effects and in our ability to ascertain the nature or extent of the effects associated with specific exposures." ([9], p. 11).

There are many ways to conceptualize uncertainty. Although from a technical risk assessment point of view the distinction between variability and uncertainty is essential [3], social science researchers introduce other aspects. Smithson [10], for instance, distinguishes between randomness of events, delay in consequences or outcomes of acts, and absence or lack of clarity in information. Another taxonomy of uncertainty distinguishes between distortion and incompleteness [11], still others – for instance, Ravetz [12] – highlight the difference between known uncertainties (knowing that we do not know) versus unknown uncertainties (not knowing that we do not know). A further interesting way to classify uncertainties is suggested by Politi *et al.* [13]. They differentiate between five types: (1) uncertainty about future outcomes, (2) uncertainty about the strength of evidence about a risk, (3) uncertainty about the personal relevance of a risk, (4) uncertainty arising from complexity of information, and (5) uncertainty resulting from ignorance. The first type of uncertainty is worth emphasizing because it is often neglected. Politi *et al.* [13] argue that risk estimates have at best a limited applicability at the individual level because they describe the aggregated outcomes in a population, but do not reduce the uncertainty at the individual level.

Taken together, a vast amount of literature exists, differentiating between various kinds of uncertainty [14–18]. Unfortunately, it remains still a challenge to integrate the variety of concepts, accounts, and theories on uncertainty in one overarching framework.

There is one distinction, however, that is almost entirely missing in these discussions of aspects of uncertainty – a distinction that is, in our opinion, of major importance for the communication of uncertainty in cancer risk assessment. It is the distinction between *hazard* and *risk* that is fundamental for risk assessment (see, for example, Refs [19, 20] and Chapter 19). Hazard means an "Inherent property of an agent or situation having the potential to cause adverse effects when an organism, system, or (sub)population is exposed to that agent" ([20], p. 12), while risk refers to the "probability of an adverse effect in an organism, system, or (sub)population caused under specified circumstances by exposure to an agent" ([20], p. 13). Much what is discussed under the label "risk assessment" is actually hazard assessment. For instance, the well-known "*IARC Monographs* evaluate cancer hazards (despite the historical presence of the word 'risks' in the title of each volume of *Monographs*) . . ." ([21], p. 107; also see Chapter 4). This distinction between hazard and risk is not well observed by the public and in risk communication [22]. We will address this specific challenge for risk communication in Section 20.6.

20.3 Reasons for Communicating Uncertainties

As can be easily seen from an acute glance at the work of IARC [23], uncertainty is the rule rather than the exception in cancer risk assessment (see Table 20.1). Since 1970, more than 900 agents have been evaluated by IARC, of which approximately only 12% have been identified as carcinogenic to humans. The remaining ones are classified as probably carcinogenic or possibly carcinogenic to humans, not classifiable, or probably not carcinogenic to humans.

From a psychological point of view, the crucial question is whether people react differently with respect to different kinds and levels of uncertainty [11, 24]. Therefore, one important question is how to communicate the uncertainties in risk assessment from science to decision makers and to the general public. More

Table 20.1 Overview of the IARC evaluation of carcinogenic risks to humans.

IARC classification	Number of agents classified by IARC monographs	
Carcinogenic to humans	107	12%
Probably carcinogenic to humans	58	6%
Possibly carcinogenic to humans	249	27%
Not classifiable as to its carcinogenicity to humans	512	55%
Probably not carcinogenic to humans	1	0%

Source: http://monographs.iarc.fr/ENG/Classification/index.php (accessed July 19, 2010).

specifically, and from a risk communication perspective, we are interested in the following questions:

- Should uncertainties in risk assessments be disclosed?
- What types of uncertainty should be addressed?
- In which format should uncertainties be described?
- In what detail should uncertainties be revealed?

From a normative point of view, it seems quite clear that uncertainties should be reported. In its handbook on risk characterization, the US EPA [8] underlines that transparency is a principal value; from this it follows that describing uncertainties is an essential criterion for good risk characterization. This view is advocated also in other fields of risk assessment. For instance, global climate change researchers – for good reasons, as the so-called "climategate" (see Ref. [25]) affair has shown – stress the need to communicate existing uncertainty beyond the scientific community [26]. In the same direction points a paper published by the Interdepartmental Liaison Group on Risk Assessment [27]. It refers to the application of the precautionary principle (see Chapter 21) and underlines the need for disclosing uncertainties:

> ... where the precautionary principle is invoked and applied, openness becomes critically important in achieving an outcome that stakeholders regard as valid. Openness demands candour in exposing, for example:
>
> - the information on which risk assessment was undertaken;
> - the scientific uncertainties and reasoning for invoking the precautionary principle, and any uncertainty factors already built into the risk assessment;
> - the assumptions made in establishing credible scenarios; and the many factors that influence the choice of risk management measures ([27], p. 10).

With regard to the question, which types of uncertainties should be addressed and especially at what depth, the case is less clear. It is sometimes argued that reporting too many details may impair clarity of the communication [28–30]. For instance, Kloprogge *et al.* [31] introduced the idea of progressive disclosure of uncertainty information, that is, the uncertainty information should be offered gradually without overwhelming the reader with all possible information on uncertainty. They suggest that in simple version, the main purpose of reporting uncertainties is to put the uncertainty in context and indicate its implications for policy. In the more elaborated version, the main purpose of reporting uncertainty is explaining the nature, extent, and sources of uncertainties.

Even more difficult to answer is the question, in which format uncertainty should be reported. It poses questions of message design, that is, the choice of risk indicators, and the selection of numerical, verbal, or graphical descriptors. Here, risk communication research offers some insights, however more or less related to the issue how the magnitude of a risk may be conveyed [1, 2]. Furthermore, it refers mostly to a small range of issues, such as the interpretation of numerical probability information or the interpretation of verbal expressions like "probable." Almost no research exists with respect to how uncertainty in risk assessment should be communicated.

20.4 Findings on Communicating Uncertainties

The most important studies on communication of uncertainty related to risk assessments are listed in Table 20.2. As seen, the range of topics is rather broad and therefore we will discuss the available empirical evidence in several topical sections.

20.4.1 Preferences with Respect to Being Informed about Uncertainty

The first question is whether nonexperts wish to get information about uncertainties. Lion *et al.* [38] examined the information preferences of laypersons regarding risks. It was shown that laypersons were particularly interested in the aspects of "How is one exposed to the risk?", "What are the consequences?", "What does the risk mean?" and

Table 20.2 Important studies on reporting uncertainty.

Author(s)	Focus	Topic
Han *et al.* [32]	Risk	Risk perception, perceived sources of uncertainty
Han *et al.* [33]	Risk	Risk perception, perceived credibility of risk information
Johnson and Slovic [34]	Risk	Information preferences, risk perception, and perceived sources of uncertainty
Johnson and Slovic [35]	Risk	Information preferences, perceived credibility of risk information
Johnson [36]	Risk	Information preferences, perceived causes of uncertainty
Kuhn [37]	Risk	Risk perception
Lion *et al.* [38]	Risk	Information preferences
Miles and Frewer [39]	Risk	Risk perception
Morss *et al.* [40]	Risk	Information preferences
Smithson [41]	Risk	Perceived credibility of risk information
Viscusi *et al.* [42]	Risk	Interpretation of uncertainty
Viscusi [43]	Risk	Interpretation of uncertainty
Wiedemann *et al.* [30]	Hazard, Risk	Risk perception, perceived expert competence, and comprehensibility as well as clarity of the information

"What is the probability of possible negative consequences?" It seems that information about uncertainties does not belong to the topics about which people prefer to get information. However, because Lion *et al.* [38] used an open response format in their focus group study and did not address uncertainty in its follow-up questionnaire study, no firm conclusions can be drawn.

Johnson and Slovic [34] report that a substantial proportion of the participants of their studies (up to 20%) had difficulties in understanding uncertainty. Results regarding the preferences people have concerning the reporting of uncertainty in risk assessment are mixed. Johnson and Slovic [35] found that while a majority of the participants of their study preferred being informed about uncertainties, about one third rather wanted a single and concrete risk number. On the other hand, Johnson [36] found that the majority of the respondents in his study prefers a simple evaluative statement, telling them whether a situation is safe or not, over the presentation of quantitative risk information.

Another aspect of presenting uncertainty was addressed by Smithson [41], who could show that information expressing uncertainty is preferred and judged more credible than conflicting information from two equally believable sources, each expressing certainty. Smithson interprets his study results as evidence that conflict aversion is stronger than uncertainty aversion.

20.4.2
Interpretation of Uncertainties in Risk Assessment

Uncertainties in risk assessment are usually reported by referring to ranges of risk estimates, for example, lifetime risk is between 1/10 000 and 1/100 000. Of course, these uncertainties may have different reasons, such as the statistical precision of the risk estimate or expert disagreements about risk estimates. Sometimes these reasons are reported, sometimes not. How do people interpret such uncertainties, how do they give meaning to them? And does reporting the reasons for uncertainty matter for their interpretation?

In an experimental study, Viscusi *et al.* [42] asked their participants to imagine that they have to decide changing their residence to one of two areas A or B. These two areas were described as differing only with regard to the risk of contracting a particular disease. First, one group of participants ($N = 65$) was given information about the risk of contracting the disease at area A in terms of ranges for risk estimates, that were based on two studies each providing a different risk estimate (study I: 150 cases per 1 million population, study II: 200 cases per 1 million population). The participants were asked to indicate what precise level of risk in area B they would consider as equivalent to risk of area A. The average estimate for this "certainty equivalent" was 178.35, which is somewhat above the midpoint (175) of the two estimates. The second group of participants ($N = 58$) was given the same task, but received risk information for area A with a larger spread (study I: 110 cases per 1 million population, study II: 240 cases per 1 million population). Again the midpoint of the two risk estimates is 175, but this time the certainty equivalent was 191.08. These results show that people do not weigh differing risk estimates equally, but tend to place greater weight on the higher risk estimate – and this tendency seems to become the more pronounced, the broader the range is.

Further experiments conducted by Viscusi *et al.* [42] indicate that the order of presenting the two risk estimates may play a role in the interpretation of uncertainty in risk estimates. The effect of focusing on the higher risk estimate is more pronounced when it is the second estimate presented (e.g., 110, 240). This effect was even stronger when the participants were informed that the second risk estimate was on a scientific study that had been conducted later than the study for the first risk estimate.

Viscusi *et al.* [42] also investigated whether the asymmetry of the range of risk estimates matters for the interpretation of uncertainty. In this experiment, one group of participants was informed that the average risk – based on a number of studies – for area A was 130 in 1 million population for contracting the disease. In addition, they were given the information that the lowest risk estimate in these studies was 125 and the highest was 155. Again participants were asked for their certainty equivalent for the risk of area B. The mean certainty equivalent was 134.90. The second group was given the same task and was also told that the average risk for area A was 130 in 1 million population, but now the lowest risk estimate was 105 and the highest was 135. Now the mean certainty equivalent was 130.38. Obviously, both experimental conditions differ only with regard to the asymmetry of the spread around the average risk estimate. When the spread was skewed in a manner that indicates that the risk may be much higher than the average risk, participants on average tended to give higher certainty equivalents for area B compared to the average risk that had been given for area A. Interestingly, however, when the skewness indicated the opposite – that the risk may be much lower than the average risk – this did not lead to a lower certainty equivalent. The average certainty equivalent for area B was basically the same as the average risk estimate that had been provided for area A.

In another experimental study, Viscusi [43] used basically the same design as Viscusi *et al.* [42], albeit with one modification. While Viscusi *et al.* did not specify the source of the differing risk estimates, the participants in this experiment were informed that the risk estimates were based on either government studies or industry studies. Low and high estimates were combined with each source (government and industry), resulting in different experimental conditions, for example, low: government, high: industry; low: industry, high: government. The results of this experiment confirm the findings of Viscusi *et al.* [42]: participants again focused on the high risk estimate. This strong weighting on the high risk estimate was particularly pronounced when the information sources differed (e.g., low: industry, high: government).

Taken together, these experimental results indicate that when people are informed about uncertainty in risk assessment by giving ranges of risk estimates, they tend to focus on the "worst case." These results from experimental studies are complemented by results from surveys and focus group studies that point in the same direction.

Johnson and Slovic [35] conducted a survey in which participants ($N = 280$) received information about a cancer risk estimate from a (fictitious) government study on contaminated drinking water. The information presented included the most likely risk (1 : 1 000 000), a lower bound estimate (0) and an upper bound estimate

(10 : 1 000 000). About 53% estimated the risk to the community to be between 1 : 1 000 000 and 10 : 1 000 000, and 15 gave even higher risk estimates, whereas only 13% chose the most likely risk (1 : 1 000 000) and 12% gave even lower risk estimates.

In an online survey with a large US sample ($N = 1520$), Morss *et al.* [40] examined how laypersons interpret and evaluate uncertainties about weather predictions, in particular whether they trust the predictions, and which format of uncertainty description they prefer. It turned out that the respondents chose point estimates over interval estimates. However, it was observed that the respondents appreciated to get interval estimates if the reasons for uncertainty were explained. This means that explaining why uncertainties do exist is honored. Moreover, it was shown that simple statements of uncertainty were preferred compared to more informative but complex statements (e.g., temperature between 70 and 85 °F and not 80% chance of 85 °F and 20% chance of 70 °F).

In a survey by Johnson [36], participants were asked why experts "would give a range of risk numbers, rather than a single number?" ([36], p. 786). This was an open-ended question and many respondents gave multiple answers. The most common answers referred to experts' lack of knowledge (44%), deception by industry in order to make the risk seem smaller (32%), and data analysis problems (15%). Nine percent stated that "the experts offered a range to protect themselves from being held accountable" ([36], p. 787). Another question in that survey asked the participants what they considered to be the reasons why experts differ in their risk estimates. While almost half of the respondents could not name any, the other half mentioned "Incompetence of the experts" (3%), "The organization for which the experts work" (18%), "Limited scientific knowledge about how risk might occur" (14%), and "All of the above" (55%). These answers indicate that lay people often see the reasons for uncertainties in risk assessment not only in the nature of the issue but also in social factors.

Johnson [36] also tried to find out whether people's perception of uncertainty could be explained by their attitude toward industry. Indeed, many of the respondents in his survey seemed to be pretty skeptical about the industry, as nearly half of them stated that industry only communicates a risk if it is actually significant. So this might be informative for understanding people's perception of uncertainty. Based on survey questions regarding attitudes toward uncertainty and toward industry, Johnson constructed two scales: an attitude toward uncertainty scale and an attitude toward industry scale. A regression analysis with uncertainty attitude as the criterion variable and industry attitude and sociodemographic factors (e.g., gender, age, education) as predictors showed that in fact the negative attitude toward industry was statistically significant ($\beta = 0.34$, $p < 0.001$) as well as age ($\beta = -0.23$, $p < 0.01$). However, the explained variance of this regression model was only 11%. In addition, the scales Johnson constructed based on the survey questions regarding attitudes toward uncertainty and attitudes toward industry had very weak psychometric properties. The attitude toward uncertainty scale had a Cronbach's alpha of 0.51; the attitude toward industry scale had a Cronbach's alpha of 0.69. This renders the analysis almost worthless.

Han *et al.* [32] conducted a focus group study to explore how lay people respond to uncertainty in risk estimates. In eight focus groups, they presented their participants (hypothetical) individualized risk estimates for colorectal cancer. Participants were informed that the data were based on a new risk prediction model. Risk estimates were given as both a numeric point estimate (9%) and a range (5–13%). Participants were then asked to discuss their interpretations of these estimates. With regard to potential sources of the uncertainty, a number of aspects were brought up: missing data in the risk prediction model, the model's reliability or accuracy, the credibility of the developers of the model, and expert disagreement about the relevant scientific evidence. While the participants' evaluations of the comprehensibility and the accuracy of the point estimate versus the range did not differ much, a majority indicated that they would be worried more and feel at greater risk when receiving the risk estimate as a range.

20.4.3 Effects of Reporting Uncertainty of Risk Estimates on Risk Perception and Credibility

Johnson and Slovic [34] report the results of three experimental studies that investigated the influence of information about uncertainty in risk assessments on risk perception and trust in the information source. Presenting uncertainty information for risk estimates led to an increase in perceived risk in two of the three experiments reported. One experiment also found that informing about uncertainty increased judgments regarding the honesty of the information source and did not reduce perceived competence of the information source, compared to the experimental group that did not receive the uncertainty information.

Johnson and Slovic [35] also investigated the effects of informing about uncertainty on credibility. About 46% of the participants disagreed with the statement that a discussion of the range of possible risk levels makes the information source seem more honest, compared to 45% who agreed. And 57% disagreed that a discussion of the range of possible risk levels makes the information source seem less competent, whereas 33% agreed.

The study by Johnson [36] found similar results with regard to the effects of presenting uncertainty information on perceptions of honesty, but less positive effects with regard to competence. In this study, about 43% of the respondents concurred with the statement that a discussion of the range of possible risk levels makes the information source seem more honest. A slightly higher percentage (45%) disagreed with that statement. And 46% agreed that a discussion of the range of possible risk levels makes the information source seem less competent, whereas 41% disagreed.

In an experimental study, Kuhn [37] investigated the effects of different ways of describing uncertainty of risk estimates, including verbal acknowledgment that uncertainty exists, a numerical range around the estimate, or two estimates from different sources (government scientists versus scientists hired by an environmental action group), on risk perception. Kuhn found that the different uncertainty formats had little effect on perceived risk. However, when the environmental attitudes of the participants were taken into account, an effect on perceived risk was found for one way of presenting uncertainty information: when uncertainty was described as resulting from two different estimates, one given by the government scientists and the other by scientists hired by an environmental action group, those with high environmental concern perceived much higher risk than those with low environmental concern. This suggests that people interpret uncertainty information in risk assessments in line with their attitudes, if they have reason to believe that the risk assessments are biased, for example, due to different underlying interests.

Han *et al.* [33] report two (Web based) experiments investigating the effects of uncertainty communication on risk perception, cancer-related worry, and perceived credibility of risk information. The first experiment tested the effects of uncertainty information provision and the representational format. Participants ($N = 240$) received (hypothetical) individualized colorectal cancer risk estimates either as a point estimate ("Your chances of developing colon cancer in your lifetime are 9%") or as a range ("Your chances of developing colon cancer in your lifetime are between 5 and 13%"). The respective risk estimates were presented either as in text format or as a bar graph. Han *et al.* found a statistically significant effect of uncertainty information on worry. When the risk estimate was presented as a range, worry was higher than when it was presented as a point estimate. No main effects were found with regard to perceived risk and perceived credibility of risk information. The representational format showed no main effect at all. There was, however, a statistically significant interaction effect between uncertainty information provision and representational format. Presenting the risk estimate as a bar graph led to increased perceived risk for the point estimate; conversely, presenting the risk estimate in text format led to increased perceived risk for the range estimate. There was also some indication for a moderating effect of a personality-related variable: participants with high scores on a dispositional optimism scale expressed much more cancer-related worry than those with low scores when the risk information was given as point estimate; however, when the range of estimates was presented, the effect of dispositional optimism was much smaller. Interestingly, another variable that had previously been found to influence risk comprehension, numeracy [44], had no effect on risk perception, cancer-related worry, and perceived credibility of risk information. In the second experiment ($N = 135$), addressing the question whether different ways of graphically representing uncertainty influence risk perception, cancer-related worry, and perceived credibility of risk information, no effects were found.

While the studies discussed so far mostly investigated only one type of uncertainty in risk assessment, namely, ranges for risk estimates, Miles and Frewer [39] investigated the effect of different types of uncertainty on risk perception. In a survey ($N = 260$), participants were asked to evaluate 12 types of uncertainty (e.g., who is affected, scientific disagreement, size of risk) with regard to risk perception (perceived seriousness of risk) for five different potential food risks (e.g., BSE, high fat diet). Hardly surprising, the perceived seriousness of risk varied for the different food risks. And although the mean ratings of perceived risk also varied for the 12 types of uncertainty, a closer inspection of the correlations of risk perception for the various types of uncertainty by means of principal components analysis revealed that much of the variance (74.4%) could be explained by only one component. Obviously, the participants did not really differentiate between the different types of uncertainty. Rather, their responses were largely based on uncertainty *per se*. Thus, these results indicate that type of uncertainty does not strongly influence risk perception.

The studies reviewed so far all deal with the uncertainty of risk estimates and not with the uncertainty regarding the existence of a hazard. A study by Wiedemann *et al.* [30] also includes this latter aspect. Wiedemann *et al.* conducted three experiments, each addressing a particular type of uncertainty regarding (1) the existence of a hazard, (2) the magnitude of a risk, and (3) the sufficiency of risk management measures. Using a 2×2 factorial design, the authors investigated the effects of two factors (uncertainty and explanations of reasons for uncertainty) on risk perception, worry, perceived expert competence and comprehensibility, and clarity of the information given. In each experiment, participants read short descriptions of the available scientific evidence relevant for the respective issue (hazard, risk, and risk management). The scientific evidence was described in a manner expressing either certainty or uncertainty whether a hazard exists, the risk has a certain magnitude, and the risk management measures are sufficient (factor 1). In addition, there was either an explanation given for the reasons for the certainty/uncertainty or not (factor 2). Only a few statistically significant results were found: In the first experiment, which was about the existence of a hazard, factor 1 (uncertainty) had an effect on two variables: when uncertainty was reported, skepsis regarding expert competence increased and the clarity of the information given was evaluated less favorably. In the second experiment focusing on the magnitude of a risk, the experimental manipulation had no effect at all. In the third experiment addressing the sufficiency of risk management measures, statistically significant effects were found for two variables. When uncertainty was reported, worry increased and text comprehensibility was evaluated less favorably. In addition, there was also a statistically significant effect of factor 2 (explanation). When an explanation was given, text comprehensibility was also evaluated less favorably. In all three experiments, risk perception remained unchanged, regardless of whether uncertainty was reported or not.

20.5
Explaining Inconclusive Evidence

A central problem of communicating uncertainties, not in cancer risk assessment only, is how to inform about *inconclusive* scientific evidence. Although many of the underlying effects of communicating uncertainty have been outlined above, this topic has been only very rarely the direct focus of research. Inconclusive evidence can be due to lack of evidence or contradictory evidence. As mentioned above, Smithson [41] found that people tend to discredit contradictory evidence; they rather prefer information expressing uncertainty – although the level of uncertainty is the same in both types of information. So the question is how to present inconclusive evidence to nonexperts in a way that fosters the understanding of the reasons for the inconclusiveness? This question is particularly important for hazard assessment, because here the problem is whether an agent poses a health threat or not. As pointed out above, with regard to cancer hazard assessment, it is often not possible to provide a clear-cut answer.

The "standard" format of presenting evidence in science (as well as in hazard or risk assessment) is the systematic or narrative review, sometimes supplemented with classifying statements that summarize the evidence (such as IARC's grading scheme; see Chapter 4). While these reviews expound in detail whether evidence is lacking or contradictory, these presentations will be usually too demanding for nonexperts. On the other hand, the classifying statements alone (e.g., *not classifiable as to its carcinogenicity*, see Table 20.1) convey no information about the reasons for inconclusiveness.

Recently, Schütz *et al.* [45] have proposed so-called evidence maps as a tool for summarizing and communicating evidence specifically for the field of risk assessment. The theoretical basis of the evidence map approach is Toulmin's argumentation model [46, 47]. According to this model, an argument typically includes at least three components: claim, data, and warrant. The starting point is the *claim* that needs to be justified. This is usually done by referring to some kind of *data*. The *warrant* serves as the logical bridge between the claim and the data. Additional components of the argumentation model, which are invoked, if necessary, are a *backing* that provides support for the applicability of the warrant, a *qualifier* that indicates the strength of the argument (e.g., by saying that a claim is, for instance, "probably" true), and a *rebuttal* that indicates conditions for which the claim is not valid.

Toulmin's argumentation model provides a useful conceptual tool for analyzing scientific argumentation, and it also has occasionally been used in a risk assessment context [48, 49]. However, to clarify the specific structure of inconclusive evidence, the model needs to be extended. In particular, the two-sided pro- and contra-structure that reflects the available positive and negative evidences needs to be included, as it is the case in weight-of-evidence approaches [50–52]. This is in fact a central feature of the evidence map approach: a graphical representation of the arguments, which are based on the respective (positive or negative) findings, as well as the conclusions that are drawn. Figure 20.1 displays the structure of an evidence map.

Preliminary results from recent experimental studies [53, 54] suggest that lay people evaluate the evidence map as a more useful tool for communicating controversial risk information compared to narrative evidence characterization.

20.6 Conclusions

20.6.1 Should Uncertainties in Cancer Risk Assessments Be Disclosed?

From an ethical point of view, there is no doubt that uncertainties should be disclosed. However, informing about uncertainty is a quite sensitive process. It requires considerable care in terms of presenting and explaining the uncertainty information. In undertaking such a communication policy, one must also recognize the cognitive

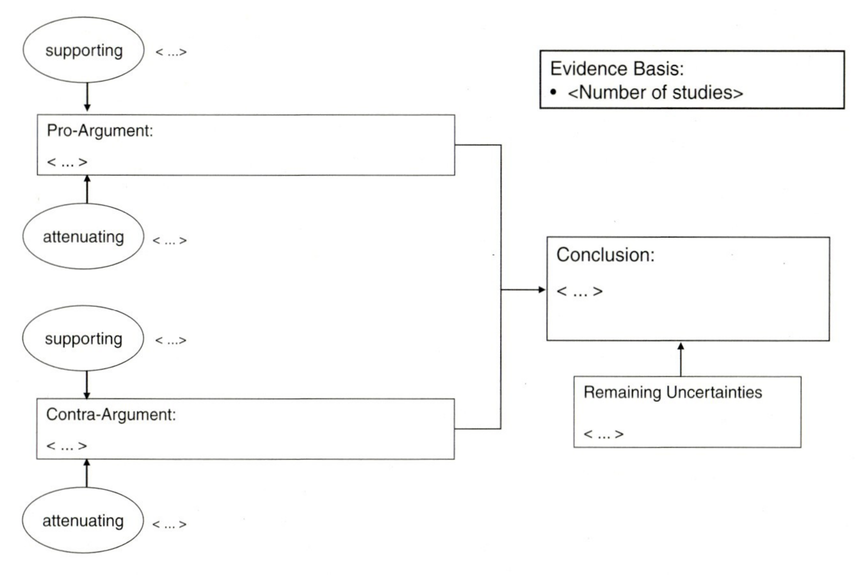

Figure 20.1 Structure of an evidence map.

limitations that nonexperts have in processing uncertainty information as well as the effects this information will have on risk perception and credibility, which may be quite different from those intended by the information provider. With regard to risk perception, there is by now no empirical evidence that clearly speaks in favor of communicating uncertainty to lay people. The effects of reporting uncertainty are less clear in regard to the credibility of the information source. Some evidence is indicating a positive effect on credibility, others point to a negative effect.

20.6.2 What Types of Uncertainty Should Be Addressed?

Basically, two types of uncertainty play a significant role in informing about cancer risk assessment: first, uncertainty with respect to the existence of a hazard and, second, with respect to the magnitude of a risk. Particularly with regard to hazard uncertainty, there is only little empirical evidence available so far; this evidence suggests that lay people tend to attribute uncertainty to a lack of competence of the risk assessors. With regard to risk uncertainty, the available evidence indicates that communicating this uncertainty may well increase perceived risk. People tend to focus on the highest risk estimate provided, that is, on the "worst case."

20.6.3 In Which Format Should Uncertainties Be Described?

With regard to explaining the magnitude of a risk, graphical representations have been found useful, whereas verbal characterizations of risk (e.g., "unlikely"), which are often considered to be more easily understood by lay people than the numerical information, seem to be of little benefit [1, 2]. Whether this also holds for the communication of uncertainty in hazard or risk assessments is an open question, as the empirical evidence about which formats are effective is rather sparse. For risk uncertainties, the available evidence does speak for a benefit of using graphical aids to explain ranges of risk estimates. With respect to uncertain hazards, it seems that the above-mentioned evidence map is an appropriate tool for reporting existing uncertainty in hazard identification.

20.6.4 In What Detail Should Uncertainty Be Revealed?

With regard to the question how detailed the explanation of uncertainties in hazard and risk assessment should be, the available evidence is mixed. Some evidence suggest that people want to be informed about uncertainties, while others indicate that people prefer simple information about whether there is a risk or not.

A crucial point that is relevant for this as well as the other questions and which has not been addressed in any of the studies reported above is the mode of information processing. This has been found to be of major importance for the effect information has on the recipient. One prominent distinction with regard to the mode of information processing is the one between *heuristic* and *systematic* processing of information [55, 56]. Systematic processing occurs when people carefully examine the information given. This mode of information processing requires both the ability and the motivation to understand the information. In contrast, when people process information heuristically – either because they are not able or not motivated to consider the information carefully – they use simple "rules of thumb." For instance, when being informed about a government-sponsored risk assessment, a person may use the heuristic "never trust a risk assessment sponsored by government," regardless of whether there is actually any indication for a bias in the risk assessment or not. The above-reported tendency to focus on the "worst case," when different risk estimates are presented, might also be considered as a kind of heuristic that people may use. This tendency has also been observed in other contexts of risk communication. For instance, Siegrist and Cvetkovich [57] found in an experiment that (hypothetical) studies reporting that a risk exists were given a higher weight by participants than (hypothetical) studies reporting that there was no risk. Some researchers have argued that this "negativity bias" represents a general human characteristic [58]. This obviously poses a problem for informing about uncertainties in hazard or risk assessment.

The general implication of the different modes of information processing is that informing about hazard or risk uncertainty can be expected to be effective only if people process the information systematically. The obvious question then is how to foster systematic information processing. Unfortunately, the answer to this question is neither obvious nor easy. The only promising approach we can think of is to design communication strategies based on valid theory and then test these strategies empirically. This is, of course, a thorny path to go – a path that risk communication research has only recently begun to follow seriously.

References

1 Lipkus, I.M. (2007) Numeric, verbal, and visual formats of conveying health risk: suggested best practices and future recommendations. *Med. Decis. Making,* **27**, 696–713.

2 Visschers, V.H.M., Meertens, R.M., Passchier, W.W.F., and de Vries, N.N.K. (2009) Probability information in risk communication: a review of the research literature. *Risk Anal.*, **29**, 267–287.

3 National Research Council (1996) *Understanding Risk. Informing Decisions in a Democratic Society,* National Academy Press, Washington, DC.

4 WHO (2005) *WHO Outbreak Communication Guidelines (WHO/CDS/2005.28)*, World Health Organization, Geneva.

5 Holmes, B.J., Henrich, N., Hancock, S., and Lestou, V. (2009) Communicating with the public during health crises: experts' experiences and opinions. *J. Risk Res.*, **12**, 793–807.

6 Wibeck, V. (2009) Communicating uncertainty: models of communication and the role of science in assessing progress towards environmental objectives. *J. Environ. Policy Plann.*, **11**, 87–102.

7 Gross, M. (2007) Communicating ignorance and the development of post-mining landscapes. *Sci. Commun.*, **29**, 264–270.

8 United States Environmental Protection Agency (2000) *Risk Characterization Handbook*, United States Environmental Protection Agency, www.epa.gov/spc/pdfs/rchandbk.pdf (retrieved July 26, 2010).

9 National Research Council (1983) *Risk Assessment in the Federal Government: Managing the Process*, National Academies Press, Washington, DC.

10 Smithson, M. (2008) Psychology's ambivalent view of uncertainty, in *Uncertainty and Risk: Multidisciplinary Perspectives* (eds G. Bammer and M. Smithson), Earthscan, London, pp. 205–218.

11 Smithson, M. (2008) Social theories of ignorance, in *Agnotology: The Cultural Production of Ignorance* (eds R. Proctor and L. Schiebinger), Stanford University Press, Stanford, CA, pp., 209–229.

12 Ravetz, J.R. (1993) The sin of science: ignorance of ignorance. *Knowledge*, **15**, 157–165.

13 Politi, M.C., Han, P.K.J., and Col, N.F. (2007) Communicating the uncertainty of harms and benefits of medical interventions. *Med. Decis. Making*, **27**, 681–695.

14 Finkel, A.M. (1990) *Confronting Uncertainty in Risk Management: A Guide for Decision-Makers*, Center for Risk Management, Resources for the Future, Washington, DC.

15 Funtowicz, S.O. and Ravetz, J.R. (1990) *Uncertainty and Quality in Science for Policy*, Kluwer, Dordrecht.

16 Krayer von Krauss, M.P. (2005) Uncertainty in policy relevant sciences. PhD thesis. Department of Environment and Resources, Technical University of Denmark, Copenhagen.

17 Morgan, M.G. and Henrion, M. (1990) *Uncertainty. A Guide to Dealing with Uncertainty in Quantitative Risk and Policy Research*, Cambridge University Press, Cambridge.

18 Walker, W.E., Harremoës, P., Rotmans, J., van der Sluijs, J.P., van Asselt, M.B.A., Janssen, P., and Krayer von Krauss, M.P. (2003) Defining uncertainty a conceptual basis for uncertainty management in model-based decision support. *Integr. Assess.*, **4**, 5–17.

19 U.S. EPA (2005) *Guidelines for Carcinogen Risk Assessment (2005)*, U.S. Environmental Protection Agency, Washington, DC, EPA/630/P-03/001F, http://oaspub.epa.gov/eims/eimscomm.

getfile?p_download_id=439797 (retrieved August 2, 2010).

20 International Programme on Chemical Safety (2004) *Risk Assessment Terminology: Part 1 and Part 2*, World Health Organization, Geneva, Switzerland.

21 Cogliano, V.J., Baan, R.A., Straif, K., Grosse, Y., Secretan, M.B., and El Ghissassi, F. (2008) The IARC monographs' approach to characterizing evidence, in *The Role of Evidence in Risk Characterization. Making Sense of Conflicting Data* (eds P.M. Wiedemann and H. Schütz), Wiley-VCH Verlag GmbH, Weinheim, pp., 101–109.

22 Wiedemann, P.M., Schütz, H., and Spangenberg, A. (2010) *Evaluation of Communication on the Difference Between "Risk" and "Hazard"* (eds E. Ulbig, R.F. Hertel, and G.-F. Böl), Bundesinstitut für Risikobewertung, Berlin.

23 International Agency for Research on Cancer (2006) *Preamble to the IARC Monographs. IARC Monographs Programme on the Evaluation of Carcinogenic Risks to Humans*, International Agency for Research on Cancer, http://monographs.iarc.fr/ENG/Preamble/CurrentPreamble.pdf (retrieved July 26, 2010).

24 Wiedemann, P.M., Schütz, H., and Thalmann, A. (2008) Perception of uncertainty and communication about unclear risks, in *The Role of Evidence in Risk Characterization. Making Sense of Conflicting Data* (eds P.M. Wiedemann and H. Schütz), Wiley-VCH Verlag GmbH, Weinheim, pp. 161–183.

25 Wikipedia Climatic Research Unit email controversy, http://en.wikipedia.org/wiki/Climatic_Research_Unit_email_controversy (retrieved July 26, 2010).

26 Wengen Conference 2000, http://www.unige.ch/ia/climat/EVENTS/Wengen/00/Wen00sum.html (retrieved July 26, 2010).

27 ILGRA (2002) The precautionary principle: policy and application. Interdepartmental Liaison Group on Risk Assessment, http://www.hse.gov.uk/aboutus/meetings/committees/ilgra/pppa.pdf (retrieved July 26, 2010).

28 Magat, W.A., Viscusi, W.K., and Huber, J. (1988) Consumer processing of hazard warning information. *J. Risk Uncertainty*, **1**, 201–232.

29 Viscusi, W.K. and Zeckhauser, R.J. (1996) Hazard communication: warnings and risk. *Ann. Am. Acad. Pol. Soc. Sci.*, **545**, 106–115.

30 Wiedemann, P.M., Löchtefeld, S., Claus, F., Markstahler, S., and Peters, I. (2010) Laiengerechte Kommunikation wissenschaftlicher Unsicherheiten im Bereich EMF. Abschlussbericht zum BfS Forschungsprojekt StSch 3608S03016.

31 Kloprogge, P., van der Sluijs, J., and Wardekker, A. (2007) Uncertainty communication: issues and good practice, Department of Science Technology and Society (STS), Copernicus Institute for Sustainable Development and Innovation, http://www.nusap.net/download.php?op=getit&lid=46 (retrieved July 26, 2010).

32 Han, P.K.J., Klein, W.M.P., Lehman, T.C., Massett, H., Lee, S.C., and Freedman, A.N. (2009) Laypersons' responses to the communication of uncertainty regarding cancer risk estimates. *Med. Decis. Making*, **29**, 391–403.

33 Han, P.K.J., Klein, W.M.P., Lehman, T., Killam, B., and Massett, H., and Freedman, A.N. (2010) Communication of uncertainty regarding individualized cancer risk estimates: effects and influential factors. *Med. Decis. Making*. doi: 10.1177/0272989X10371830

34 Johnson, B.B. and Slovic, P. (1995) Presenting uncertainty in health risk assessment: initial studies of its effects on risk perception and trust. *Risk Anal.*, **15**, 485–494.

35 Johnson, B.B. and Slovic, P. (1998) Lay views on uncertainty in environmental health risk assessment. *J. Risk Res.*, **1**, 261–279.

36 Johnson, B.B. (2003) Further notes on public response to uncertainty in risks and science. *Risk Anal.*, **23**, 781–789.

37 Kuhn, K.M. (2000) Message format and audience values: interactive effects of uncertainty information and

environmental attitudes on perceived risk. *J. Environ. Psychol.*, **20**, 41–51.
38 Lion, R., Meertens, R.M., and Bot, I. (2002) Priorities in information desire about unknown risks. *Risk Anal.*, **22**, 765–776.
39 Miles, S. and Frewer, L.J. (2003) Public perception of scientific uncertainty in relation to food hazards. *J. Risk Res.*, **6**, 267–283.
40 Morss, R.E., Demuth, J.L., and Lazo, J.K. (2008) Communicating uncertainty in weather forecasts: a survey of the US public. *Weather Forecasting*, **23**, 974–991.
41 Smithson, M. (1999) Conflict aversion: preference for ambiguity vs. conflict in sources and evidence. *Organ. Behav. Hum. Decis. Process.*, **79**, 179–198.
42 Viscusi, W.K., Magat, W.A., and Huber, J. (1991) Communication of ambiguous risk information. *Theory Decis.*, **31**, 159–173.
43 Viscusi, W.K. (1997) Alarmist decision with divergent risk information. *Econ. J.*, **107**, 1657–1670.
44 Reyna, V.F., Nelson, W.L., Han, P.K., and Dieckmann, N.F. (2009) How numeracy influences risk comprehension and medical decision making. *Psychol. Bull.*, **135**, 943–973.
45 Schütz, H., Wiedemann, P.M., and Spangenberg, A. (2008) Evidence maps: a tool for summarizing and communicating evidence in risk assessment, in *The Role of Evidence in Risk Characterization. Making Sense of Conflicting Data* (eds P.M. Wiedemann and H. Schütz), Wiley-VCH Verlag GmbH, Weinheim, pp. 151–160.
46 Toulmin, S. (1958) *The Uses of Argument*, Cambridge University Press, Cambridge.
47 Toulmin, S., Rieke, R., and Janik, A. (1984) *An Introduction to Reasoning*, Macmillan Publishers, New York.
48 Driedger, S.M. and Eyles, J. (2001) Organochlorines and breast cancer: the uses of scientific evidence in claims making. *Soc. Sci. Med.*, **52**, 1589–1605.
49 Mackay, J., Schulz, P., Rubinelli, S., and Pithers, A. (2007) Online patient education and risk assessment: project OPERA from cancerbackup – putting inherited breast cancer risk information into context using argumentation theory. *Patient Educ. Couns.*, **67**, 261–266.
50 Krimsky, S. (2005) The weight of scientific evidence in policy and law. *Am. J. Public Health*, **95** (Suppl. 1), S129–S136.
51 Linkov, I. and Satterstrom, F.K. (2006) Weight of evidence: what is the state of the science? *Risk Anal.*, **26**, 573–575.
52 Weed, D.L. (2005) Weight of evidence: a review of concept and methods. *Risk Anal.*, **25**, 1545–1557.
53 Börner, F., Schütz, H., and Wiedemann, P. (2009) Evidence maps as a tool for risk communication. SRA 2009 Annual Meeting: Risk Analysis – The Evolution of a Science, Baltimore, MD.
54 van Wingerden, A.M. (2010) The effect of information lay-out on risk comprehension, risk perception, perceived reader friendliness and preference: the evidence map as a tool for risk communication. Diploma thesis. University of Twente, Enschede.
55 Chen, S. and Chaiken, S. (1999) The heuristic–systematic model in its broader context, in *Dual Process Theories in Social Psychology* (eds Y. Trope and S. Chaiken), Guilford Press, New York, pp. 73–96.
56 Trumbo, C.W. (1999) Heuristic–systematic information processing and risk judgment. *Risk Anal.*, **19**, 391–400.
57 Siegrist, M. and Cvetkovich, G. (2001) Better negative than positive? Evidence of a bias for negative information about possible health dangers. *Risk Anal.*, **21**, 199–206.
58 Rozin, P. and Royzman, E.B. (2001) Negativity bias, negativity dominance, and contagion. *Pers. Soc. Psychol. Rev.*, **5**, 296–320.

21
The Precautionary Principle and Radio Frequency Exposure from Mobile Phones

Gary E. Marchant

21.1
Introduction

The precautionary principle and the risk management of radio frequency electromagnetic fields (RF EMF) from mobile phones and other handheld wireless communication devices challenge each other. On the one hand, the precautionary principle challenges the relatively light-handed risk management and regulation of RF EMF to date, suggesting that the inherent uncertainties about the risks from RF EMF requires consideration of more stringent controls on RF EMF exposures from mobile phones. On the other hand, the overwhelming consumer demand for mobile phones, notwithstanding the uncertainties about their risks, challenges the vitality, meaning, and assumptions of the precautionary principle.

The conflation of these reciprocal challenges presents two questions that are examined in this chapter. First, is precaution, and more pointedly the precautionary principle, an appropriate concept for guiding or contributing to the risk management of potential carcinogens, including RF EMF from cellular phones? Second, to the extent that precaution or the precautionary principle should or does apply, how should the precautionary principle be construed and applied to a particular concern such as RF EMF? These two questions are examined after first briefly providing some background on the precautionary principle, including its history, purpose, and definition.

21.2
Background on the Precautionary Principle

The precautionary principle is a relatively new concept in risk management, but one that has proliferated very quickly around the world over the past two decades. Despite its adoption by many legislators, regulators, and jurists, the precautionary principle remains the subject of considerable controversy in at least some jurisdictions and sectors, even if broadly accepted in others. The core concept behind the precautionary principle is quite straightforward and unobjectionable – it is better to be safe than

Cancer Risk Evaluation: Methods and Trends,
Edited by Günter Obe, Burkhard Jandrig, Gary E. Marchant, Holger Schütz, and Peter M. Wiedemann.

sorry. Operationalizing this simple concept into specific language and criteria proves, however, to be much more challenging. One of the first attempts to give concrete expression to the precautionary principle was the Rio Declaration on Environment and Development issued in 1992 by the United Nations Conference on Environment and Development (UNCED), which defined the principle as follows: "Where there are threats of serious and irreversible damage, lack of full scientific certainty shall not be used as a reason for postponing cost-effective measures to prevent environmental degradation." [1]. A subsequent definition, known as the Wingspread Statement, was offered by a prominent group of supporters of the principle: "When an activity raises threats of harms to human health or the environment, precautionary measures should be taken even if some cause and effect relationships are not fully established scientifically." [2].

The precautionary principle has been incorporated into a number of international agreements and national laws, albeit usually without reference to a specific definition. Europe has been at the forefront of adopting and applying the precautionary principle, and the European Community (EC) has formally committed to implementing environmental policy in conformity with the precautionary principle in the 1992 Maastricht amendments to the EC Treaty. In 2000, the European Commission issued a 28-page Communication that provides the most comprehensive official guidance to applying the precautionary principle to date [3]. The Communication made it clear that the precautionary principle applies to risk management rather than risk assessment, and requires measures based on the precautionary principle to be "*proportional* to the chosen level of protection; *non-discriminatory* in their application; *consistent* with similar measures already taken; *based on an examination of the potential benefits and costs* of action or lack of action (including, where appropriate and feasible, an economic cost/benefit analysis); *subject to review*, in the light of new scientific data; and *capable of assigning responsibility for producing the scientific evidence*." ([3], p. 4). Notwithstanding this helpful contribution to articulating the practical meaning of the precautionary principle, the Communication failed to provide a definition of the precautionary principle or to provide any specific criteria for when it applies or what it requires [4].

The precautionary principle has been applied to a growing number of potential risks, but of particular relevance here, it has been suggested by some as providing an appropriate risk management approach for regulating exposure to RF EMF. For example, the UK Independent Expert Group on Mobile Phones ("Stewart Report") recommended in 2000 a "precautionary approach" to RF EMF exposure and in particular recommended restrictions on the use of cellular phones by children [5]. Some countries, including Switzerland, Italy, and Belgium, have reduced allowable exposure levels to RF EMF based on the precautionary principle.

21.3 Pros and Cons of the Precautionary Principle

A vigorous debate exists within government, public policy, and academic circles over the merits and value of the precautionary principle. The need to apply some measure of precaution in risk management decisions by government, companies, and individuals is beyond dispute. Moreover, there is little disagreement that the application of precaution in the past has not been as transparent and explicit as it should be, and thus a more open and considered approach to the application of precaution would be useful. That is where consensus ends, however, as there is strong disagreement about how much precaution is appropriate in a given situation and whether the precautionary principle serves a useful function in raising and applying precaution.

Proponents of the precautionary principle are of the view that past regulatory efforts have generally been underprotective for new or developing risks. A prominent resource supporting this perspective was a EU-commissioned study entitled "Late Lessons from Early Warnings" that documented 14 examples such as asbestos, chlorofluorocarbons (CFCs), and lead where government, industry, and society generally failed to adequately heed indications of an unacceptable health risk, to the ultimate detriment of public health [6]. The lessons drawn from these examples are that we need to apply precaution more proactively and aggressively, hence the precautionary principle.

While the argument for the precautionary principle is quite straightforward and concise, the argument against the principle is more complex and multifaceted. Key arguments against the precautionary principle are summarized as follows:

- **Ambiguity**: There is no single agreed definition of the precautionary principle, but rather numerous different versions with subtle but important distinctions [7]. For example, some versions (such as the UNCED version cited above) apply only to "serious" or "irreversible" risks, while others (such as the Wingspread version also cited above) are not so limited and appear to apply to any risk. Every version is ambiguous as to the nature and certainty of the evidence for a risk that is needed to trigger the principle. For example, what if there is some scientific evidence of a potential risk, but the overall body of available evidence weighs against the existence of a significant risk? There is no consistent answer as to what, if anything, the precautionary principle commands in such circumstances. Similarly, every version of the precautionary principle is ambiguous as to what quantum of safety data a technology developer must provide to satisfy the precautionary principle. Given that complete safety is impossible to prove, and zero risk is impossible to achieve, what level of risk, if any, is acceptable to allow the product to proceed under the precautionary principle? Can the benefits of the product, and the economic and social costs of restricting it, be weighed against the risks in deciding an appropriate course of action? Some versions and interpretations of the precautionary principle seem to allow or require this, while others are silent or disfavor such balancing.
- **Arbitrariness**: Perhaps as a consequence of its inherent ambiguity, the precautionary principle is prone to arbitrary application. With no clear meaning or criteria for its application, the precautionary principle is frequently invoked for reasons other than the likelihood or potential severity of risk. Rather, inappropriate factors such as bias, political pressure, self-interest, competitive advantage,

or protectionism often explain why the precautionary principle is being applied to some problems but not others of equal or greater potential indications of risk. These types of improper motivations can lead to extreme and absurd applications of the precautionary principle, such as the attempt by Norway to ban corn flakes cereal fortified with vitamins because extra vitamins could be harmful, by France to ban caffeinated energy drinks so that pregnant women would not consume caffeine (yet not banning coffee), by Denmark to prohibit cranberry juice cocktails because the extra vitamin C could conceivably harm some individuals, and by Zaire to ban US food aid to its starving population because some of that food aid might have contained low levels of genetically engineered corn eaten every day by millions of Americans with no apparent adverse health effects [4]. All these decisions were based primarily on the precautionary principle.

- **Overprotection**: Although the precautionary principle is aimed at preventing underregulation of technologies, application of the precautionary principle, particularly in its more draconian forms or interpretations, has the potential to overly restrict new technologies. For every false negative (i.e., technology incorrectly assumed to be safe) that the precautionary principle will protect people, there will be a false positive (i.e., technology incorrectly assumed to be dangerous) that will be unnecessarily regulated by the precautionary principle. Examples of such false positives whose risks were overestimated include saccharin, silicone breast implants, and even coffee. Thus, the precautionary principle has the potential to suppress innovation by unreasonably and unnecessarily delaying or even preventing the introduction of many socially beneficial products. New technologies and new products provide many benefits to the public, often including improved health and safety. History demonstrates that while some technologies and products have had unfortunate adverse consequences, the net effect of the introduction of new technology and products has been to increase the overall health and welfare of consumers [8]. By placing a roadblock in the path of new products and technologies, the precautionary principle may contribute to technological and economic stagnation and threaten the economic and health well-being of citizens.

In sum, there are competing arguments for and against the precautionary principle. The case for the precautionary principle is strongest when the principle is approached as a general philosophy or normative principle that provides relevant background for a regulatory decision but does not specify a specific decision rule or legal requirement. In this context, the ambiguity and uncertainties about the meaning of the precautionary principle are less important because the principle is serving as a general guide rather than dictating a specific result, where more precision would be needed. If, however, the precautionary principle is adopted as a legal requirement or binding obligation, the ambiguity of the precautionary principle would be more problematic because it would be vulnerable to manipulation and arbitrary and capricious decision making [9]. As Daniel Bodansky noted when the precautionary principle was first receiving widespread attention two decades ago, the principle may be "too vague to serve as a regulatory standard [10]."

21.4 Applying the Precautionary Principle to Radio Frequency Electromagnetic Fields

If the precautionary principle has been adopted in a given jurisdiction, two questions are raised with respect to its application to RF EMF from mobile phones. First, assuming that the application of the precautionary principle is limited to some subset of products and technologies based on risk, uncertainty, or other factors, is cell phone RF EMF a potential hazard to which the precautionary principle should apply? Second, assuming the precautionary principle is applied to these fields, what does it require? These two questions are examined sequentially next.

21.4.1 Should the Precautionary Principle Apply to Radio Frequency Electromagnetic Fields?

As with the arguments for and against the precautionary principle generally, there are arguments both for and against the application of the precautionary principle specifically to RF fields from mobile phones. There are two primary arguments in favor of application of the precautionary principle to this potential hazard. First, there is significant inherent uncertainty about the health effects, if any, of RF EMF from cell phones, although the extent of this uncertainty is itself contested [11]. Safety testing of RF EMF in animals and *in vitro* systems may not be representative of human risks, and the significant latency period expected for any brain tumors caused by RF EMF suggest that epidemiological studies may not be capable of detecting any increased risk yet [12]. Thus, even if most studies conducted to date have failed to document any human health risks, these results available to date cannot definitely establish that RF EMF is safe. Given that the primary purpose of the precautionary principle is to bridge uncertainty, the greater the uncertainty, arguably the greater the need for the precautionary principle.

The second primary argument for application of the precautionary principle is the widespread human exposure to RF fields from mobile phones. A very large percentage of the population now uses a cellular phone or similar wireless communication device, resulting in extensive RF EMF exposure. There are an estimated 4.6 billion cell phone subscribers worldwide [13]. Accordingly, if RF EMF emitted by cell phones does, in fact, increase the risk of cancer or other adverse health effects, the population risk could be quite large, even if the individual risk is relatively small. The large exposure associated with RF EMF thus argues in favor of application of the precautionary principle.

While uncertainty and large exposure weigh in favor of more precaution, the relevance of the existing set of health studies published to date is more equivocal and could be argued from either side. While most animal, *in vitro*, and human studies to date have not shown any increased risk of brain cancer or other serious health effects, proponents of the precautionary principle can point to some relatively rare findings buried in the large data set that are arguably consistent with an increased risk. These few positive findings arguably could be used to justify application of the precautionary principle, reflecting the ambiguity about the triggering criteria for the precautionary principle in general (i.e., should the precautionary principle nevertheless apply when the weight of scientific evidence and expert opinion suggest there is no risk, but a minority of the data and experts lean toward a risk).

The leading argument against application of the precautionary principle to RF EMF is that the clear weight of available evidence supports the absence of any significant risk [14]. The few contrary results are, under this view, inevitable blips that would be expected to occur by chance alone in any large-scale toxicological data set. For example, the largest study (INTERPHONE) reported to date on the association between cell phone use and brain cancer concluded:

This is the largest study of the risk of brain tumours in relation to mobile phone use conducted to date and it included substantial numbers of subjects who had used mobile phones for 10 years. Overall, no increase in risk of either glioma or meningioma was observed in association with use of mobile phones. There were suggestions of an increased risk of glioma, and much less so meningioma, at the highest exposure levels, for ipsilateral exposures and, for glioma, for tumours in the temporal lobe. However, biases and errors limit the strength of the conclusions we can draw from these analyses and prevent a causal interpretation ([15], p. 688).

A survey by the US National Cancer Institute found no increase in the incidence of brain tumors or other cancers of the nervous system in the period from 1987 to 2007, notwithstanding the dramatic increase in cell phone use during that period [16].

The lack of any concrete evidence of actual health problems occurring to date, combined with predictive models and theoretical considerations (e.g., nonthermal radio frequency unable to break chemical bonds) supporting the safety of cellular phones, is the central argument against application of the precautionary principle to restrict RF EMF exposure from cellular phones and related handheld communication equipment. Those formulations of the precautionary principle that are limited to "serious" or "irreversible" risks would seem to be most inapplicable. Because RF EMF exposure has generally produced no adverse effects in the battery of traditional toxicology tests in which many other types of products have produced positive effects, it would rank low on the priority list of products to which the principle potentially should be applied under the EU Commission's nondiscriminatory criteria for application of the precautionary principle.

A second argument against applying the precautionary principle to RF EMF is that manufacturers of mobile phones already meet regulatory limits on RF EMF exposure. These limits were established with a large margin of safety with respect to the known (i.e., thermal or heating) effects of RF EMF exposure [17]. By adhering to these stringent limits, manufacturers have arguably already taken protective action sought by the precautionary principle.

Another factor weighing against application of the precautionary principle is the benefits of, and consumer demand for, mobile phones. The various versions of the precautionary principle are inconsistent with respect to the consideration they give, if any, to the benefits of a product (and thus the socioeconomic costs of their prohibition or delay). Nevertheless, most versions of the principle contain some element of proportionality, as does the EU Commission's Communication on the precautionary principle, whereby the precautionary action warranted must be determined relative to the benefits of the product that may be foregone. In other words, the greater the benefits provided by a product, the less justified are precautionary actions that preclude the realization of those benefits. This factor weighs against application of the precautionary principle to mobile phones because of the substantial benefits in the form of convenience and security that these products provide to the public. These benefits are demonstrated by the dramatic growth in the public demand for and use of mobile phones and other handheld wireless communications equipment in the past decade.

In addition to convenience, mobile phones provide substantial safety benefits to the public. They are frequently used to make distress calls to report accidents, to call for emergency assistance, and to report crimes, fires, and other emergencies. For example, mobile phones are used to make more than 290 000 calls to 911 emergency services *each day* in the United States [18]. This results in significant saving of human lives and injuries. These substantial health and safety benefits could be jeopardized by restraining the use of mobile phones through implementation of the precautionary principle. Such an action would almost certainly have the net effect of doing more harm than good to public health and safety, in direct conflict with the protective intent of the precautionary principle.

Finally, according to its proponents, a core principle underlying the precautionary principle is that the public voice be given considerable weight in applying precaution [2]. In the case of mobile phones, consumers appear to have spoken based on their widespread adoption of wireless communication equipment, undoubtedly due to the convenience and benefits that such products provide. One could argue that the public would be more concerned, and less willing to purchase and use such products, if they understood that there are potential risks associated with RF EMF exposure. Yet, the weight of evidence does not support such a finding of risk at this time, and the public seems complacent about and willing to accept the limited evidence of a hazard that has surfaced to date. Any attempt to restrict the rights of the public to use this equipment or to buy new models would be extremely unpopular, throwing into question the relevance of precautionary restrictions in this context.

In sum, the question of whether the precautionary principle should apply to RF EMF from cellular phones depends in large part on what version and interpretation of the precautionary principle one applies. Versions that emphasize proportionality, nondiscrimination, and democratic choice would be less applicable than more stringent versions. Of course, application of the precautionary principle need not be an all-or-none proposition, and so perhaps the key question is what action the precautionary principle would point to if applied to RF EMF from mobile phones.

21.4.2
What Action Does the Precautionary Principle Recommend for Radio Frequency?

If the precautionary principle is applied to RF EMF fields from cell phones, the requirements imposed by the precautionary principle are ambiguous and open to argument. Some more extreme formulations of the precautionary principle appear to require that a product not be placed on the market unless and until all uncertainties have been resolved in favor of complete safety. Such a requirement would effectively lead to a ban on cellular phones (and virtually every other product) since it is not possible either to eliminate all uncertainties about risk or to prove the absence of any risk.

More reasonable constructions of the precautionary principle implicitly anticipate a spectrum of potential precautionary actions other than simply a ban, although the range of available options along the spectrum and the decision criteria used to determine which option should apply to a particular technology are rarely articulated explicitly. Nevertheless, the precautionary principle could be construed to suggest a range of possible precautionary actions for RF EMF from mobile phones, including the following:

1) **Ban or moratorium**: The most stringent option would be a permanent ban or temporary moratorium on some or all cellular phones unless or until more conclusive data on their safety were available. One problem with this requirement is that it may never be possible to prove the products are safe unless they are permitted to be used. Of course, this then opens the door to the argument that consumers are being used as guinea pigs, but the reality is that for products, probably including RF EMF, that can only be shown to be safe using human data, there may be no alternative other than to allow people to use the product and then monitor them for any adverse effects.
2) **Partial restrictions**: A somewhat less stringent option would be to prohibit cell phones in certain contexts, presumably those that present the greatest risks. For example, the use of cell phones by or sale of cell phones to children could be prohibited. Alternatively, the siting of cell phone towers could be restricted from certain areas, such as near schools.
3) **Performance standards**: Cell phones must meet performance standards that limit the strength of the radio frequency field they can generate. These standards are based on protecting people from thermal effects, with a margin of safety built in. The precautionary principle could be used to justify more stringent performance standards with an additional margin of safety.
4) **Warnings**: Various types of warnings could be required, either as a label on the phone, in the phone manual, or in another format. These "warnings" could range from a neutral conveyance of information (e.g., power rating) with no specific recommendation to a more active warning that recommends limiting exposure to radio frequency from cell phones. For example, the City of San Francisco recently mandated that retailers must post information on power levels next to every phone they sell [19].
5) **Avoidance measures**: Individual mobile phone users could take various actions to reduce their exposure to RF EMF. Examples include the use of a headset, holding the phone away from the head, and limiting the duration and frequency of cell phone calls.
6) **Public education**: Another possible action would be some form of public education campaign alerting consumers to potential risks from cell phones and/or recommending certain prudent actions (e.g., use a headset, avoid long-duration exposure, etc.) with respect to the use of cell phones. The campaign could be financed and conducted by governmental entities, private industry, advertising councils, or some combination of the above.

7) **Research**: The precautionary principle could be used to justify additional safety studies to further understand any risks from cell phone radio frequency. Of course, this would raise questions about what types of studies, who conducts the studies, who funds the studies, and similar practical issues.

From this list, it is clear that a menu of possible precautionary actions is available, assuming that a determination has been made to apply the precautionary principle to cell phone RF EMF fields. The precautionary principle provides little guidance (if any) on which of these options should be selected for a particular problem. Since the purpose of the precautionary principle is to better protect human health and the environment, potential risk–risk trade-offs is one important factor to consider before taking precautionary action. As discussed above, cell phones not only provide convenience and enhanced communication opportunities for users but also provide important safety benefits, such as being able to call for help in an emergency, reporting being lost or threatened, or communicating with children to help them make safe choices. While these health and safety benefits have not been quantified, they are both real and significant. Given that any health harms from cell phones are speculative and at present go against the weight of scientific evidence, it would likely be contrary to public health to prohibit or significantly restrict the availability of cell phones. These risk–risk trade-offs present a challenge to the precautionary principle – it is often not clear in such contexts whether the precautionary principle should prohibit or require technologies presenting such risk trade-offs [20].

Similarly, concerns can be raised against at least some types of warning or public education campaign. To the extent that such warnings deter use or ownership of cell phones, they might have a net detrimental effect on public safety. Moreover, invoking the precautionary principle appears to increase rather than decrease public anxiety about a technology [21, 22]; so, public communication campaigns might backfire if they increase public anxiety while providing negligible or even negative public health impacts.

One advantageous aspect of application of the precautionary principle to cell phone RF fields is that there is the opportunity for individualized choice rather than the need to impose a single decision on entire society. Many other risk decisions potentially involving the precautionary principle, such as whether to allow a particular ingredient into the food supply or determining the allowable level of emissions from power plants, impose a societal level of risk and precaution that do not enable individual departures or opt-outs from the societal decision (however made). With respect to decisions about RF EMF exposure from cell phones, however, each individual can make his or her own decision on how much exposure (and hence maybe risk) to accept notwithstanding the uncertainty, by deciding whether to use a cell phone at all, whether to use a headset, or whether to limit the amount of time they use a cell phone. Such options are referred to as prudent avoidance strategies [23] and can be made at the level of the individual consumer. Given that risk preferences and precautionary attitudes differ within the population, the opportunity for individual citizens to make their own choices on prudent avoidance options is an attractive dimension in applying the precautionary principle to this technology. To empower such choice, it is important that people have access to objective and credible information about what is known and not known about the risks from RF EMF, as well as options that a person can take to reduce exposure (e.g., use a headset).

In addition to empowering personal choice about protective measures, another sensible application of the precautionary principle in this context would be to continue to research the risks of RF EMF. Given that the latency period for any tumors that may be caused by RF EMF is likely to exceed 10 years, it is important to maintain active vigilance for any early indicators of risk. It would also be important, and again something that could be cast as an application of the precautionary principle, to ensure that any early indications of risk are not suppressed or ignored, as occurred in previous technologies such as smoking and asbestos.

21.5 Conclusions

Precaution is an important and essential element of any risk decision. The precautionary principle represents the most organized and sustained effort to date to make the role of precaution in risk decision making more explicit and relevant. The principle has gained significant momentum in recent years as it has been incorporated into a growing number of international environmental agreements and national legislative enactments and regulatory programs. Yet, despite its rhetorical appeal, the precautionary principle remains problematic in its practical application, in large part due to the ambiguity and arbitrariness of the principle. Nevertheless, if the precautionary principle were applied to RF EMF from cell phones, there would be a range of possible policy interventions. Some of the more stringent options, such as bans, moratoria, or warnings, would likely do more harm than good to public health, and thus are not justifiable. Other more modest interventions, such as informing people of what is known (and unknown) about RF EMF risks, empowering individual decisions on prudent avoidance options available for reducing exposure and thus any possible risks, and continuing safety research and population monitoring with safeguards in place to ensure transparency, likely represent the most sensible application of precaution to mobile phone RF EMF at present.

References

1 United Nations Conference on Environment and Development (1992) Rio Declaration on Environment and Development. UN Doc. A/CONF. 151/5/Rev. 1.

2 Raffensperger, C. and Tickner, J. (eds) (1999) *Protecting Public Health & the Environment: Implementing the Precautionary Principle*, Island Press, Washington, DC.

3 Commission of the European Communities (2000) Communication from the Commission on the Precautionary Principle. Brussels, COM(2000) 1.

4 Marchant, G.E. and Mossman, K.L. (2004) *Arbitrary & Capricious: The Precautionary Principle in the European Union Courts*, AEI Press, Washington, DC.

5 Independent Expert Group on Mobile Phones (2000) Mobile Phones and Health. Available at http://www.iegmp.org.uk/report/text.htm (accessed August 16, 2010).
6 Harremoës, P., Gee, D., MacGarvin, M., Stirling, A., Keys, J., Wynne, B., and Vaz, S.G. (2001) Late Lessons from Early Warnings: The Precautionary Principle 1896–2000, European Environmental Agency Environmental Issue Report No. 22. Available at http://www.eea.europa.eu/publications/environmental_issue_report_2001_22/Issue_Report_No_22.pdf (accessed August 16, 2010).
7 Sandin, P. (1999) Dimensions of the precautionary principle. *Hum. Ecol. Risk Assess.*, **5** (5), 889–907.
8 Huber, P. (1983) The old–new division in risk regulation. *Va. Law Rev.*, **69** (6), 1025–1107.
9 Marchant, G.E. (2003) From general policy to legal rule: the aspirations and limitations of the precautionary principle. *Environ. Health Perspect.*, **111** (14), 1799–1803.
10 Bodansky, D. (1991) Scientific uncertainty and the precautionary principle. *Environment*, **33** (7), 4–5, 43–44.
11 Leszczynski, D. and Xu, Z. (2010) Mobile phone radiation health risk controversy: the reliability and sufficiency of science behind the safety standards. *Health Res. Policy Syst.*, **8** (1), 2 (http://www.health-policy-systems.com/content/8/1/2).
12 Saracci, R. and Samet, J. (2010) Commentary: call me on my mobile phone ... or better not? A look at the INTERPHONE study results. *Int. J. Epidemiol.*, **39** (3), 695–698.
13 International Telecommunication Union (2010) Measuring the Information Society. International Telecommunication Union, Geneva. Available at http://www.itu.int/ITU-D/ict/publications/idi/2010/Material/MIS_2010_without%20annex%204-e.pdf (accessed August 28, 2010).
14 Dolan, M. and Rowley, J. (2009) The precautionary principle in the context of mobile phones and base station radiofrequency exposures. *Environ. Health Perspect.*, **117** (9), 1329–1332.
15 The INTERPHONE Study Group (2010) Brain tumour risk in relation to mobile telephone use: results of the INTERPHONE international case–control study. *Int. J. Epidemiol.*, **39** (3), 675–694.
16 Altekruse, S.F., Kosary, C.L., Krapcho, M., Neyman, N., Aminou, R., Waldron, W., Ruhl, J., Howlader, N., Tatalovich, Z., Cho, H., Mariotto, A., Eisner, M.P., Lewis, D.R., Cronin, K., Chen, H.S., Feuer, E.J., Stinchcomb, D.G., and Edwards, B.K. (eds) (2010) *SEER Cancer Statistics Review, 1975–2007*, National Cancer Institute, Bethesda, MD, http://seer.cancer.gov/csr/1975_2007/ (based on November 2009 SEER data submission posted on the SEER web site).
17 Grandolfo, M. (2009) Worldwide standards on exposure to electromagnetic fields: an overview. *The Environmentalist*, **29** (2), 109–117.
18 CTIA-The Wireless Association®. U.S. Wireless Quick Facts. Available at http://www.ctia.org/media/industry_info/index.cfm/AID/10323 (accessed August 19, 2010).
19 May, P. (2010) San Francisco's Cell Phone Safety Ordinance Stirs Controversy. *San Jose Mercury News*, August 8, 2010.
20 Sunstein, C.R. (2002) *Risk and Reason: Safety, Law, and the Environment*, Cambridge University Press, New York.
21 Wiedemann, P.M. and Schütz, H. (2005) The precautionary principle and risk perception: experimental studies in the EMF area. *Environ. Health Perspect.*, **113** (4), 402–405.
22 Barnett, J., Timotijevic, L., Vassallo, M., and Shepherd, R. (2008) Precautionary advice about mobile phones: public understandings and intended responses. *J. Risk Res.*, **11** (4), 525–540.
23 Morgan, M., Florig, H., Nair, I., and Hester, G. (1987) Power-frequency fields: the regulatory dilemma. *Issues Sci. Technol.*, **3** (4), 81–91.

Index

Cancer Risk Evaluation: Methods and Trends,
Edited by Günter Obe, Burkhard Jandrig, Gary E. Marchant, Holger Schütz, and Peter M. Wiedemann.

d